Transdisciplinary Perspectives on Complex Systems

Franz-Josef Kahlen • Shannon Flumerfelt •
Anabela Alves
Editors

Transdisciplinary Perspectives on Complex Systems

New Findings and Approaches

Springer

Editors
Franz-Josef Kahlen
Kahlen Global Professional Solutions
Gronau
Germany

Department of Mechanical Engineering
University of Cape Town
Cape Town
South Africa

Shannon Flumerfelt
Department of Organizational Leadership
School of Education and Human Services
Oakland University
Rochester, MI
USA

Anabela Alves
Department of Production and Systems
School of Engineering
University of Minho
Campus of Azurém
Guimarães
Portugal

ISBN 978-3-319-38754-3 ISBN 978-3-319-38756-7 (eBook)
DOI 10.1007/978-3-319-38756-7

Library of Congress Control Number: 2016947786

Printed on acid-free paper

This Springer imprint is published by Springer Nature
The registered company is Springer International Publishing AG Switzerland

Foreword

A few years ago, a group of colleagues representing different parts of the aerospace industry were lamenting the current state of our development programs and the prevalent trend of rising costs and significant schedule delays across the industry. We agreed that increasing complexity was a consistent characteristic for many of our under-performing programs. Being the good engineers that we were, we first looked at the technical root causes leading to the increased complexity. We specifically looked at how we, as an industry, approached Systems Engineering thinking and accountability. We thought we would find that we were not effectively addressing the complexity of our systems or that we were unwittingly introducing complexity into our systems. What we found was it was a bit of both in addition to contributors outside of our control.

To start this discussion, we decided to host a Complex Aerospace Systems Exchange (CASE), sponsored by the American Institute of Aeronautics and Astronautics (AIAA) to engage in this investigation and determine where in the Systems Engineering approach we may be failing. We explored examples within the design, integration, and verification phases of programs to examine those that were successful and those that were not. We also reached out to some of our nontechnical peers to see if there were other contributing factors. What we found was that this was not going to be an easy problem to solve.

The first issue we found was that the definition of complexity was not consistently understood. Many people used *complexity* synonymously with *complicated.* Complicated is generally defined as something with many intricate parts that can be difficult to understand, but all the parts can still be identified. While a system being complicated is definitely a factor when determining complexity, others asserted that a system could be complex due to the attribute of emergence. Although there is some debate about whether to call it emergence or ignorance. In either case, this attribute of emergence is used to describe phenomena that result from the interacting parts of the system. In the end, it was safe to say that most of those participating in the conversation agreed with the latter interpretation.

The second issue was that, while we tried to take a narrow look at design, integration, verification, or even how we manage programs, the topics kept getting intertwined. We quickly concluded that the emergent behaviors resulted in unintended consequences within our programs. Some programs incorporated technology that was not designed for how it was being used, resulting in unintended consequences that presented themselves during different stages of the program. Some programs had requirements to use multiple suppliers, levied upon them from either internal or external sources, resulting in varied or unanticipated performance during integration.

The third issue that also continued to creep into the discussions was the human element. How we, as people, interact and engage in the execution of a program greatly influences the success of the program. We are facing an increasing global and distributed environment in which we work to produce products. This requires us to look beyond the engineering and technical aspects of development programs and to consider the impact of cultural and regional value systems on policy and decision making.

Another aspect of the human element is related to organizations and the roles within those organizations. Originally, CASE targeted an audience that tended to be in the more senior positions, but what we found was that the topic of complexity attracted individuals from all levels of the organization. Those in more senior positions were there to share their experiences and insights particularly around creating work environments and establishing a culture that can effectively address the emergent behavior. Those in entry to mid-level positions were engaged to help understand and recognize the emergent behaviors and what actions they can take to mitigate the side effects of complexity. Even those at more of an entry-level position can play an important role in responding positively to mitigate the side effects of complexity. A common denominator was the recognition that experiential learning was an important factor in dealing with complexity and that the way projects organize can be critical to their success.

What you will find in the pages ahead are some insightful discussions about complexity, how to recognize it, and approaches to mitigate the impact it might have on our systems. We know it is unrealistic to eliminate complexity, and we will likely be exposed to even more emergent behavior in the future. It is imperative for us to consider this topic much more aggressively. Through the discussions presented in this volume and in the discussions referenced earlier, we are trying to raise the awareness of complexity and share possible approaches for dealing with it successfully. The hope is that by exposing engineers and other stakeholders to these discussions, they will begin to look for and recognize these potentially weak signals and emergent behaviors and start to address the problem of complexity.

St. Louis, MO Tsutsumi Sophia Bright

Preface

Complex systems play an increasingly important role in the advancement of numerous technologies and even in the conduct of our daily lives. The development and support of these systems is challenged by several factors. We continue to push the limits of hardware and software to deliver greater performance, sometimes at the risk of failures that can quickly become catastrophic. New technical disciplines, such as cyber security, are being integrated into legacy and emerging systems, while subsystems are gaining in both complexity and interdependency. Local control systems interact directly, and sometimes autonomously, with global command and control centers. High tech sensors and feedback loops are embedded in all levels of components to enhance capability, but their effects are not always fully understood or characterized by the product designers.

Traditional systems engineering approaches provide a general methodology, but often break down when applied to systems so complex that it is not possible to fully characterize all of the subsystems under all potential conditions. Contributing elements include the remoteness and extremity of the actual operating environment, the inability to integrate with critical infrastructure during development due to considerations for safety and security, the unpredictability of human interfaces, or just the inexhaustible number of circuit and software logic paths of the system. Even as we bring on more high power computing and continue to advance modeling and simulation capabilities, there is no substitute for testing, early and often, but that approach often has limitations.

When failures occur or out of phase technical issues are discovered, the subsequent root cause analyses and corrective actions are often accompanied by investigations to determine where the process went wrong. Inevitably, breakdowns in the execution of the traditional systems engineering process discipline are identified, to which we respond by embedding more checklists, reviews, approvals, and analysis tools, as well as organizational changes, better communications, etc., all wrapped up with the promise to do better next time. The findings are real and there are always opportunities to improve our systems development; however, we should be open to new techniques that supplement our conventional systems engineering methods to address the challenges of complex systems.

Not all of the opportunities will lie in traditional engineering disciplines. Research to replicate the evolution of biological systems has led to real applications for technical design. Embracing the power of social media through crowdsourcing is also proving to be a viable model for design development. There are many examples that demonstrate the innovations that can come from opening up our engineering aperture and looking to other fields of study for potential solutions.

This book casts a wide net in seeking out current work from multiple disciplines, each focused on advancing the state of complex systems development or management. There is growing consensus on the definitions and attributes, but not complete agreement. A number of researchers use the term "complex" to indicate a system that has, or is likely to have, emergent behavior, while there are still many practitioners who use "complex" in the generic English sense to mean "complicated" or "multifaceted." Approaches to addressing system complexity range across technical and sociological foundations, resulting in numerous candidate models and methods—each offering an opportunity for solutions and increased understanding. Value comes from the open-minded consideration of these distinctly different perspectives. By presenting this wide scope of approaches, the editors hope to foster conversation, additional research, and overall advancement of the science of complex systems.

Tucson, AZ Laura J. McGill

Contents

Mathematical Characterization of System-of-Systems Attributes

W. Clifton Baldwin and Wilson N. Felder

1 Introduction

Our intention for this chapter is to provide a cross-disciplinary vocabulary designed to ease communication about complexity among practitioners in the fields of systems engineering, management, and philosophy. Our specific goal is to inform the practice of systems engineering, while being consistent with the scholarship in these other disciplines. The definitions in this vocabulary have mathematical characterizations in order to be as precise as possible. The structure we propose is intended to facilitate further investigation of complexity in a broad subset of engineered systems. It is neither intended to serve as a universal lexicon, nor is it intended to stifle the wider discussion around issues of complexity that is based on broader, semantic definitions.

The very act of creating mathematical definitions limits the scope of the problem space open to discussion. As eloquently summarized by John L. Casti [1], this is an inevitable result of important findings in the theory of formal logic by Bertrand Russell, Kurt Gödel, and Walt Chaitin. It is perhaps because of this limitation that there is such a great reluctance to attach mathematical definitions to complex system concepts; by staying clear of such definitions, we preserve the unlimited scope of the conversation. On the other hand, by not restricting the field of study in such a way, we also limit our ability to measure the results of our actions. Our project, then, (as the philosophers like to say) will be to develop a useful test of complexity for systems, with the understanding that we fall short subject to the limitations imposed by our approach. The ultimate goal remains elusive, but the results to this point are encouraging.

W.C. Baldwin (✉) • W.N. Felder
School of Systems and Enterprises, Stevens Institute of Technology, Hoboken, NJ, USA
e-mail: wbaldwin@stevens.edu

F.-J. Kahlen et al. (eds.), *Transdisciplinary Perspectives on Complex Systems*,
DOI 10.1007/978-3-319-38756-7_1

This chapter represents the latest accomplishment of a decade-long effort to find a simple, straightforward mathematical test that could be applied to any given system and would establish whether or not that system had the architectural predisposition to unpredictable behavior as the result of complexity. Our wish to find this test—we cannot really call it a goal because we are not by any means confident that it can be achieved—is driven by our combined personal experience with the systems engineering, design, development, test, and operation of the U.S. National Airspace System (NAS), arguably one of the largest, most convoluted, and complex systems of systems (SoS) ever built. Repeatedly since roughly 1990, efforts to modernize this system have come up short in various ways on numerous occasions despite the hard work of smart, well trained and highly motivated engineers, operators and managers.

As we have pointed out previously, [vide 2–6], systems like the NAS exhibit features of complexity that cause the fundamental tools of classical systems engineering to be inadequate. As Keating [7] has so eloquently pointed out, this insufficiency is because the usual way of tackling system problems is to break the system down into manageable chunks, and then reassembling it during integration. Unfortunately, a complex system like the NAS is defined by the existence of system results that emerge from the interactions among its constituent parts. Looking at such a system in its piece-wise form cannot reveal the source of unpredicted and perhaps detrimental behaviors.

As we design and test incremental improvements to an overall SoS like the NAS, we attempt to imagine as many system states as we can, and devise ways to simulate the system's response to these scenarios. But we have no way of knowing (1) whether there will be unpredicted emergent behaviors with bad consequences; nor (2) whether we have tested a change sufficiently to prove that it will not have any latent problems.

Complex behavior in a SoS is built on a series of precursor attributes, and so our work has focused on understanding those attributes in preparation for tackling the higher level question of system complexity. Based on a comprehensive review of over 40 published papers on SoS, Boardman and Sauser [8] collated the various identified features into five characteristics, which they named autonomy, belonging, connectivity, diversity, and (in some cases) emergence. In reaching their conclusions, Boardman and Sauser leaned strongly on previous efforts along these lines by Ackoff [9], who coined the phrase "system of systems," and Maier [10], who first proposed a set of relevant system attributes. Baldwin and others have offered a framework for organizing our knowledge about SoS behavior using the Boardman and Sauser attributes, and have provided some preliminary ideas on how to characterize those attributes [2].

We are now ready to formalize the definitions of four of the five attributes: autonomy, diversity, connectivity, and most importantly, belonging. These four will provide the foundation for a formal definition of emergence as we continue to solidify our understanding of complexity as it relates to the engineering of complex systems. In addition to reviewing a set of mathematical expressions suitable for the characterization of the above system attributes, we also provide a simple exemplar based on the air traffic control system that demonstrates the use of the proposed belonging definition. Although we suggest that these attributes serve to distinguish various types of systems from each other, including SoS, we stop short of offering a mathematical definition of system complexity. Instead, we propose one out of a number of possible lines of investigation as the one with the highest likelihood of resulting in such a characterization.

Our goal is to be able to unequivocally determine, based on a given system architecture, whether a particular SoS is or is not likely to exhibit emergent behavior (or to encourage emergent behavior in the socio-technical system to which it belongs). If a system has this potential, then we are in a position to design and engineer it in a way that acknowledges the inevitable, even though unpredictable, potential of an emergent result resulting in system failure. If we have a mathematical test that confirms a system as complex according to our taxonomy, then we should not expend the effort to find the failure modes of the system, regardless of its criticality, since such a system has an operationally infinite number of possible system states, and we can never be sure that we have explored them all.

We know the people who were engaged in the development of the Federal Aviation Administration's (FAA) air traffic control systems to be intelligent, highly trained, dedicated and competent engineers and managers. Some of these systems are the Advanced Automation System, the Standalone Terminal Replacement System, and the En Route Automation Modernization (ERAM) program. The problem with taking a purely heuristic view of system failure—of falling back on the socio-technical context of which the engineered system is merely a part—is that it may lead to the conclusion that the failure of the system was a function of human incompetence. But we know that to be false, at least in the above referenced cases. In fact, there is something going on in the engineering of the system, something that flows directly from the attributes of the architecture and causes the unpredicted behavior.

2 Background: Review of Attributes and Taxonomy

Baldwin et al. [2] offer a taxonomy of system descriptions based on observed system behaviors, and tied that taxonomy to the five attributes described by Boardman and Sauser [8]. The attribute of autonomy has been defined as the ability of a system to achieve its own goals [11], and diversity ensures there are different goals among the constituent systems. The characteristic of connectivity is more than having a connection since a system with any usefulness has some type of connection to the outside world. In this case, connectivity refers to a dynamical nature of information flow in that the flow is stochastic, dynamic, event-driven, unpredictable, or asynchronous. On the other hand, the lack of connectivity indicates either information flow that is static, predetermined, or orderly; or no connection at all. The goals of a system of systems (SoS) are attained through the

Table 1 System attributes and taxonomy (adapted and extended from [2])

System Type	Characteristic Attribute(s)	Description	Mathematical expression(s) (See following text for explanation of notation)	Example (Biological)
Component	Existence	Physical entity	$\exists x$	Vein, artery
Sub-system	Process	Transforms inputs into outputs via a process	T: $x \rightarrow f(x)$	Circulatory System
Simple system	Autonomy	Able to meet a stated system goal without outside help	A: $\min_{\epsilon_i \geq 0} \epsilon_i : f_i(x) + \epsilon_i = G_i(x), \forall S_i \in SoS$	Organism
Composite SoS	Diversity, Connectivity, Belonging	Able to exchange information to provide mutual support; outputs predictable	D: $\exists S_i \neq S_j, \{S_i, S_j\} \in SoS : G_i(x) \neq G_j(x)$; C: $C^* = \begin{bmatrix} c_{11} & c_{12} & & c_{1n} \\ c_{21} & c_{22} & \dots & c_{2n} \\ \dots & \dots & & \dots \\ c_{n1} & c_{n2} & & c_{nn} \end{bmatrix}$; B: $\min_{\epsilon^* \geq 0} \epsilon^* \cup f_i(x_k) + \epsilon^* = G^*$	Family Group
Complex SoS	Emergence	Unpredictable outputs	TBD	Schools, herds and the like
Adaptive SoS	Adaptability	Able to change architecture of system in response to outside pressures	TBD	Species

contributions of the constituent systems, since basically the SoS is nothing more than the union of its constituent systems. Therefore it is more important for systems to interact and contribute appropriately to each other rather than simply possess some cluster of connections. Belonging, in this context, is the ability to collaborate on goals. For ease of reference, the 2011 taxonomy is replicated in a slightly different format in Table 1. The systems we are interested in are those described by von Bertalanffy [12] as "open," (that is, they perform work). Note that there is a hierarchical relationship between "systems", as we use the term, and subsystems, which also perform work, but are not capable of achieving a goal autonomously. Please note that the description of the notation used will be introduced in a later section as we discuss each attribute in detail.

Thus an organism is a system, whose goal is famously to convert energy into replicas of itself [13]. An organism is not, however, a complex system nor is it even a SoS according to our taxonomy, because (1) no purported "component system" of an organism is capable of achieving its own goal independently, and (2) an organism does not produce results that cannot be predicted solely from a full understanding of the functioning of the organism's constituent parts. Think of an organism with a cardiovascular system. Despite its name, the cardiovascular system cannot function outside of the organism, and the heart, lungs, arteries, and other related parts dictate how the cardiovascular system will work in the organism.

Arguably, however, a species of organisms is indeed a complex system, and an adaptive one, at that, as argued by Minai et al. [14]. It is complex because the richness of combinations among members of the species is operationally infinite, the dynamic connectivity among members of the species is asynchronous, random, and frequent, and the resulting character of the species taken as a whole is unpredictable given a complete understanding of the functioning of the individual organisms within it. The species is an adaptive system because as a result of the emerging behavior, which flows from the connectivity among species members, a species can change its character in response to changing environmental conditions, transforming *Tyrannosaurus rex* into a Ruby Throated Hummingbird (*Archilochus colubris*) and tiny vole-like thecodonts into Blue Whales (*Balaenoptera musculus*). A species of organisms fills all the definitional requirements of the systems taxonomy proposed in Baldwin et al. [2]. In addition to adaptability and complexity, as just described, it possesses diversity in that every individual member of the species differs, sometimes quite significantly, from every other. It possesses autonomy, since each individual organism in the species has the ability to fulfill the basic goal of turning energy into potential offspring. Of course, in species that reproduce sexually, it has to connect to at least one other member of the species in order to fully achieve its larger goal, but the point remains. It has belonging because every individual organism gets something from being a member of the species (if nothing else than a chance for sharing DNA) and every organism contributes (strongly!) to the species goal of survival over time.

It is hard to think of a better exemplar of a successful complex adaptive SoS than a species! Indeed, if we believe that dynamic connectivity is a defining attribute of a complex system, and the mechanism whereby complexity is mediated, then a species is the perfect example for this argument. Imagine a species without the attribute of dynamic connectivity. What would *that* look like? Well, to satisfy the definition of static connectivity, each organism in a species would be so constructed as to only be able to mate with one, two, or at most a highly limited number of others. Suddenly, the wild and crazy plethora of possible system states would reduce to just a few. Complexity would almost completely be eliminated. And the necessary precondition for adaptability would vanish. And guess what? In the marvelous laboratory of real life, we have more than one existence proof of this happening. When a species is reduced to a small number of individuals, the connectivity in the system becomes "essentially" static, and almost always in these conditions, barring some "zookeeper *ex machina*" intervention, the species becomes extinct!

The biological species is not only a great example of a complex SoS, but it is also a great example of a case in which complexity is a good thing. Too often we tend to see complexity in socio-technical systems as a problem, potentially causing jet aircraft to fall out of the sky and nuclear power plants to melt down. But biology certainly does not share that view. In fact, we can very safely say that biology as we know it could not exist without complexity. Hopefully this description of species biology helps to put in perspective the reality that there are different kinds of systems out there, and they are different because of the attributes they do or do not possess.

The question is then how do these system characteristics express themselves when what we are talking about is an aircraft, an algorithm or a socio-technical system such as the National Airspace System? We are professionally interested in the answer to that question, but we also firmly believe that before we tackle it, we need to be very clear in our definitions. And that means bringing mathematics into the room.

3 Discussion: Development of System Attribute Characteristic Expressions

3.1 Objectives of the Present Development

Baldwin et al. [3] described the so-called belonging behavior of collaborative systems of systems (SoS) but did not rigorously define the differentiating features. In Baldwin et al. [2] we described the role of certain characteristics in differentiating systems from more complex composite systems, but we did not rigorously define the characteristics mathematically. This chapter provides a mathematical framework for the identifying features of a SoS. Although we will derive these characteristics from first principles, we use the naming conventions set forth by Boardman and Sauser [8] and demonstrate how these characteristics describe the functionality of any SoS.

3.2 Premise

A SoS exists as the union of its constituent systems; there is no SoS entity apart from its constituent systems. Keeping that in mind, consider the system elements. A SoS consists of some number, *n*, of constituent systems. Collectively there are multiple, *m*, goal-directed outputs from the various systems. Each system has a goal set, which must be proper subsets of the *m* goal-directed outputs. For those interested in emergence, note there is no restriction limiting the SoS to the set of *m* goal-directed outputs. Emergence has been defined as global behavior which cannot be traced to any individual component [inter alia 15, 16]. In essence, the theory of emergence declares that the whole is greater than the sum of its parts and manifests as a non-additive property [17]. Accordingly, a result of the theory of emergence is to limit the ability to predict [18]. Therefore it describes behaviors of the SoS as a whole and cannot be predicted based on the attributes of the individual constituent systems [summarized in 4].

The elements of a system are subsets within the system set. It is convenient to use matrices to work with these sets. Potentially, many of these matrices will be multi-dimensional. For ease of comprehension, we discuss each matrix in the context of two dimensions, although any given element may represent a multi-dimensional matrix.

3.3 Principles of Systems Theory

In its most basic form, a system, S_i, takes input x and transforms it via a set of processes into output, $f_i(x)$, to attain some specific goal, G_i (Fig. 1) [12]. The links to the external world are the system's connections.

Thus, we can describe a system by $S_i = \{F_i, X_i, G_i, C_i\}$. The set of the system's functions are represented by F_i, which form the system's processes. These processes transform inputs, X_i, into outputs that attempt to achieve the system's goal set, G_i. Since a goal may be an abstract entity, a specific input, x_i, is transformed by the system's processes into an output, $f(x_i)$, which is a measurable result of a system attempting to achieve its goals. The subset of connections, C_i, or interfaces, allows an output from one system to be an input to another one.

Our notation here differs slightly from that used previously to describe a system using set theory [3]. In that case, the inputs X_i were represented by A_i or the action of a function on an input, and the subset of connections was represented within a subset E_i, which included any additional elements not represented by functions, activities, and goals. Using the earlier notation, $S_i = \{f_i, A_i, G_i, E_i\}$, which is identical in meaning to the expression used here.

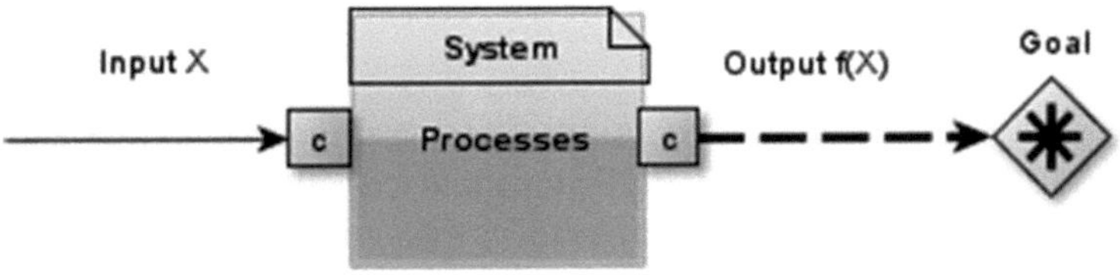

Fig. 1 Basic system principles: the Input/output model of system functionality

3.4 Autonomy

From first principles, a system attempts to achieve its own goals. For example, these goals may be a set with some "weight" goal, some "speed" goal, some "cost" goal, and potentially many other objectives. Therefore each goal set G_i is actually a vector of specific goals, where the size of this vector will differ by system. A system that exhibits autonomy can fulfill its own goals [11]. Let us define ϵ_i to indicate the difference between what the system does, $f_i(X_i)$, and its goal set, G_i. These goals may be quantified with utility functions [3], but for simplicity we will refer to the utility function of each goal as the actual goal. For a system to be truly autonomous, the difference between output and goal must be vanishingly small and ideally be equal. It is important to note, that in order for a system to exhibit this characteristic, it need only be self-sufficient *in theory*. There are many examples of systems that fail to meet their goals in reality, often for the very reasons of complexity that are the subject of this book. Systems within a SoS exhibit autonomy by definition, although they may have to sacrifice some actions in order to belong to the SoS [8]. In any case, the system's functions attempt to satisfy its goals, although they may fall short by some amount ϵ_i [Eq. (1)]. In order to exhibit autonomy while making these sacrifices, the constituent systems strive to minimize the error ϵ_i. Alternatively, if ϵ_i has a significant value, then the system does not possess autonomy (and consequently the composite system fails to be a true SoS). What the specific amounts are, as well as what qualifies as significant, are yet to be determined. For now, we can assume ϵ_i is determined qualitatively.

$$\min_{\epsilon_i \geq 0} \epsilon_i : f_i(x) + \epsilon_i = G_i(x), \forall S_i \in SoS \text{ (Autonomy)} \tag{1}$$

The result of this discussion is summarized in Table 1, which includes the important conclusions from Baldwin et al. [2], extended according to the analysis above. Table 1 also includes material discussed below, including the important conclusions about belonging.

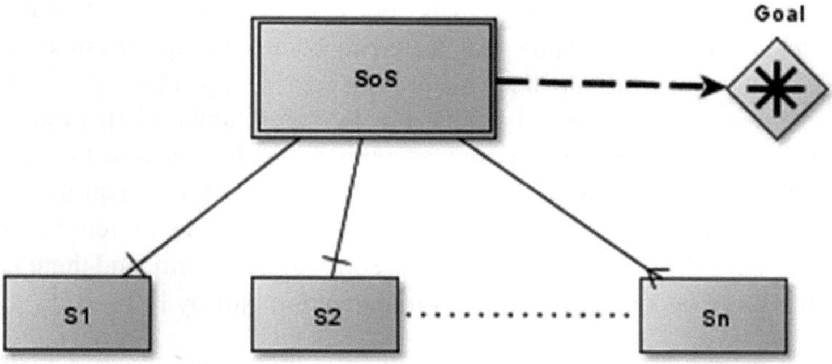

Fig. 2 Basic composite system principles

3.5 Systems of Systems

At a minimum, a SoS is a composite system comprised of multiple autonomous systems that as a whole has its own goal set, G*, achieved through the actions of the constituent systems (Fig. 2).

We will write *Composite System* $= \{S_1, S_2, \ldots, S_n, G^*\}$, where S_i are the constituent systems and G^* is the overall composite system goal. Based on this description, there are certain restrictions when describing the functionality of a SoS. The constituent systems must perform all functions since the SoS is just the composition of individual systems. The SoS goal, G^*, may be known to the individual systems, but otherwise all SoS behavior is in some way the direct or indirect result of the constituent systems.

We will further require that a mathematical model of a SoS must address the constituent systems as independent entities and address the collective nature of constituent systems. As we will show, autonomy and diversity will be framed as the qualities of the independent constituent systems. It is easy to recognize that autonomy applies to individual systems. Although diversity cannot be determined without comparing systems within a SoS, the given description requires that it be based on the qualities of the constituent systems. Likewise we describe connectivity and belonging in terms of the collective behavior since these characteristics are meaningless if applied to any one system alone.

3.6 Diversity

Diversity is a critical attribute of every SoS, since without it we just end up with a larger version of the component system! Without diversity, the collection of constituent systems cannot really do anything different than any one constituent system, just more of it. So diversity is the attribute that allows a SoS to be a totally different entity. Stated another way, diversity is a characteristic that ensures some constituent systems have different goals [Eq. (2)]. Of course a SoS may include some systems that share the same goal. Referring back to the species of an organism example, some organisms may be twins or even copies, depending on the species in question. However there must exist at least two systems (and probably more) within the SoS that do not have the same goal, as mathematically stated in Eq. (2). On the other hand, a composite system without diversity has all constituent systems with the same (or basically the same) goals. Note, that a composite system all of whose components have the same goals has been described as a swarm, and there is a large body of literature on swarm theory as summarized in part by [19].

$$\exists S_i \neq S_j, \{S_i, S_j\} \in SoS : G_i(x) \neq G_j(x) \text{ Diversity} \tag{2}$$

Although Eq. (2) provides a suitable definition of diversity, we note that in practical terms there is a difference between a system's goals and the outputs of the system. In accordance with the expected-utility maximization theorem [20], natural numbers can be assigned to utility functions for each of a system's goals. Since a system produces outputs to attain its goals, we can measure this output as a representation of the goals. The utility functions of the goals are assigned the quantity of the output. Although the output is not synonymous with the system's goals, it is a sufficient representative, but we need to remember that the desired output is technically a product of the system while the goal is an ideal purpose for the system. Furthermore, there exists a preference relationship among the different goals with some providing more benefit than others [3]. Due in part to this diversity, systems within a SoS will contribute differently towards the SoS goal. As noted in Eq. (3), system i may contribute via its goal(s) to the SoS more than or at least as much as system j contributes via its goal(s), and so on until the system that contributes the least, system n. However, despite the preferences of goals, it is the combination of all the constituent systems' contributions that produce the unique behavior of the SoS.

$$G_i(x) \succeq G_j(x) \succeq G_n(x), \; \{S_i, S_j, S_n\} \in SoS \tag{3}$$

3.7 *Connectivity*

The concept of connectivity is both extremely simple and very complicated. To support our discussion of the belonging attribute to follow, we will limit our definition of connectivity here to the most basic observation that every system has a subset of connections, C_i. Although the information flow may be static or dynamic, we will treat the connections, at least initially, as being equivalent. We use an adjacency matrix from graph theory to represent those systems that have the capability to connect to other systems [21]. We define a simple connection matrix

C* for the SoS such that each row is each system's vector of connections, C_i, to the other systems within the SoS. Then a connection, of whatever sort, from system i to system j is represented in the matrix C* at position c_{ij}. Continuing our theme of simplicity, we define $c_{ij} \equiv 1$ if there is a connection from system i to system j, and 0 otherwise.

$$C^* = \begin{bmatrix} c_{11} & c_{12} & & c_{1n} \\ c_{21} & c_{22} & \cdots & c_{2n} \\ \cdots & \cdots & & \cdots \\ c_{n1} & c_{n2} & & c_{nn} \end{bmatrix} \tag{4}$$

For our purpose here, the matrix C* as defined in Eq. (4) simply provides an accounting of the potential locations for information transfer among constituent systems. It is obvious that the quantities c_{ij} could take a wide variety of forms, with attendant complexity in the definition. One advantage of defining the connections of a SoS in this way is that we could easily calculate how many paths there are between any two systems. From graph theory, we know the number of k-step connections between system i and system j is the (i, j) entry of $(C^*)^k$. Therefore $(C^*)^2$ gives all the two-step connections and $(C^*)^3$ gives the three-step connections. This representation allows the entries in C* to be directional. If there exists a connection from *i* to *j* but not the other way, then $c_{ij} = 1$ but $c_{ji} = 0$. Although not important for the present discussion, this observation points to potentially fruitful areas for examination in the future.

Although each of the attributes described above could be examined in greater detail, we have limited our exploration of diversity and connectivity to those aspects necessary to support our development of a suitable characterization of belonging as a separate attribute.

3.8 Belonging

Baldwin et al. [3] provide a mathematical description of the decision behavior for a system to belong to a SoS. Yet there is no mathematical description of what belonging does for a SoS. The goals of a SoS are attained through the contributions of the constituent systems. Therefore it is more important for systems to contribute appropriately to each other rather than just have a connection.

Belonging is the ability to accept and contribute goal-directed actions with respect to another entity, namely the SoS [11]. A SoS with belonging has its goal achieved within some margin of error, $\epsilon^* \geq 0$, and each constituent system is responsible for some part of the SoS goal, G*. Arguably some systems may not contribute anything, which is permitted by the mathematics here. Since each system's output is depicted by f(x), there is some combination of the $f_i(x)$ to attain the contributions to the SoS, which allows it to attain its goals. Let the vector G* represent the output of the SoS. Then $\cup\, f_i(x_k) \rightarrow G^*$. Ideally, a proper SoS minimizes the error between the SoS goal and the sum of the individual systems' contributions.

$$\min_{\epsilon^* \geq 0} \epsilon^* \cup f_i(x_k) + \epsilon^* = G^* \text{ (Belonging)} \tag{5}$$

Although Eq. (5) effectively describes the formal aspects of belonging, it is of limited practical value. A more useful formulation can be developed if we consider the role that connectivity plays in the functioning of SoS. Such an expression links the concepts of belonging and connectivity allowing us to assess the manner in which the individual component systems support each other (and the overall SoS goal G*). Connectivity enables the output of one system to be utilized by another one. Therefore connectivity provides a function to the SoS such that the constituents systems' outputs can be used to attain the SoS outputs, G*. We can rewrite $\cup f_i(x_k) \to G^*$ as $\mathbb{C}(f_i(x_k)) \to G^*$. This expression suggests that connectivity is an essential precondition for belonging, prompting us to use the connectivity matrix C* as an additional clarification of the belonging attribute.

In the previous section, we defined the connectivity matrix, C*, such that entry (i, j) = 1 if there is a connection from system i to system j and 0 otherwise. This convention allows an easy mechanism for computing the number of connections between systems. However, there exist in the realm of Network Theory many algorithms for computing paths between many types of graphs. These algorithms go by the collective name of shortest path algorithms [21]. If we define what we are measuring between systems, we can allocate that measurement to the matrix and then select the appropriate algorithm to determine the contribution of each system to the others. Although we do not explore the use of this property here, we note that it could play a role in future expansions of this concept.

We now define a matrix B* derived from the matrix C* with the entries replaced by the contributions of each system to each other. Let b_{ij} represent the contribution of system i to system j. Note $f_i(x) \to b_{ij}$ but b_{ij} may be only a portion of $f_i(x)$, which is why we cannot simply insert $f_i(x)$ for b_{ij}.

$$B^* = \begin{bmatrix} b_{11} & b_{12} & & b_{1n} \\ b_{21} & b_{22} & \cdots & b_{2n} \\ \cdots & \cdots & & \cdots \\ b_{n1} & b_{n2} & & b_{nn} \end{bmatrix} \tag{6}$$

A SoS has the columns sum to the goals of each system, $\sum_k^n b_{ki} + \epsilon_i = G_i$. Also a SoS has the sum of the row sums equal the SoS goal, $\sum_i^n \left(\sum_k^n b_{ki} + \epsilon_i \right) + \epsilon^* = G^*$.

The drawback from this matrix method is we rarely, if ever, can fully quantify in practice b_{ij} or even $f_i(x)$. Yet, it should be possible to identify when an output of one system is used as an input to another system. Although we would prefer well-defined utilities to populate the B* matrix, we can again use an agency matrix where $b_{ij} = 1$ if the output of system i is an input to system j and 0 otherwise. In the following section, we will consider a practical application of the definitions described so far.

4 Discussion: Belonging in the National Airspace System: A Practical Example

Needless to say, mathematical descriptions are all well and good, but as W. A. Baldwin [22] has suggested, practical application is what validates the worth of a theory. Accordingly, we now offer an application of the above concepts to the National Airspace System (NAS). Let us be very clear about one thing: the NAS is certainly a complex system under anyone's definition of the term! Well, a complex socio-technical system, anyway. Just one look at the performance of the NAS on a bad weather day should provide a good example of emergent behavior! But, if we were to take the humans out of the loop, then what is it? And, why do we care, anyway? Perhaps what we should be more concerned about is has the NAS been designed so that when the inevitable but unpredictable results happen, it still hangs together and functions even with the numerous, asynchronous and unpredictable interactions mediated by the humans involved? Actually, why should we be surprised by the emergent results? Although our analysis of collections of organisms (such as humans) was based on their reproductive behavior, organisms are hard-wired to exhibit complex behavior. So, applying the rigorous definitions, is the NAS a complex SoS, or is it not?

The NAS is composed of, among other things, radar systems and automation systems. In the traditional arrangement, the radar was attached in a one-to-one relationship to the processing and display function (see Fig. 3). Even when all elements of this architecture were made digital, the relationship held. But eventually, more than one radar system became connected to a single display. In the beginning, this was done by segmenting the geographic coverage into polygons, with only one radar assigned to each polygonal area. So, architecturally, we were still dealing with one radar for one display. Today, inputs from multiple radars are fused using, perhaps, Kalman filter logic. Oddly, the result is the same. The nature of the Kalman filter algorithm is such that it is wholly deterministic. Given a set of inputs, the result will always be the same. The filter is designed to accept multiple information flows and to produce consistent outputs (of a higher quality than is possible with a single incoming data flow, by the way). The connectivity in this (small) SoS is static, and there is no possibility of emergence from the flow of information (even though autonomy, diversity and for all intents and purposes belonging exist).

The radar systems interconnected in this arrangement carry the information from the secondary transponder beacon aboard the aircraft, and the computed azimuth

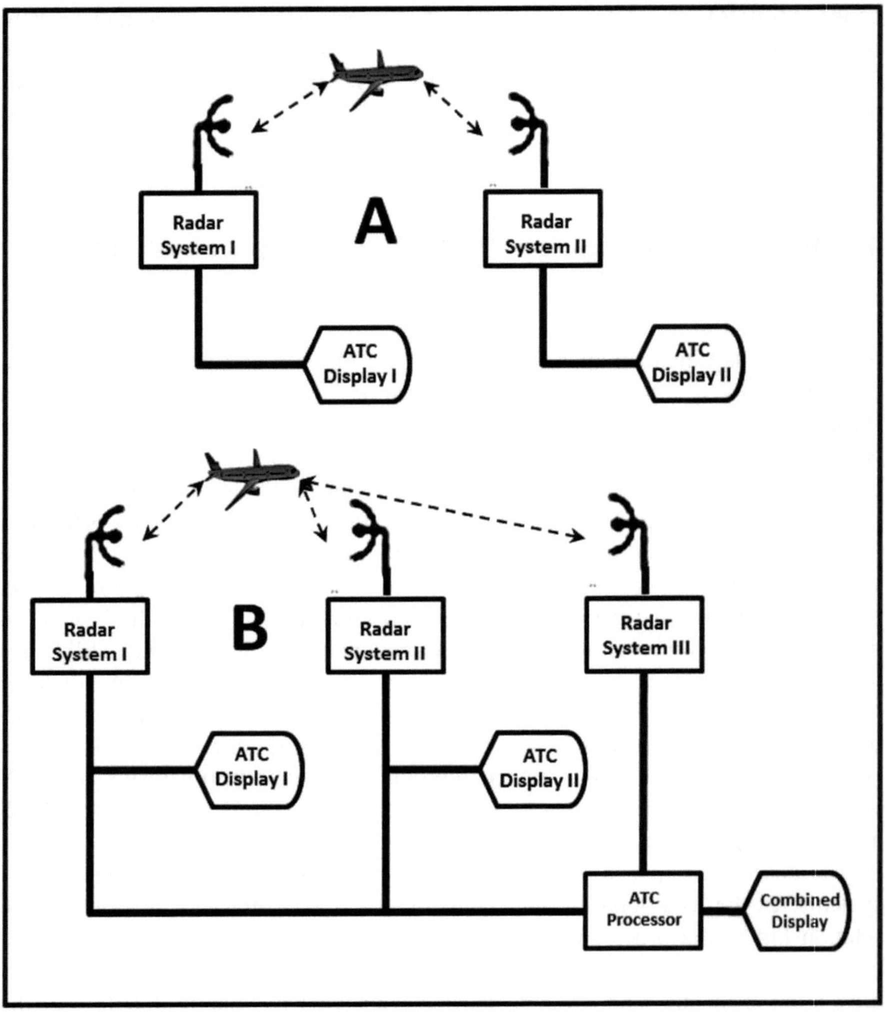

Fig. 3 ATC Examples: (**a**) early stage radar system showing only autonomy; (**b**) later stage automation system with diversity, connectivity and belonging

and slant range to that transponder, from the radar's secondary antenna to the radar display. Each radar system has the full ability to carry out its mission of providing an aircraft location—"NAS to glass"—as air traffic control jargon has it. The radar system possesses autonomy as we have defined it. Note that the supposed similarity between the radar system and an aircraft propulsion system is fallacious since although the radar system "needs to be attached" to the air traffic control center in order to do its job, this situation is qualitatively different than that of an aircraft propulsion system that "needs to be attached" to the aircraft in order to fulfill its goals. In the latter case, the propulsion system, properly called a subsystem, fails the autonomy test because the critical nature of its goal is to impart motion to the thing it is attached, while the radar system in the NAS is trivially connected to the building, and need not be attached to succeed in its purpose. We can write down the B* matrix [from Eq.(6)] for this small SoS:

$$B^* = \begin{bmatrix} 1 & 0 & b_{13} \\ 0 & 1 & b_{23} \\ b_{31} & b_{32} & 1 \end{bmatrix} \tag{7}$$

Where b_{13} and b_{23} are, very concretely, the inputs on aircraft position to the ATC processor system contributed by radar system 1 and 2 respectively. The net Circular Error Probability (CEP) metric for aircraft position (the G* output of the total SoS) is demonstrably improved through the contributions of radar systems I and II to the overall SoS in this simple example. On the other hand, the quantities b_{31} and b_{32} represent information about aircraft position returned from the automation system back to the individual displays of the separate radar installations. Although this information is not currently used operationally in the NAS, it is clear that the better location information produced by the automation system could improve the information at each constituent system, and thereby contribute to the individual system goals G_i. One could continue this sort of thought experiment with the other systems currently deployed in the NAS. All would end up looking like "composite SoS" from a definitional point of view (see Table 1). Even though they are pretty darned complicated in their interrelationships, none currently meet our taxonomic definitions of "complex SoS" since their outputs are predictable. We must conclude this result is due to the static connectivity.

What about some of the eventual evolutions of the NAS as part of the ongoing NextGen program, which intends to modernize the NAS? Although the early stages do not contemplate a direct connection from the ground control system to the onboard flight management systems of aircraft, such a connection is almost inevitable in time. If as a result of this initiative, a flight plan change were calculated by algorithms in the ERAM software, communicated to the onboard flight management system via some data communication system, and executed by the aircraft's flight control system automatically, then it is relatively easy to imagine that behavior could emerge in unpredictable ways. It is important to note that the *predictable* behavior expected of this arrangement: that aircraft 4-D trajectories are deconflicted in a smooth, energy and pollution reducing, and efficiency increasing manner, is highly desirable. The *unpredictable* behavior that might ensue from this scheme, and which cannot be a priori determined because of the astronomically large number of possible variations in system state that can happen, could involve catastrophic consequences as the systems involved attempt to function, perhaps under high workload conditions, and with added factors such as weather avoidance becoming part of the equation.

A data communication system is just one example of a NextGen system that could lead to emergence in the NAS. Another is the use of System Wide Information Management (SWIM), which enables the sharing of information between diverse systems. With SWIM, the best and most readily available data is used to make decisions, such as how to sequence aircraft for arrival at a given airport or navigational fix, and the result would be different each time it is calculated, based on what data were available from the SWIM network.

To summarize, Eq. (4) allows us to calculate the benefit obtained through participation in a SoS, in this case the exact improvement to the CEP of the aircraft position as a result of the information flows b_{12} and b_{13} in Eq. (7). For more complicated situations, more information flows come into play and include bidirectional ones, in which individual systems more clearly receive benefits from belonging to the SoS. For example, it is clear that in the SoS described by DeLaurentis et al. [23] involving naval force collaboration based on the Littoral Combat Ship scenario, information flow among the units could be characterized in the form of Eq. (6) and the specific performance benefits accruing to the overall task force and each participant as a result of membership could be readily assessed.

We now have a complete set of expressions describing the behavior of the four fundamental attributes that make up a SoS. It is on this foundation that we will have to build the analysis of complexity as a system attribute.

5 Foundations for Future Work: Our Understanding of Complex Systems

So far, we have demonstrated that we can write down a set of precise (although admittedly limited) and internally consistent set of descriptions sufficient to characterize the most salient features of a system of systems (SoS). We have said nothing, however, about the characteristics of a *complex* system. What we need to do now is ask the question: “what is a complex system?” and relate complexity of systems to more general concepts of complexity. We will ask the question in the context of the attributes described so far.

In Table 1, we have indicated that complex systems are associated with the characteristic of emergence, a property for which we have as yet no mathematical definition. In this section we will examine the features of emergence as a property of complex systems, and suggest ways in which we might be able to construct a mathematical expression that serves as a test of complexity.

5.1 The Critical Role of Emergence

We begin by noting that emergence is an inextricable feature of complex systems; you cannot have a complex system without emergence. Any systems engineering definition of complexity, therefore, needs to take this into account.

The strongest and most consistent set of conclusions about complexity comes out of the field of physics, as articulated by Casti, Crutchfield, and others.

According to this view, complexity, as reviewed by [24], is defined as the emergence of chaos out of determinism. As Crutchfield has put it elsewhere [25] "One recent and initially baffling example of emergence is deterministic chaos. In this, deterministic equations of motion lead over time to apparently unpredictable behavior. When confronted with chaos, one question immediately demands an answer—Where in the determinism did the randomness come from? The answer is that the effective dynamic, which maps from initial conditions to states at a later time, becomes so complicated that an observer can neither measure the system accurately enough nor compute with sufficient power to predict the future behavior when given an initial condition. The emergence of disorder here is the product of both the complicated behavior of nonlinear dynamical systems and the limitations of the observer."

Anderson [15] seems to have been the first author to recognize that the new, emergent properties are not directly determined by the physics of the lower level systems (e.g. in a SoS, although he was not specifically dealing with SoS). Keating [7] has provided a particularly lucid discussion of the issues surrounding the conundrum of emergence in complex systems. While we do not agree with one of his assumptions (namely, that all SoS are necessarily complex) that disagreement does not have any bearing on the overall framework for complexity that he describes.

Putting the same idea another way, Weaver [26] has observed that organized complexity within systems is a unique state, differentiated from simplicity and disorganized complexity through the evolution of an emergent whole. Simplicity describes the "easy to solve" problems from science involving just two or maybe three variables. Disorganized complexity is the opposite extreme of scientific problems requiring millions of variables, which can be addressed with probability theory or statistical mechanics. Organized complexity, on the other hand, involves the problems with a sizable number of variables but not enough for statistical methods to adequately address. This is precisely the type of system of interest to us here.

The philosophical literature identifies six properties associated with emergence [27, 28]. The first property, in order of appearance, is radical novelty, where a new feature is observed in the system. The second property, coherence/collective/wholeness, is the characteristic of an emergent feature that integrates the lower-level components into some whole with its own sense of identity. Global is the property whereby emergent phenomena occur at the macro-level. The emergent phenomena arise over time and are therefore labeled as dynamical. The ostensive property describes the fact that the emergent features are recognized once they appear. The sixth property, self-transcending constructions (STC), expresses the dual nature of emergence and applies more to the process of emergence than the emergent phenomena. This property describes emergent features as both following from and being discontinuous with the parts from which they emerge. In addition, Goldstein [27] offers four characteristics as necessary preconditions for emergence: nonlinearity, self-organization, beyond equilibrium, and attractor. Nonlinearity includes small causes leading to large effects. Self-organization is described as the adaptability-seeking behavior of a complex system and not a process for constituent systems coming together to form a SoS. Beyond equilibrium means a system does not reach a steady state or tend to a final state of equilibrium. The final characteristic of attractors indicates new qualitative levels of the emergent feature. Although we have not argued for these attributes in a complex SoS, it is not difficult to see how they may be present. Nonetheless, these are elements of emergence worthy of analysis in the future.

Incidentally, the species example we gave in our earlier discussion about system attributes also illustrates the different types of emergence that have been identified in the literature: the first expressed by the goals of the SoS being more than the sum of the goals of the individual constituent systems [17], the second being characterized by the patterns of organization among the SoS' constituent systems [25] in that organisms dynamically connect to form reproductive pairs, family groups, herds or pods or schools, races, and subspecies, and sometimes the subspecies break off to form entirely new species. Species also exhibit the third type of emergence, in which the complex SoS develops behaviors or results that are not predicted by the behavior of the constituent systems [18, 29]. In this case, we say that the species is "evolving."

Having established the criticality of emergence to the definition of complexity in systems, we additionally note that emergence appears to be a consequence of the interactions among systems, and indeed, that the existence of emergence is the result of particular types of interactions: asynchronous, transient, and situation dependent.

Thus, Casti [30] makes several assertions about complex systems that we believe are central to the successful pursuit of our work, namely: (1) emergent behavior is the defining evidence of a complex system (as just presented); (2) the action of connections within a system is key to the emergence; (3) complex problems (read "systems") are irreducible, the essential properties of the complex system residing in the whole. In a similar vein, Rouse [31] summarizes the nature of complex systems by identifying four characteristic parameters (paraphrased here): multi-level architecture, uncertain state transitions, non-linear dynamics, and agent based emergence. The last three reflect different forms of emergence, while the first is captured implicitly in our diversity and autonomy Eqs. (1) and (2) in Table 1. The summary views given by these authors are consistent with each other and will serve as a foundation for our further approach to the definition of emergence.

It would appear that the behavior of static versus dynamic information flow via connectivity, and the relationship between the forms taken by the quantities c_{ij} in the matrix C* of Eq. (4) are key to characterizing emergence. There is a tantalizing sense that the existence of dynamic information flow in a system is what creates the environment in which emergence can occur, as suggested by Baker [32]. There seems to be something about the dynamic, asynchronous and unpredictable flow of digital data across the interfaces among systems (and even among the subsystems inside a sufficiently rich digital system) that triggers unforeseen behavior. It appears that the dynamic flow via connectivity may be capable of accounting for all the different types of emergence described in the literature.

Based on the above discussion, we expect that the proposed mathematical expression for emergence (hence complexity) will be based on the previously given Eq. (6). That expression captures the Rouse [31] requirement for multilevel architectural structure (recall that we have said the individual cells in the matrix B could in and of themselves be matrices). In addition, the form of the matrix itself captures the dynamic nature of connection among the component systems, allowing us to describe the uncertain transition states proposed by Rouse [31], particularly if B is instantiated as B(t). Finally, by making the individual b_{ij} non-linear, perhaps through one of the techniques suggested in Baldwin et al. [3], we can also capture the non-linear dynamic nature of Rouse's [31] third criterion. We expect that a successful implementation of this strategy would result in emergent outcomes if, for example, a system were modeled using such an instantiation of Eq. (6) as the foundation for the model.

This expectation is in part because conclusions about complexity in models, as described in Edmonds [33], are fully applicable to systems since models are representations of the phenomena they model. Edmonds has spent a couple of decades struggling with the concept of emergence in complex systems, mostly from a philosophical point of view. His conclusions regarding modeling of complex systems are right on point when it comes to understanding the functioning of these systems. Indeed, the conclusions from his 2013 paper should serve as a basis for further work in the development of systems engineering theory, since physical theories are in essence contextual models of systems. At least Edmonds' conclusions about models are *prima facie* applicable to systems themselves.

5.2 Epistemology Versus Ontology

It remains an open question whether emergence can actually exist in a system made up entirely of engineered components. Edmonds [34] has eloquently posed this as the question of emergence versus ignorance. In other words, is an unpredictable result truly a feature of the complex system, or merely an artifact of our lack of understanding? Rouse and Bodner [35] restate this principle as the idea of "perceived" versus "real" complexity. As we learn more about complex systems, in some cases, they become less complex.

This question is the classical philosophical problem of ontology versus epistemology, and it has dogged the concept of emergence from the beginning. Baker [32] has suggested that a fundamental linkage between complexity and connectivity can be found in the philosophy of structural realism [vide 36], a line of inquiry that traces back to Poincaré and thence to Immanuel Kant. It has its roots in the question of whether we can ever actually know anything about the "real" world, which is just another way of stating the epistemology versus ontology question. The potential application of this suggestion to our study of complex systems is attractive because it would provide the necessary theoretical link among the concepts of complexity, emergence and connectivity.

In McIntyre [37], a choice is presented between an ontological and an epistemological perspective on complexity. McIntyre chooses the epistemological view, and makes a compelling case. However, it is possible that EITHER choice, on the surface of it, is wrong. Why? Because complexity, at least when we are looking at complex systems, is a feature resident in the architecture of the system itself, and in this sense is patently ontological (i.e. there is a real, physical "cause" for the complexity). However, to view the problem as ontological, in McIntyre's view, is to say that we will never be able to understand it! On the other hand, to accept the contrary result that complexity is primarily an epistemological phenomenon forces us to conclude that complexity is only "perceived" but not "real."

The underlying assumption is that the word "complex" describes something "too complex to understand." However, there is a different potential point of view in that the features of complexity that irks us, namely the emergent behaviors that are not explicable at the level of the individual component parts, are in fact resident in the reality of the system. In other words, they are ontological phenomena. On the other hand, we do have a problem of epistemology in the sense that we do not yet have a theory strong enough to explain the observed phenomena of complexity. The problem is similar to the fluid dynamics problem of modeling turbulent flow. Although the flow is "real" enough, it can never be explicitly described on a particle by particle basis. In the end, we may just be stuck with a similar sort of conclusion with regard to emergent behavior in complex systems.

We note here that the literature relevant to the application of complexity theory to management, as exemplified by e.g., Jackson [38], Johnson [39], and San Miguel et al. [40] derives from many of the same sources discussed here and in Rouse and Bodner [35]. Since these practitioners are more interested in the application of system theoretic concepts to management rather than in the derivation of new fundamental concepts, it is small wonder that they find the concepts of complexity as found in the philosophy of science literature of little value. By the same token, the concepts useful to the management literature have limited utility for our purposes.

6 Conclusions

To this point, we have developed an internally consistent set of descriptions that can be used to analyze the characteristics of a system of interest in order to determine whether or not it is a simple system, a composite system of systems (SoS), or a complex SoS. We know that if the latter is true, then we need to take specific design steps to mitigate the impact of the eventual, unpredictable but inevitable emergent behavior. For clarity, we define a complex system as equivalent to (the same thing as) a complex SoS and we define a "complex" system as one that exhibits emergence in the third sense (that is, not as overarching goals, and not as patterns of connectivity, but only as unpredictable results). A collection of systems that does not exhibit these traits, we will call a "composite" system. There is a time dimension to all of this, which we have not addressed in this paper. One of the issues alleged with complex systems is that SoS goals, G*, change over time as a consequence of emergence. Although we have discussed the ways in which emergence may be characteristic of complex system behavior, we leave the development of this idea for a future project.

In extending our previous work on the taxonomy of SoS types, we have shown that we have been able to construct a set of four set theoretic equations expressing the essential attributes used in constructing the taxonomy. These mathematical expressions are complete; they follow from the accepted verbal descriptions commonly used to denominate system autonomy, diversity, connectivity, and belonging. They are consistent with the set of attributes evolved by Boardman and Sauser [8], and together they provide a complete description of the taxonomy of systems up to but not including complex SoS (i.e. those that we have defined as expressing the attribute of emergence). Furthermore, the four expressions follow logically from each other, are internally consistent as a group, and are based on the same mathematical field (i.e. set theory). The use of set theory to describe system attributes is also consistent with the work of others, and shows promise as a tool for the description of emergent system features. Finally, we have used the expression for belonging to analyze a case involving a realistic (though simplified) air traffic control system example. While the example is simple, it is not simplistic. The results from this analysis are realistic, and provide insight into the manner in which constituent systems collaborate to meet SoS level objectives in a composite SoS. Additionally, it is clear from this first use of the tool that there are a number of potentially fruitful elaborations of the approach that may result in interesting conclusions about how emergence is enabled by the character of the information exchanged among constituent systems in a SoS over time. We look forward to continuing the evolution of this line of inquiry, with the expectation of further interesting developments.

In a particularly insightful article published in *Nature* in 2001, John Casti has made the point that "…one can only wonder when the… informal idea of 'complex' will be formalized into a rich mathematical theory of complex systems" [1].

This statement is a call to action which has yet to be satisfied and which has been forcefully advocated on numerous occasions by former NASA Administrator Mike Griffin, who has said that "System Engineering is not a discipline: it's not a discipline because we have no theory," [41] and more recently, that we must pursue "the mathematization of relationships among disciplines in complex systems," [42]. We look forward to the continued pursuit of this worthy goal.

7 Addendum: An Additional Example from the First Responder System

In the Air Traffic Control example given earlier, we stress the importance of connectivity and belonging, but let us look at one additional example to demonstrate how an understanding of the four described characteristics improves our understanding of the complexity of a SoS, or rather the presence of emergence. A generic first responder system is a composite system, with the goal of managing emergency incidents in order to preserve lives and property during a crisis [43]. According to the U.S. Federal Emergency Management System (FEMA), the goals of a first responder system are "to prevent, protect against, respond to, recover from, and mitigate the effects of incidents, regardless of cause, size, location, or complexity, in order to reduce the loss of life and property and harm to the environment" [44].

Consider three constituent systems of a first responder system, which represent law enforcement, a fire fighting department, and an emergency medical service. Autonomy can be tested for any one of these systems. For example, law enforcement performs several actions to accomplish their goal of preserving the peace. These actions include maintaining order, providing medical assistance, and enforcing applicable laws. Since the goal is completely accomplished by the actions, the law enforcement system is autonomous. Similar arguments can be made for the fire department and emergency medical service. Belonging can be tested accordingly as these three constituent systems contribute to each other's goals as well as the overall first responder system's goals in exchange for assistance with their goals. For example, one goal of the first responder is to reduce the loss of life. In that case, law enforcement supports the first responder goal by offering basic medical assistance, and the first responder system saves property from disaster, which helps law enforcement maintain order.

Connectivity among the constituent systems takes many forms, not the least of which is communication. First responder systems work hard to ensure valuable information is dynamically communicated to the appropriate systems. The communication may be person-to-person, direct via radio, indirect via a dispatcher on the radio, or even some acoustical system such as whistles or horns.

Each constituent system has its own individual goals, such as preserving the peace for law enforcement and mitigating hazardous situations for fire fighters. This uniqueness of goals qualifies the constituent systems as diverse. Hence, a first responder system has autonomous and diverse constituent systems that exhibit belonging and connectivity via sufficiently dynamic information exchange. The result is a first responder system with emergent capabilities to reduce the loss of life, loss of property, and harm to the environment during any emergency. The emergent features of the SoS cannot be traced to constituent systems. For example, every constituent system in the first responder SoS contributes to the preservation of life, property, and the environment, but no one system alone achieves these objectives, except in very limited ways. The overall effect comes from the interaction among the constituents in a way that cannot be allocated to the individual systems.

References

1. Casti, J. L. (2001). Formally speaking. *Nature, 411*, 527.
2. Baldwin, W. C., Felder, W. N., & Sauser, B. J. (2011). Taxonomy of increasingly complex systems. *International Journal of Industrial and Systems Engineering, 9*, 298–316.
3. Baldwin, W. C., Ben-Zvi, T., & Sauser, B. J. (2012). Formation of collaborative system of systems through belonging choice mechanisms. *IEEE Transaction on Systems, Man, and Cybernetics Part A: Systems and Humans, 42*, 793–801.
4. Baldwin, W. C., Hostetler, S., & Felder, W. N. (2013). *Mathematical models of emergence in complex systems-of-systems*. 23rd Annual INCOSE International Symposium. INCOSE, Philadelphia, PA, pp. 1–15.
5. Felder, W. N., & Collopy, P. (2012). The elephant in the mist: What we don't know about the design, development, test and management of complex systems. *Journal of Aerospace Operations, 1*, 317–327.
6. Felder, W. N. (2013). *Interactions among components in complex systems*. AIAA Complex Aero-space Systems Exchange.
7. Keating, C. (2009). Emergence in system of systems. In M. Jamshidi (Ed.), *System of systems engineering: Innovations for the 21st century* (pp. 169–190). Hoboken, NJ: Wiley.
8. Boardman, J., & Sauser, B. (2006). System of systems—The meaning of of. In *Proceedings of the 2006 IEEE/SMC International Conference on System of Systems Engineering*. IEEE, Los Angeles, CA, pp. 118–123.
9. Ackoff, R. L. (1971). Towards a system of systems concept. *Management Science, 17*, 661–671.
10. Maier, M. W. (1998). Architecting principles for system-of-systems. *Systems Engineering, 1*, 267–284.
11. Baldwin, W. C., & Sauser, B. (2009). *Modeling the characteristics of system of systems*. 2009 I.E. International Conference on System of Systems Engineering (SoSE). IEEE, Albuquerque, NM, pp. 1–6.
12. Von Bertalanffy, L. (1950). An outline of general systems theory. *British Journal for the Philosophy of Science, 1*, 139–164.
13. Stein, S. (1993). *Noah's garden*. New York: Houghton Mifflin Company.
14. Minai, A. A., Braha, D., & Bar-Yam, Y. (2006). Complex systems engineering: A new paradigm. In D. Braha, A. A. Minai, & Y. Bar-Yam (Eds.), *Complex engineered systems: Science meets technology* (pp. 1–22). New York: Springer.
15. Anderson, P. W. (1972). More is different. *Science, 177*, 393–396.
16. Miller, J. H., & Page, S. E. (2007). *Complex adaptive systems: An introduction to computational models of social life*. Princeton, NJ: Princeton University Press.
17. Ablowitz, R. (1939). The theory of emergence. *Philosophy of Science, 6*, 1–16.
18. Pap, A. (1952). The concept of absolute emergence. *The British Journal for the Philosophy of Science, 2*, 302–311.
19. Dorigo, M. (2007). Editorial. *Swarm Intelligence, 1*, 1–2.
20. Myerson, R. B. (1997). *Game theory: Analysis of conflict*. Cambridge, MA: Harvard University Press.
21. Newman, M. (2010). *Networks: An introduction*. Oxford: Oxford University Press.
22. Baldwin, W. A. (1903). *Industrial-social education*. Springfield, MA: Milton Bradley Company.

23. DeLaurentis, D. A., Marais, K., Davendralingam, N., Han, S. Y., Uday, P., Fang, Z., et al. (2012). *Assessing the impact of development disruptions and dependencies in analysis of alternatives of system-of-systems*. Systems Engineering Research Center.
24. Crutchfield, J. P. (2012). Between order and chaos. *Nature Physics, 8*, 17–24.
25. Crutchfield, J. P. (1994). *The calculi of emergence: Computation dynamics, and induction*. Santa Fe, NM: Santa Fe Institute.
26. Weaver, W. (1948). Science and complexity. *E:CO, 6*, 65–74.
27. Goldstein, J. (1999). Emergence as a construct: History and issues. *Emergence, 1*, 49.
28. Goldstein, J. (2013). Complexity and philosophy: Re-imagining emergence: Part 1. *Emergence: Complexity and Organization, 15*, 78–104.
29. Pepper, S. C. (1926). Emergence. *The Journal of Philosophy, 23*, 241–245.
30. Casti, J. L. (1986). On system complexity: Identification, measurement, and management. In J. L. Casti & A. Karlqvist (Eds.), *Complexity, language, and life: Mathematical approaches* (pp. 146–173). Berlin: Springer.
31. Rouse, W. B. (2003). Engineering complex systems: Implications for research in systems engineering. *IEEE Transactions on Systems, Man, and Cybernetics Part C: Applications and Reviews, 33*, 154–156.
32. Baker, A. (2013). Complexity, networks, and non-uniqueness. *Foundations of Science, 18*, 687–705.
33. Edmonds, B. (2013). Complexity and context-dependency. *Foundation of Science, 18*, 745–755.
34. Edmonds, B. (1999). *Syntactic measures of complexity*. Dissertation, Manchester, UK: University of Manchester.
35. Rouse, W. B., & Bodner, D. A. (2013). *Multi-level modeling of complex socio-technical systems—Phase 1*. Hoboken, NJ: Stevens Institute of Technology.
36. Ladyman, J. (1998). What is structural realism? *Studies in History and Philosophy of Science Part A, 29*, 409–424.
37. McIntyre, L. (1997). *Complexity: A philosopher's reflections*. Santa Fe, NM: Santa Fe Institute.
38. Jackson, M. C. (2003). *Systems thinking: Creative holism for managers*. Chichester: Wiley.
39. Johnson, J. (2010). The future of the social sciences and humanities in the science of complex systems. *Innovation: The European Journal of Social Science Research, 23*, 115–134.
40. San Miguel, M., Johnson, J. H., Kertesz, J., Kaski, K., Díaz-Guilera, A., MacKay, R. S., et al. (2012). Challenges in complex systems science. *European Physical Journal-Special Topics, 214*, 245–271.
41. Griffin, M. D. (2011). *Keynote address*. AIAA Complex Aerospace Systems Exchange 2011. Pasadena, CA.
42. Griffin, M. D. (2014). *Address to the school of systems and enterprises advisory board, Stevens Institute of Technology*. Washington, DC: School of Systems and Enterprises Advisory Board.
43. Baldwin, C., Li, Q., Sauser, B., & DiMario, M. (2010). *Simulating a first responder scenario*. 8th Annual Conference on Systems Engineering Research. Stevens Institute of Technology, Hoboken, NJ, pp. 465–473.
44. Homeland Security, U.S. (2008). *National incident management system*. http://www.fema.gov/emergency/nims/

So It's Complex, Why Do I Care?

Steven Holt, Paul Collopy, and Dianne DeTurris

1 Introduction

The subject of this chapter is systems engineering and how it is impacted by complexity. Systems engineering is a transdisciplinary process that engineering organizations use to develop aircraft, spacecraft, radars and other similar large systems. Prior to systems engineering, an informal process existed for designing engineered objects, and the process worked well for individuals or teams who designed light bulbs, electric motors, and even ocean liners. However, space launch rockets and jet propelled aircraft could not be designed by individuals, or even small teams. These are systems with many components, and different engineering specializations are necessary to design different components. Designing a modern aircraft requires whole organizations of engineers that includes many teams.

Systems engineering seemed effective in the 1960s. Apollo landed on the moon. The Concorde was designed in the late 1960s, and eventually carried passengers at Mach 2 across the Atlantic Ocean. The Boeing 747 and the Mach 3 Lockheed SR-71 were also designed in the late 1960s. Even the sky was no limit.

However, the record of systems engineering achievement from 1990 to the present has been less impressive. Although the Curiosity rover has actively explored Mars and the Hubble Space Telescope has set records for generating

S. Holt (✉)
The Boeing Company, Seattle, WA, USA
e-mail: steven.c.holt@boeing.com

P. Collopy
Department of Industrial & Systems Engineering & Engineering Management, University of Alabama in Huntsville, Huntsville, AL, USA

D. DeTurris
Aerospace Engineering Department, Cal Poly State University, San Luis Obispo, CA, USA

F.-J. Kahlen et al. (eds.), *Transdisciplinary Perspectives on Complex Systems*,
DOI 10.1007/978-3-319-38756-7_2

terabytes of scientific data, and even though the safety record of modern jet airliners is the envy of the transportation industry, today's process for developing large engineered systems is problematic. Cost overruns are so common that they are actually expected, and the average overrun is around 50 %. Most projects finish years late, if they finish at all. One third of large US defense projects (from 1990 to 2007) and one half of NASA human space exploration projects (from 1991 to 2011) have been cancelled.

Why has systems engineering become so difficult? The usual explanation is that modern engineered systems are complex. This is descriptively informative, but the practicing systems engineer needs a pragmatic guide to assessing complexity, and based on that assessment, taking an appropriate approach to reduce, mitigate, adapt to, or exploit complexity and bring large projects to successful conclusions.

So, why do I care? One must care whether the system is complex, because complex systems are not predictable. Contemporary systems engineering is founded on requirements, which are essentially predictions of the outcome of the system design and development process. The difficulties in large system development programs today are that essentially, the system fails to meet its requirements. Cost and schedule overruns are mere symptoms, as programs attempt to recover or patch over the gap between what the system was predicted to do and what it actually does. In this sense, the root of system development difficulties is the prediction, via requirements, of the outcome of a process that is fundamentally not predictable.

In this chapter, the management of complexity is explored as a transdisciplinary element of system design. That is, complexity does not belong to mechanical design, software engineering, or human factors. Instead, a design can be characterized as complex or not complex from any or all of these perspectives.

At the same time, management of complexity is not itself a hard science like dynamics or electromagnetics. It is barely quantitative, and perhaps only qualitative. But even as a qualitative measure, appreciation of the complexity of a system provides the design organization with a warning of approaches to avoid and perhaps some positive steps toward more successful outcomes.

For example, this chapter recommends the Cynefin Framework as an approach to system development that will accommodate emergent behaviors which are characteristic of complex systems. Small, safe to fail experiments are suggested for learning about a system that defies the best attempts to predict behavior.

The purpose of this chapter is to develop the beginnings of a prescriptive approach to systems engineering for complex engineered systems. It shows that complexity inherently means system behaviors will occur that cannot be predicted. The standard, rational approach of taking actions that leads to the desired outcomes will not work on complex systems, because it is not clear where actions lead. This inability to predict system attributes and behavior necessitates a re-examination of the fundamentals of systems engineering practice. This re-examination ultimately must entail not only complexity science, but also cognitive science, social psychology, organizational sociology, and the physics of the designed artifact.

1.1 *The Rising Cost of Aerospace Systems with Complexity*

The design of large engineered systems today is increasingly difficult because of the level of complexity in the systems that are designed. For example, in the spring of 1939, the aircraft manufacturer Consolidated received a contract to develop a prototype aircraft for what became the B-24 Liberator long range bomber. The aircraft began service in 1941. In the fall of 1986, Boeing and Lockheed received contracts to develop competing prototypes for the aircraft eventually designated the F-22 fighter. The aircraft began service in 2005. The progression of development time increased from 2 years to 19 years between these programs. The F-22 development cost is roughly 1000 times the comparable B-24 cost. So what changed? The usual answer is that the F-22 is more complex.

The historical trend of increasing cost with increasing complexity can be seen in Fig. 1, developed by Paul Eremenko at the Defense Advanced Research Projects Agency (DARPA). This historical schedule trends chart compares the linear growth of military aerospace development time with complexity with the flat trend of automobiles and the declining trend in the semiconductor industry. The inset box on the chart also notes the large exponential growth in development cost, doubling every 7 years in real dollars, compared to automotive development where real cost doubles every 20 years, and semiconductors, where real cost has not grown at all [1]. Something about the design of aerospace systems causes system development

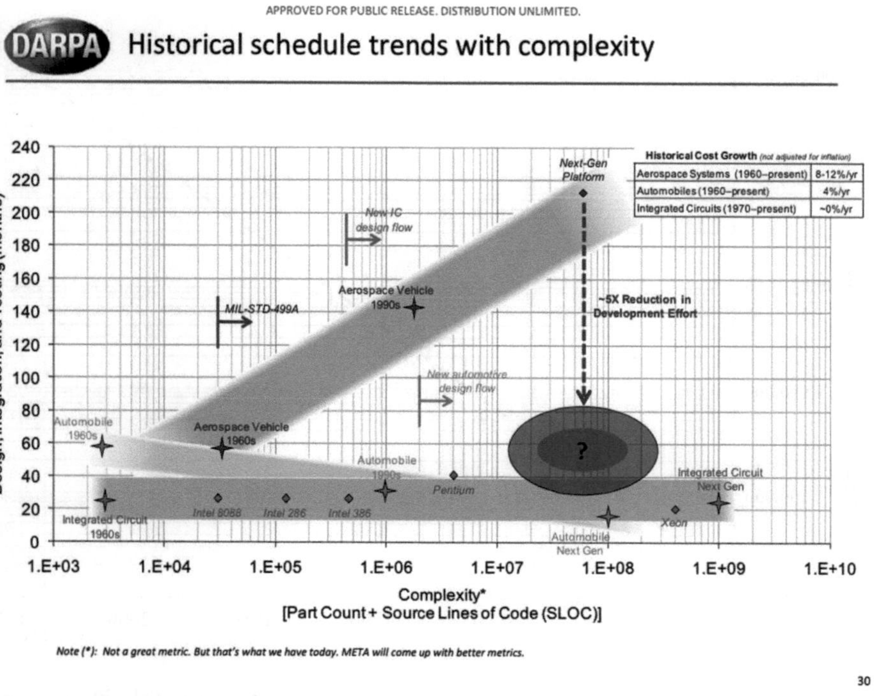

Fig. 1 Historical schedule trends with complexity for aerospace, automobiles and integrated circuits [1]

to become catastrophically more difficult (as reflected by growth in cost and development time) as the systems themselves become more complex.

The growth of development cost parallels the growth in production cost. Norm Augustine's 16th Law puts this very provocatively,

> In the year 2054, the entire defense budget will purchase just one aircraft. This aircraft will have to be shared by the Air Force and the Navy 3.5 days per week, except for leap year, when it will be made available to the Marines for the extra day. [2]

This exponential cost growth, first observed in 1983 but updated in Fig. 2, is a phenomenon that has continued to the present.

Why did Augustine's graphs not have the ultimate effect of changing the status quo? Perhaps the answer lies in the complex nature of modern systems. The complexity slows down the project and increases the costs until it gets to a point that no amount of development time can pull it off. A good example of what can happen is the Army's Future Combat System (FCS), the largest and most ambitious planned acquisition program in the Army's history, which was cancelled early in development [3]. The system of systems approach to FCS included challenging software, integration, and life-cycle components [3].

FCS is not an outlier. In fact, the cancellation is an outcome we have come to expect. A value analysis of large Department of Defense development programs measured the combined impact of overruns, delays and cancellations as a loss of $200 million per day on average over the past decade [4]. Accounting for complexity is a way to address these issues.

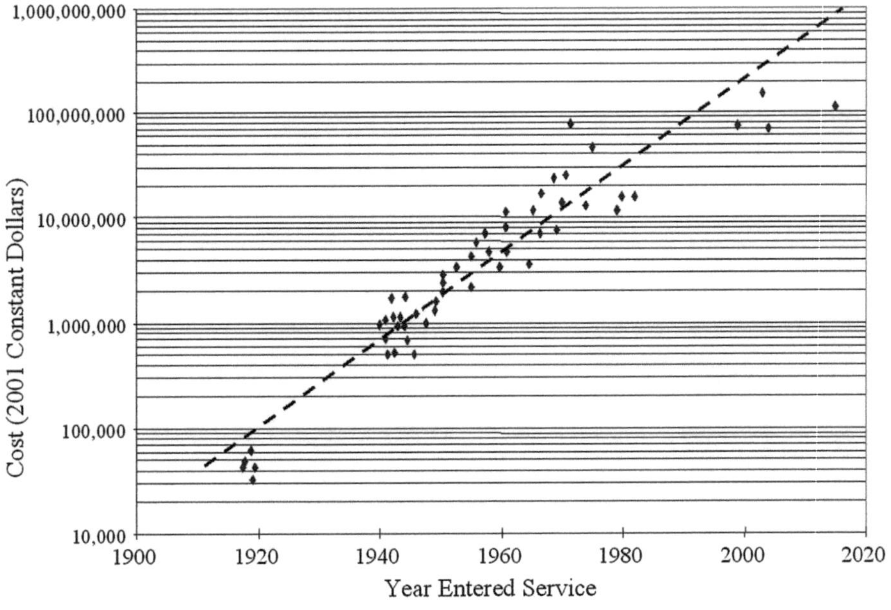

Fig. 2 Augustine's trend of fighter aircraft production cost versus time, updated with contemporary aircraft, and adjusted to use constant dollars. Augustine observed similar relationships for other military aircraft, ships and ground vehicles, and commercial aircraft

1.2 *What to Do About Complexity*

The view that systems engineering is very difficult on modern large systems because they are so complex seems reasonable. However, the practicing systems engineer needs more than a truism. He or she needs a pragmatic guide to assessing complexity, and based on that assessment, an appropriate approach to reduce, mitigate, adapt to, or (better yet) exploit complexity and bring large projects to successful conclusions. A complete roadmap from complexity to assessment to approach does not yet exist, but this chapter will discuss the beginnings of such a strategy. Note that some of this roadmap will lie outside of the realms of study normally considered to be the Systems Engineering Body of Knowledge. Consequently, the final product will certainly be transdisciplinary. Several of the methods discussed here are examples of that.

Discussions of complexity often revolve around emergent properties or behaviors of the system. Emergence is considered by many theorists to be the defining characteristic of complexity. There is considerable debate as to whether such emergent properties are objective or subjective. Are we surprised by system behaviors (and therefore label them as emergent) because we do not know the cause of the behaviors, or because the cause is somehow unknowable? This chapter takes the position that it does not matter. Whether system behavior is unpredictable because of ignorance or some deeper nature of the system does not matter—what does matter to the designer is the unpredictability itself, and the way unpredictability interferes with traditional system design and development.

2 Complexity

Before ways to deal with complex systems can be suggested, it is important to survey how complexity and complex systems have been defined and what techniques and methods have been tried to cope with it. The techniques and methods referenced here are not the result of rigorous analysis; they are heuristics developed over time, based on what seemed to be effective. In that regard these approaches tend to be non-hypothesis based methods.

2.1 *A Survey of Complexity Definitions*

Complexity lacks a single definition. In fact, it has a multitude of definitions. Seth Lloyd came up with forty or so [5]. Melanie Mitchell describes hosting a panel discussion between esteemed experts for the Santa Fe Institute's Complex Systems

Summer School in 2004. The first question to the panel was, "How do you define *complexity*?" None of the panelists' definitions matched and some of them disagreed with each other [6]. The definitions that exist range from the highly formal to the informal, from the mathematical to the non-mathematical, from those based on trying to measure the amount of complexity to those who see it as unpredictable and, hence, unmeasurable. And yet, several categories of definitions exist.

Scott Page defines complexity as a collection of diverse, connected interdependent entities whose behavior is determined by rules, and which may adapt, but need not. The interactions of these entities often produce phenomena that are more than the sum of the parts. These are called emergent phenomena, and most complex systems are therefore not predictable [7].

One characteristic of complexity is that it lies Between Order And Randomness (hence, the BOAR complexity model), and another is that it cannot be easily Described, Evolved, Engineered or Predicted (the DEEP complexity model) [7].

Melanie Mitchell extends the definition beyond "complex" or "complexity", suggesting that a complex system is one that exhibits nontrivial emergent and self-organizing behaviors [6].

A valuable adjunct to our definitions of complexity and complex systems is a description of the behaviors or attributes of complex systems since these are what we normally encounter when dealing with complexity. David Snowden and Mary Boone [8] describe complex systems as having the following characteristics:

- Large number of interacting elements
- Non-linear interactions such that seemingly minor changes can produce significant consequences
- Solutions emerge from dynamic circumstances, that is, they are emergent
- Elements of the system evolve together in irreversible ways
- Hindsight cannot lead to foresight because conditions and the system constantly change
- The system and the agents operating within it constantly operate on each other in ways that are unpredictable

Many of these attributes contribute to the unpredictable characteristics of complex systems. Large numbers of interacting elements means it is difficult or impossible to predict how a change in one element will impact another. And, the interactions can be highly non-linear; a small action may have a large effect and vice versa. These interactions vary with time so semi-stable states or solutions will often evolve and emerge over time in ways that could not be predicted in advance. And, once they have emerged, things will not go back to the starting position.

Paul Cilliers uses the wonderful example of making mayonnaise when he said, "a jumbo jet is complicated, but a mayonnaise is complex" [9]. That is, once built, each part of an airplane can be taken apart, analyzed, and then put back together. The result will still be an airplane. A person confronted with the constituent ingredients of mayonnaise would be hard pressed to predict what the result of beating them all together would be and, once created, the mayonnaise cannot be "unassembled."

Cilliers' use of the words "complicated" to describe an airplane and "complex" to define mayonnaise also highlights one of the challenges of creating specific definitions for words that are in common English usage. Unfortunately, the words "complex," "complicated," and "chaotic" are often used as synonyms in common conversation. "It's complex," "It's complicated," and "It's chaotic" can all describe the same situation. If you ask someone to define either complex or complicated you'll find that it's often difficult for them to define complex without using the word complicated and vice versa. This will be described more fully in the section covering the Cynefin Framework, but for our purposes, a complex system is inherently unpredictable while a complicated system could be predicted through analysis.

Complex systems are highly dependent on even tiny variations in starting conditions; consequently, you can never exactly replicate the same results. Consider, for example, a baseball game. The same two teams could play games on consecutive days and the games will not be identical even if it's the same players on the same field in the same weather conditions using the same equipment and game strategies. It's the inherent complexity that makes sports popular. If every game were completely predictable, sports would be dull.

2.2 *The Consequences of Complexity*

Part of the reason that so many definitions of complexity have been suggested is that there are many and varied aspects that appear to exist in complex systems. Consider, for example that complexity descriptions can include:

1. Technical complexity: Technologies can interact and integrate in ways that are difficult to determine in advance.
2. Business system complexity: Large organizations with multiple interrelated organizations or complex international supply chains can create complex relationships.
3. Managerial or organizational complexity: Similar to business systems complexity. A single entity developing a product with their own funds is significantly different from a project with a mix of private and government funding from multiple sources that may have different (and possibly competing) goals and objectives.
4. Scale complexity: Sometimes new ideas are piloted successfully in small experiments and yet fail when scaled up to a large organization or enterprise.
5. Coupling complexity: Describes the degree in which things are related to each other. The greater the number of interacting agents the more difficult it is to predict the results.

6. Interactive complexity: The amount of time or slack between interactions of elements in a system. A trigger mechanism for chaos is the inability to have time to reflect; the plan is moving forward all the time, people don't stop to look at what happened or go to another route.
7. Cognitive complexity: There are simply too many details and too many connections for any one person to fully understand the system.

People who must deal with complexity often focus their attention on one or more of these types of complexity. This can often be the result of the real or perceived causes of past problems. If a past program took on too much technical complexity ("invention on the critical path") for instance, a current program leader may focus significant amounts of attention on dealing with technical complexity…and completely miss a scaling or organizational problem.

Experience has taught people a series of rules of thumb, or heuristics, to try to avoid the problems often associated with complex systems development. Unfortunately, experience alone can be a notoriously poor teacher and may simply reinforce our own biases as to why things are happening and what to do about them [10]. Worse yet, our biases can decrease our ability to detect true root causes such that we continue to react inappropriately in our attempts to successfully manage complex systems.

Consider, for example, large projects of the type that are often government funded since they are so large as to all but require state-level funding. In his 2003 book *Megaprojects and Risk*, Bent Flyvbjerg describes this phenomenon by showing that huge infrastructure building projects are frequently not economically viable despite reams of initial analysis showing they were. The political nature of these infrastructure projects leads to exaggeration and distrust about the facts. Ultimately, the politicking threatens to bankrupt a company or even a country. Although projects suffer poor performance, the need for these infrastructure capabilities dictates that the projects not be canceled. Flyvbjerg suggests an approach specifically created to overcome the weaknesses of the conventional approach in megaprojects by emphasizing risk, institutional issues, and accountability [11].

Sometimes an appeal is made to decrease the complexity as a way to manage successfully. Consider an example frequently used by Dr. Eli Goldratt (creator of the Theory of Constraints). He drew the shapes shown in Fig. 3 and then asked the question, "Which is more complex?" [12].

Most people say that System B is more complex. Their sense of complexity has to do with the level of difficulty or the number of words needed to define the system (A version of Kolmogorov or Descriptive Complexity). Goldratt would say that, as a physicist, System A is more complex because it has many more degrees of freedom and it would be extremely difficult to predict how any input would influence the system since we have no information on how the circles are related. In contrast, we know exactly how any change in any element in System B would impact the rest of the system (as indicated by the arrows between the circles). This is why Goldratt maintained that any complex system must have a sense of "inherent

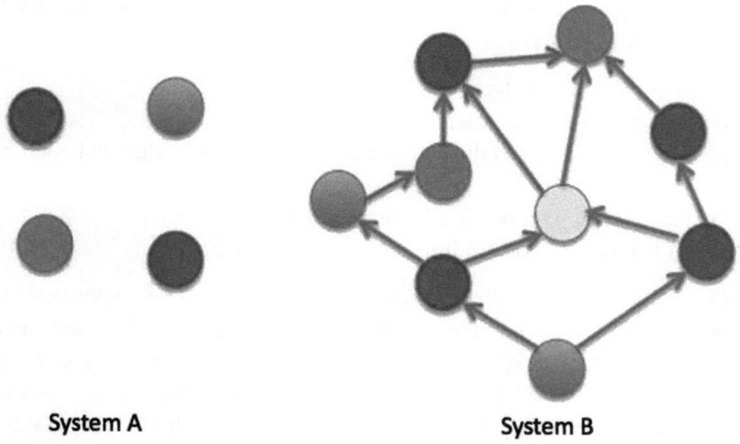

Fig. 3 Goldratt, creator of the Theory of Constraints, asks which is more complex, System A or System B? [12]

simplicity" behind it that would provide significant leverage for improvement. . .if we can find it.

A less mechanistic and more organizational approach is seen in the story of a software company in the US that set out to create a more empowered work force by flattening the organizational structure to essentially two levels: they had about a half dozen managers and 300 employees. The intent was that employees would interact with each other and self organize as required to accomplish tasks, solve problems, make improvements. What they discovered was that a new level of "phantom middle managers" began to appear. Initially these were simply people who had more knowledge about some aspect of the business than others and they gained a reputation for being helpful and informative. Essentially they became major nodes in the overall web of personal interaction. When this new phantom level began appearing, the actual management tried to break it up, hypothesizing that it would block the empowerment they were after. But they realized that this was a case where the right people, in the right positions, were being identified and organically promoted. It was essentially a successful, emergent management system that was not derived from any traditional means of designing organizations. And it fit the company and its culture.

Another way to think about adapting is through rapid, small scale testing of ideas. The approach described by Eric Ries in *The Lean Startup* [13] presents a new way of looking at the development of innovative products that emphasizes fast iteration and customer insight, a huge vision, and great ambition, all at the same time [13]. The approach can be applied both to startups and to internal teams of a parent company that act as startups. "If the fundamental goal of entrepreneurship is to engage in organization building under conditions of extreme uncertainty, its most vital function is learning" [13].

The lean startup method has five principles that accommodate the uncertainty found in complex systems:

- entrepreneurs are everywhere
- entrepreneurship is management
- validated learning happens through frequent experiments that test each element of the vision
- the experimentation is a build, measure, and learn activity
- innovation accounting, or holding innovators accountable

The Lean Startup approach is based on extensive use of experimentation to evolve products. But an organization following this approach does not do experiments just for the sake of experiments. The experiments are done by a small group for a short time and the results then inform the project. If the project team is smart about their experiments, they save months of ineffective time on assessment. It's not always known what is going to happen when the experiment is run, but the assumptions are set explicitly and then rigorously tested [13]. The intent is to maximize learning, not maximize failure as the key strategy. This sort of learning through tests and failures has often been interpreted as meaning that the goal is failures; it isn't. The goal is learning and often the most important lessons are with respect to what does not work.

3 Ideas to Build on

The issues raised in this chapter are not new, they have been puzzling the systems engineering and software engineering communities for decades. There is no final or complete answer, but a foundational approach is presented and some practical approaches to confront complexity in systems development can be useful.

Perhaps the most direct assault on the challenge of developing large, very complex systems is the Northrop Report from the Software Engineering Institute [14]. An important conclusion of Linda Northrop and her colleagues is that, when systems are very complicated, specifying requirements is no longer practical or even useful. Complex systems are defined by both the environment into which they will be inserted (or launched, or released), and the broad changes that are hoped for as the system interacts with its environment. What the system itself will be, what precisely is being built, is not knowable. It is not knowable ahead of time, and it is not precisely knowable even when the system is in the field. This unknowability is inherent when we say the system is "complex." What we need to do is face up to unknowability, and address the consequences.

To the extent that there are understood definite functions in such a system, they are conflicting and subject to rapid evolution. The components of complex systems are not a neatly ordered tree, they are heterogeneous, changeable and not-quite consistent. And, they will include fields of study beyond what is normally considered to be traditional Systems Engineering.

Felder and Collopy [15] identify more of the challenges facing systems developers today. The traditional systems engineering V or the waterfall process of functional decomposition, component design and test, and successive re-integration is impossible when systems have tens of thousands of components and the probability of any single component or interface working is less than 99 % [15].

We trusted systems in the past because we employed a strategy for verification testing that was ironclad:

- Test every system state
- Test every transition from one state into another
- Make sure each state and each transition behaves as intended

However, current systems, like the Airbus A350 or the F-35 fighter have a vast number of states and an astronomical number of state transitions. These cannot be tested. We cannot even test a representative sample of states and transitions, or test only the transitions that are most likely to occur. Instead, we are facing a future where the most likely type of system failure is a once-in-a-billion year occurrence that happens once and is never seen again. But there will be so many potential failures of this type that they can make systems utterly unreliable.

And yet, some large complex systems succeed. The landing of the Jet Propulsion Laboratory's (JPL's) rover Curiosity on Mars in August of 2012 was a triumph of modern engineering. JPL has accomplished several similarly stunning successes in the complex systems arena. Can we ask the Curiosity team how they did it, and follow their lead? Probably not. Donald Schön informs us that we will not find the answer by asking. When experts describe how they do what they do, they create an honest reconstruction of how they think their behavior must occur, but it is frequently very different than what they really do because expertise is generally unknowable to the expert [16]. Most expertise consists of tacit knowledge that is not expressed in words, concepts, or symbols. Much of what makes JPL successful is not describable by the people who work at JPL, and this means that their success cannot simply be transplanted to another organization.

One way to identify complex systems is by the nature of the developer. Individuals can develop simple and even complicated systems. A bicycle sits right on the boundary of what a person, working alone, can do well in development. If a good engineer devotes his or her career to understanding all the aspects of bicycle design, from metallurgy to tribology to dynamics to coatings, it is conceivable that they might be able to design and develop a very good bicycle all on their own.

For anything more complex than a bicycle, a team of designers is needed. No single person knows enough to design every minute part of an airplane. But the maximum effective team size is around 15 people.

Beyond 15 people, the team must be replaced by an organization, with roles and responsibilities, communication channels, and reporting relationships. This is the realm of complex systems development. Frederick Brooks managed one of the earliest very large development organizations, and wrote about it. He noted that communication presents a barrier to complexity, because once a design team requires dozens of designers, they cannot all communicate with each other—there is not enough time in the day for all that communication to take place. Therefore, people need to be partitioned into groups, and communication among the groups needs to be channeled. The result is that important ideas cannot always be adequately shared, and the prevalent fault becomes "I thought the system was being designed to behave like x, but in fact it did y" [17].

Research suggests that the optimal size for a team is around seven people [18]. An extended group can operate effectively with minimal structure as long as everyone trusts everyone else. However, the maximum size of a group where such trust can practically exist is about 15 people. Team size also has an effect on the value of transparency. In a small team (15 or less) transparency of who is working on what and their progress is a benefit since it allows frequent integration. Essentially it is the design equivalent of small batches or Single Piece Flow. Learning from each other can be rapid. But as the team grows beyond that level there is increasing loss associated with transparency. Increasing numbers of people mean increasing coordination requirements, which nearly always reduces the time available to spend on the design. And, because the trust limits have been exceeded, it means that people will be in the position of exposing their "failed" designs to people they do not know and do not trust. This makes coordination more difficult and slows down organizational learning.

The work of Dr. Robin Dunbar suggests that there is another organization size breakpoint at about 150 people. This is the approximate number of people that we can be acquainted with at one time. That is, in an organization of 150 or less, each person would know each of the other people. This is a Skunk Works type development environment, and groups like this have been very effective, successfully developing systems such as the F-117 fighter and the SR-71 very high speed reconnaissance aircraft. However, a complete modern aircraft is beyond the reach of a 150 person team. Neither the F-117 nor the SR-71 was adequately designed for mass production or military maintenance and support, and that is why both are retired today. To design a F-35 requires tens of thousands of people, not just 150.

Still, large organizations can be very effective at developing simple or even complicated projects. McDonnell Douglas developed the F-18 E/F models on time and on budget, and the derivative aircraft has been very successful. Boeing and Airbus have become very effective at developing derivatives of their existing aircraft models such as the Boeing 767–400 (a fuselage stretch of the 767–300).

3.1 Process of Design and Design of Process

Product development is a process that is filled with uncertainty. Part of that uncertainty depends on the product being designed and part depends on the process being used. One person believes that the key to a successful design is a complete set of requirements based on an upfront understanding of customer requirements so the way to start is with requirements. Another is convinced that customers don't really know what they want until they see it, so the best approach is to start designing and then share the results with customers to get their feedback. Experiential and anecdotal evidence suggests that both work sometimes, but not always. There's no way to know in advance which is the best approach in a particular case. And, no matter which approach is used, it seems that things happen that were never in the plan but must, nonetheless, be dealt with. In this situation, what's needed is a more transdisciplinary "sensemaking" approach—a way to help guide us despite having incomplete information.

The process of design can be thought of as starting with a blank slate and coming up with a product. Analysis, on the other hand, is evaluating a pre-existing idea, concept, or design. This seems to imply that analysis must start with something and that design can start with nothing. The reality is a bit different in that a designer is the product of their own background and will quite naturally use their existing knowledge and experience to craft their designs. A designer, then will not truly start from nothing, they will start, even without realizing it, from a set of baseline requirements, many of which seem obvious when stated: An aircraft must have some way of moving through the air. It likely has some sort of propulsion system, some form of guidance and control system. If it has to carry a pilot then an additional set of requirements come up. If the project is to design a commercial jetliner, then it must carry passengers, it will have jet engines, it must have a way of fitting into the transportation network, etc. Realistically this means that many design details are given, no matter what sort of commercial jetliner it is. Designers know about how big humans are, they know things about jet engines, their fuel, how to control them, etc.

Then, if we consider characteristics of our desired airplane like payload and range or which airports it is intended for, we can use a series of analytical calculations to determine design parameters. What's left is the truly unknown parts; the "blank slate" elements of design.

If we step back and consider that process, that means that there are some aspects of design that are essentially given, other aspects that can be analytically calculated and finally, some that must be developed in a sort of trial and error, iterative approach. The given aspects are the simplest, the analyzed portions the next hardest and the "invented" parts the most difficult. In an actual design effort, we may be doing parts of each of these domains simultaneously. This can lead to confusion, design churn and rework, and conflict.

Two approaches to providing some guidance in this situation are presented, the Cynefin Framework and methodology associated with Agile Software Development.

3.2 The Cynefin Framework

The Cynefin Framework (created by David Snowden, Cynthia Kurtz and others) [20] provides a sensemaking approach to help identify the degree of complexity we

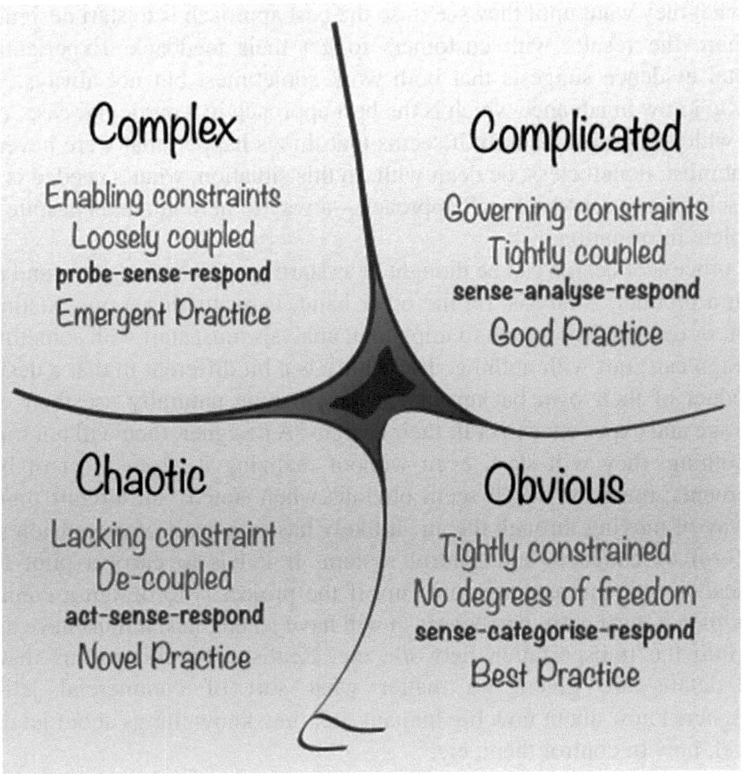

Fig. 4 The domains of the Cynefin Framework [20] (Image used with permission of David Snowden)

are operating in and determine an appropriate response. Knowing everything that is desirable to know is often not possible; sensemaking is a way to determine enough of what's going on to choose a reaction that matches the situation and increases the chance of success (or survival). As shown in Fig. 4, the Cynefin Framework includes a specific context for complex, complicated, obvious (simple) and chaotic systems.

The Cynefin framework is specifically illustrated with curved lines and no outer boundary to emphasize that it is not a 2 by 2 matrix and that the shape of the domains can change as our degree of understanding of the context changes. This is an important aspect of a sensemaking approach and is in contrast to a more deterministic categorization approach. An example of a categorization model is the 2 by 2 matrix used by Charles Perrow to illustrate Normal Accident Theory [19].

In general, the right side of the framework represents domains of Order in which cause and effect apply. The left side represents "unorder" in which cause-effect relationships are only evident in hindsight, if at all. The odd shape in the center represents Disorder, a state of not recognizing which domain you are in.

Obvious: The Obvious domain, previously referred to in literature as the Simple domain, describes a situation in which causality is clear and obvious to all observers. Consequently, the appropriate response is to follow a sense-categorize-respond approach. This is the legitimate domain of best practice [8].

Complicated: The Complicated domain applies where cause and effect still apply, but the relationship can only be determined by analysis or expert determination. Consequently the approach is to sense-analyze-respond. There is no longer a single best practice but there may well be several good practices [8].

Complex: The Complex domain describes a system in which a closed form solution is not possible. There are so many potential interactions in so many unknown ways that the observer cannot determine a "right" answer in advance. Consequently, the approach is to utilize probe-sense-respond. Interacting on the system with multiple experiments will lead to emergent practice. Note that after the fact, Complex domain interactions often appear as if they had been in the Complicated domain. This retrospective coherence, or 20/20 hindsight, frequently leads to the misperception that if only more analysis would have been done then the answer would have been determinable. This seems true, but it is not [8].

Chaotic: The Chaotic domain describes a system in which actions at the local level appear completely random. Not only is it not possible to analyze an approach (Complicated domain) but it is also not possible to develop coherent experiments (Complex domain). Consequently the appropriate approach is one of act-sense-respond which leads to novel practice. Novel practice sounds innovative and appealing until you realize that you can't plan on it ahead of time [8].

Disorder: The domain of Disorder describes the situation of not knowing which domain you are in. Within the Disorder domain, it is not uncommon for people to fall back on habitual behavior based on the domain they are most used to or most comfortable operating within. People who are experts in one problem solving method will, for example, tend to see all problems as solvable using their expertise.

The domain differences also suggest that the way that people operate most effectively within them should be different. And, as a project transitions from one domain to another the organizational structure should change to accommodate the new domain. The following are provided as general guidance for operating within the four main domains.

Obvious: For the cases in which there is one best way of doing things and everyone should be following it, the organization can have a strong, central authority overseeing the actions of hundreds or even thousands of people. Nearly everyone, for example, knows what a Stop sign looks like.

Complicated: In this domain experts and groups of experts must work together. This is where an organizational size limit of about 150 people makes sense. That allows them to all interact with each other while still having smaller groups of discipline experts.

Complex: In this domain the focus needs to be on relatively small, multi-disciplinary teams of people who trust each other, hence the limit of about 15. The group needs to be within the trust limits because emergence is expected and any of the team members might spot it first. Trust between team members is beneficial to ensure that everyone acts in the best interest of the overall team. Note that the size of sports teams and military units are usually a bit less than 15. This is primarily an experiential heuristic, but seems remarkably consistent in human activities.

Chaotic: Because operating in the Chaotic domain requires speed and aligned action the teams should be small, 5–7 people. This is approximately the number of items we can hold in short term memory and appears to be correlated to the number of people that forms the most effective crisis management team.

As an example of how to use the Cynefin Framework in deciding how to react to a given situation, consider the two modes of learning described by James March. Low-Intellect Learning describes the situation where you copy what someone else does and if it works, you don't need to know the theory behind why it works [21]. High Intellect Learning is when you do understand the theory such that you can modify the practice to match your context. When you first encounter a problem, getting a sense of which of the domains of the Cynefin Framework it falls in increases your chance of being able to use High Intellect Learning and reduces your chance of relying on the often less successful approach of Low Intellect Learning.

The first goal of sensemaking is to establish a common basis of reference; to avoid scenarios like the blind men and the elephant in which each person may have an equally valid and defensible position. Although everyone's observations are correct, their conclusions are not correct. In a complex product development system each of these observers may be in a different discipline. Ultimately the only way to recognize the whole, which is the elephant, is by taking a transdisciplinary view of it.

Although the general construct is referred to as the Cynefin Framework, when applied to a specific situation it becomes a model of that situation. The application is the model. An example of building a model from the context of a specific situation is using a technique called Cynefin Contextualization or Four Corners. Basically, a group of issues, concerns or statements describe a particular situation. It works well to have each on a separate sticky note and then have a large piece of blank paper, a tabletop, a wall, etc. Note that the temptation is to draw the domains or domain boundaries in and then place notes within the boundaries. But if the domains are drawn first and then the issues are fit into them, this becomes a categorization approach and not a sensemaking approach. It's more informative to place the issues first and then draw lines later. Instead of drawing the domains, think of the four corners of the blank paper as the extremes of the four main domains. Place the issues by picking them up one at a time and imagine that each sticky note has four rubber bands attached to it and to each of the four corners. Using the four corner conditions as guides, place the sticky note where it best seems to fit. Some will be more toward one side or one corner. Others will tend to clump near the center. Be alert for differences of opinion among team members as to where a particular note goes. If, for instance, one believes it should go in the lower right and another in the upper left, create a second note and place both of them on the page. Come back to these notes later because the difference may be a sign that people are looking at the same situation and seeing different situations. This is useful. Once all the issues have been placed, then the apparent domain boundaries can be drawn. Note that there may be more issues in one area than another, and there may be a number of issues in the center or Disorder domain, etc. Do not expect the result to be symmetric or the domains to be of uniform size. This context specific representation indicates how the domains are derived from the information (sense-making) as opposed to placing the issues into pre-selected domains (categorization). More detailed instructions can be found in Methods section of the Cognitive Edge website [22].

A related method is The Future, Backwards, which is akin to a cross between scenario planning and vision creation. It is primarily used to help learn how to recognize weak signals in a system. The process involves the creation of three simple state descriptions representing (1) the present, (2) a wonderfully positive future, and (3) a horribly negative future and then working backwards to create a stream of turning points. The turning point streams of the two potential futures intersect the turning point stream of the present in the past. This is a way of showing that all systems have already sown the seeds of their futures.

In practice, this method works well with multiple small teams working independently. It can be very effective if the members of each team are drawn from the same constituency (shared background, same organization, same role, etc.). This is ironic, since it is purposely NOT taking a transdisciplinary approach at this point as a way to call attention to the differences. Once the maps have been created, members of each group look in turn at the other team's maps to see what is the same, what is different and what is surprising about each of the other group's efforts. Frequently, people realize that the signals indicating whether an organization will tend toward a positive or a negative path are initially quite subtle. Also, it can be very informative to learn that one group's positive is another's negative [23].

3.3 Agile Development

As software became more of a part of daily life, software development programs grew increasing large and encountered complex system behavior. Due dates and budget commitments became increasingly more difficult to meet and user requirements were either difficult to determine or changed more quickly than code could be written. The Standish Group Chaos Report is often sited in industry to show that most software projects that are completed are over budget, late, and/or not what the users wanted [24]. The report also says that many projects are cancelled before completion. The initial reaction was to apply linear, deterministic systems engineering and more detailed, process-centric project management methods more rigorously: more tasks, more details, more milestones, more status reports. This led to the use of the Waterfall Method for planning and managing software development projects. Waterfall is a top down approach with a mandated series of development phases that are planned and scheduled in advance and worked sequentially. Waterfall and similar methods were called "heavyweight" methods. Despite the effort and the intent of Waterfall, the increasing complexity of software did not result in appreciable improvements.

The explosive growth of software occurred at about the same time as the growing knowledge of complexity theory and software developers began looking to complexity for answers. This created a pendulum swing, of sorts, with the perception that software creation was always on the edge of chaos, that it was unplannable and had to evolve based on empirical evidence. Software development entered an experimentation phase and saw the development of a number of alternative development methods. Some of the ideas seem to be guided by a rejection of the heavyweight methods like Waterfall. This rejection led some to assert that software projects should not have any planning, no documentation, no process descriptions, no fixed budgets or schedules, and no management. In contrast to the heavyweight methods, the new methods were called Lightweight methods and included Scrum, Extreme Programming (XP), Feature Driven Development (FDD), etc.

This proliferation of Lightweight methods led to a meeting in February 2001 at Snowbird, Utah where a group of 17 proponents met to reconcile their different methods. They came up with the name Agile as the collective term for what they did and wrote the Agile Manifesto and accompanying principles to guide further development [24–27].

The Manifesto, or, more properly, the Manifesto for Agile Software Development, has since been translated into dozens of languages and signed by thousands of people. It is both an acknowledgement of past practice and an appeal for more flexible and responsive processes as a more effective approach for dealing with complexity. For example, one of the precepts is that the signatories value responding to change over following a plan. One of the key principles is that software that does what customers want should be created and delivered quickly and frequently. This allows for rapid learning and evolutionary development as well as the ability to respond to change rapidly. Given these principles, Agile methods usually have small teams that produce working software in frequent, incremental releases. They make use of very close coordination with customers and are open to changes in requirements even late in the development cycle (For the full text of the Manifesto and the 12 Principles of Agile Software, see: http://agilemanifesto.org/).

Agile has continued to evolve and draw in useful features from other disciplines. Notable examples include the inclusion of theory and applications from Lean and Theory of Constraints. Interestingly, because of the success Agile methods have seen at dealing the complexity of software development, these combined approaches are now being fed back into the development of non-software products where they are finding utility for their ability to deal with variation and uncertainty in the development process.

3.4 Risk and Uncertainty

We care about complexity because the outcome of complex system development projects is unpredictable. This is much like saying that the outcome is uncertain, but uncertainty is a broader and more subtle term than unpredictability.

When systems engineering projects develop technology, uncertainty is unavoidable. If the outcome of development were certain, there would be no need for engineering design. Manufacturing drawings could be prepared immediately. The result would be known at the start.

But when a development project pushes technology frontiers, engineering is necessary, and therefore the outcome is uncertain. Project planning estimates include many aspects of the outcome (schedule, cost, product performance, reliability, and so on), but the actual outcome may be better or worse than the estimate.

This presents a problem for systems engineering, because the basic processes of decomposition, construction and integration assume a deterministic world. The only area of systems engineering that moves beyond determinism is risk management. Can risk management handle the uncertainty inherent in systems development?

There are two problems with using risk management to cope with the uncertainty that is endemic in systems engineering and systems development. First, risk and uncertainty differ in important ways [28]. Risk addresses bad outcomes that might occur. Uncertainty encompasses all outcomes that differ from the expected outcome, some of which are bad and some of which are good. Engineers want to mitigate or eliminate risk. However, with uncertainty, the right approach, according to decision theory, is to maximize the mean utility of the outcome. Thus, better than expected outcomes impact the overall design positively. Second, the systems engineering risk management process assumes that there are a small number of risks to be addressed—at most, several dozen. However, in a high technology project, uncertainty is ubiquitous. For a new commercial airliner, there are tens of thousands of uncertain outcomes [29]. The design of every component and part has uncertain cost, mass, development schedule, and so on. Although the magnitude of most of these uncertainties may be small, the collective effect can have a huge impact on the success or failure of the overall project.

All these uncertainties must be managed. Also, uncertainty management must be an essential element of the foundation of systems engineering, not an add-on process, which is how risk management is currently structured.

As an example of the important difference between risk and uncertainty, consider 10 tasks that are sequential. The owner of each task estimates that their task is 90 % sure of success. From a risk assessment standpoint, each would rate their task as being of very low risk. But the overall project has a probability of success of 0.9 to the tenth power, or only about 35 %. That means that the uncertainty as to when the overall project will complete is very high even though the risk that any individual task will miss their due date is very low.

4 Common Threads

We need a new paradigm for systems engineering. Michael Griffin has suggested that Systems Engineering is a collection of heuristics that lacks a foundational theory [30]. The current systems engineering V doesn't work for complex systems. We need to overcome the perception that the reason things didn't work in the past was because we didn't try hard enough, this is frequently the results of retrospective coherence as described in Sect. 3.

We don't have it yet, but here are some elements toward a solution. The ideas that have been explored in this chapter reinforce the need for change and point toward a solution. We find that there are a number of useful principles that can be applied to complex systems and hope that a better defined process can be developed in the future. The better we understand the theory of what is happening, the better we can match practice with context.

4.1 Implications of Complexity Definitions

There are ways to be successful operating within complex systems, but you have to remain flexible, there are no recipes. Fred Brooks argues that you have to design the organization that designs the system. The design of the organization has to match the product being designed, even though it is uncomfortable, or at least unusual, for engineers to look at the engineering of organizations [31].

This understanding was slow in developing and still today many Systems Engineers see the role as primarily one of technical integration. Brooks's point is that Systems Engineering must look at product development as a whole system and the role must include the full socio-technical integration.

The opinion of attendees at a National Science Foundation workshop on the Science of Systems Engineering in November 2014 was that, in addition to the technical challenges to be faced, successfully navigating a complex development project will also require defining and/or accounting for mission context, markers for success, organizational and team culture; values, assumptions, and behaviors [32]. Since a number of those roles involve fields normally found outside of Systems Engineering, such as Sociology, Psychology, Organizational Development, and Behavioral Economics; a transdisciplinary approach is necessary in order to successfully integrate large scale, complex product development projects.

4.2 Recommendations

These recommendations represent some basic heuristics that will change for each instance in different ways. The common thread is to create an approach that does not presuppose the solution in advance by employing agility, resiliency, flexibility, and constant feedback. Our recommendations can be summarized as:

- be able to react quickly to emergent risks and emergent opportunities
- use rapid, small scale, safe to fail experiments
- use small co-located teams of specialist generalists
- integration needs both lateral communication and vertical communication
- failure is a way of learning what works and what doesn't work

Being able to react quickly is one of the main attributes of Agile and is a part of the sensemaking aspect of probe-sense-respond advice from the Cynefin Framework. When applying a flexible process that accounts for emergent risks and opportunities, the normal bureaucracy will interpret the flexibility as neither ordered nor controlled. It will look like managers have no clue about how to manage, but it can't be presupposed that the system is deterministic. Elon Musk uses this approach very successfully with SpaceX. Up until now, we have been convinced at some level that the appropriate response to complexity is to impose more order on the systems engineering approach, not less. Recall from Sect. 1 how the Future Combat Systems was a failed attempt at trying to impose more order by insisting on repeated use of unrealistic timelines.

Use of rapid, small scale experiments enables insights through learning and/or failing experiences. The experiments are a way to accommodate the unpredictable aspects of the project. In some cases you need to persevere, and in some cases you need to pivot your approach to a new direction. If the experiments are truly safe-to-fail, they have the potential to spur innovation. When Nick Swinmurn wanted to test his hypothesis that people would buy shoes online, he worked with local shoe stores. He took pictures of the shoes, posted the pictures online, and when someone purchased the shoe, he bought the shoe from the store and mailed it to the customer. Yes, people will buy shoes online and that's how Zappos got started. At that time, it was not a valid business model, but his simple experiment proved that the concept was viable [33].

The more speculative the technology is, the more complex it is, the more need exists for a small team of specialist generalists that are co-located. The co-location is desirable in order to encourage rapid, frequent communication. If what is being built is a repetitive, mature product, it is easier to use a spread out organization, which is not as dependent on the need for frequent communication. Co-location works on a small scale but not on a large scale.

Vertical integration is important for new technologies because meeting customer requirements is very challenging without a carefully tailored product design. The first computers had to be built with all specially made parts, but this is not the case anymore. Now individual modules are available that just snap together and they work. Computers now use diverse supply chains, and it is possible to mix and match components. Fifty years ago, it would have taken a small team of experts to design and integrate all the systems in a computer. Now, modularity has reduced the need for that same level of lateral communication for systems integration [34].

Last, but not least, is that failure is a very useful experience. We don't learn from failing all the time, we don't learn from succeeding all the time either. Through failure, the limits of the solution are revealed.

5 Summary

A new paradigm is needed for systems engineering to account for the unpredictable nature of complexity. Standard systems engineering techniques used to create large modern products have resulted in an exponential growth in time and development cost due to complexity. Simply adding layers of modification to the existing paradigm will not work because complex problems today cannot be solved using deterministic or stochastic methods. As evidenced by the growth of integrated hardware and software for large aerospace systems, standard systems engineering is not obsolete, but it will not work for everything anymore. Of course not all systems are complex, and it is still very effective to use a closed form systems engineering solution for deterministic systems. If the final design can be accurately predicted in the early design phase, it is not a complex situation in need of an innovative solution.

When complexity is encountered, it does not need to be eradicated—it can be managed. Definitions of complexity vary, but most include emergent behavior in the system. Complex systems have a large number of interacting elements where minor design changes can have major consequences. And a solution that works in one context does not necessarily work in another. The inquiry is the same; the outcome is not [8]. Also, accounting for complexity is not just about finding a technical solution for an unpredictable system. Complexity is found in every facet, such as business, organizational, scale, coupling, interactive, and cognitive aspects of the system.

The Cynefin Framework can be employed as a sensemaking vehicle to interpret the character of the system by revealing what is complex, complicated, chaotic or obvious. Cynefin is a meta-strategy that opens a window to transdisciplinary thinking as a theoretical basis for the process of design that includes the design of the process both organizationally and technically. Agile represents an existing creative approach developed for software, which is now being introduced into hardware applications. Agility is the ability to both create and respond to change in order to profit in a turbulent business environment [35].

Cynefin works to illuminate a clearer picture of the context. Agile methods are an example of appropriate tools that fit in the appropriate context, but lack a fundamental theory to explain why they work [36]. Creating a foundational theory for both of these approaches will lead to the ability to manage effectively in complex systems.

For now, the best approach is to expect the unexpected such that the team is able to react quickly to both emergent risks and opportunities. Rapid feedback using small scale experimentation and finding out what doesn't work has also been found to be extremely useful. A small co-located team of people with diverse background works well along with effective lateral and vertical communication within the organization. For complex system development, it is best to accept that uncertainty is inherent and unavoidable, there are ways in which it can be accommodated, and it can provide significant opportunities to those who know how to take advantage of it. And that is why all engineers should care about complexity.

References

1. Eremenko, P. (2010, October). *Adaptive Vehicle Make (AVM)*. Proposers' Day Briefing. Tactical Technology Office, DARPA.
2. Augustine, N. R. (1983). *Augustine's laws* (2nd ed.). New York: American Institute of Aeronautics and Astronautics.
3. Pernin, C., Axelband, E., Drezner, J., Dille, B., Gordon, J., Held, B., et al. (2012). *Lessons from the army's future combat systems program*. Santa Monica, CA: RAND Corporation.
4. Maddox, I. D., Collopy, P. D., & Farrington, P. A. (2013). Value-based assessment of DoD acquisition programs. *Procedia Computer Science, 16*, 1161–1169.
5. Lloyd, S. (2001, August). Measures of complexity—A nonexhaustive list. *IEEE Control Systems Magazine*.
6. Mitchell, M. (2009). *Complexity: A guided tour*. Oxford: Oxford University Press.
7. Page, S. E. (2011). *Diversity and complexity*. Princeton, NJ: Princeton University Press.
8. Boone, M., & Snowden, D. (2007, November). *A leader's framework for decision making. Harvard Business Review*.
9. Cilliers, P. (1998). *Complexity and postmodernism: Understanding complex systems*. London: Routledge.
10. Kahneman, D. (2011). *Thinking, fast and slow*. New York: Farrar, Straus and Giroux.
11. Flyvbjerg, B. (2003). *Megaprojects and risk, an anatomy of ambition*. Cambridge: Cambridge University Press.
12. Goldratt, E. (2008). *The choice*. Great Barrington, MA: North River Press.
13. Ries, E. (2011). *The lean startup: How today's entrepreneurs use continuous innovation to create radically successful businesses*. New York: Crown Business.
14. Northrop, L., Feiler, P., Gabriel, R. P., Goodenough, J., Linger, R., Longstaff, T., et al. (2006). *Ultra-large-scale systems: The software challenge of the future*. Pittsburgh, PA: Carnegie Mellon University, Software Engineering Institute.
15. Felder, W. N., & Collopy, P. (2012). The elephant in the mist: What we don't know about the design, development, test and management of complex systems. *Journal of Aerospace Operations, 1*, 317–327.
16. Schön, D. A. (1983). *The reflective practitioner: How professionals think in action*. New York: Basic Books.
17. Brooks, F. P., Jr. (1982). *The mythical man-month, essays on software engineering*. Reading, MA: Addison-Wesley [Originally published in 1972].
18. Levitan, B., Lobo, J., Kauffman, S., & Schuler, R. (1999). *Optimal organization size in a stochastic environment with externalities*. Santa Fe Institute Working Paper: 1999-04-024. http://www.santafe.edu/research/working-papers/abstract/c3b337d659e39e8ef7e6e02c1f21f292
19. Perrow, C. (1999). *Normal accidents: Living with high-risk technologies*. Princeton, NJ: Princeton University Press.
20. Snowden, D. *Cynefin framework image*. http://commons.wikimedia.org/wiki/File:Cynefin_as_of_1st_June_2014.png
21. March, J. G. (2010). *The ambiguities of experience*. Ithaca, NY: Cornell University Press.

22. Cognitive Edge website. Ref: http://cognitive-edge.com/library/methods/four-tables-contextualisation-basic/
23. http://cognitive-edge.com/library/methods/the-future-backwards-basic/
24. Hibbs, C., Jewett, S., & Sullivan, M. (2009). *The art of lean software development*. Sebastopol, CA: O'Reilly Media.
25. Shalloway, A., Beaver, G., & Trott, J. R. (2010). *Lean-agile software development*. Upper Saddle River, NJ: Addison-Wesley.
26. Anderson, D. J. (2004). *Agile management for software engineering: Applying the theory of constraints for business results*. Upper Saddle River, NJ: Prentice Hall.
27. Highsmith, J. (2004). *Agile project management*. Boston: Addison-Wesley.
28. Reinertsen, D. G. (2009). *The principles of product development flow: Second generation lean product development*. Redondo Beach, CA: Celeritas Publishing.
29. Collopy, P. D. (2015, March). *Technical risk management. IEEE Aerospace Conference, Big Sky, MT*.
30. Griffin, M. D. (2010). *How do we fix systems engineering? (IAC-10.D1.5.4)*. Prague: International Astronautical Congress.
31. Brooks, F. (2010). *The design of design: Essays from a computer scientist* (1st ed.). Reading, MA: Addison-Wesley Professional.
32. Collopy, P., & Mesmer, B. (2014, November). *Report on the science of systems engineering workshop*. AIAA 2015-1865.
33. Donelan, J. (n.d.). *Do lean startup principles have a place in the enterprise?* http://thenextweb.com/entrepreneur/2013/08/06/do-lean-startup-principles-have-a-place-in-the-enterprise/
34. Christensen, C. M., Raynor, M., & Verlinden, M. (2001). Skate to where the money will be. *Harvard Business Review, 79*(10).
35. Highsmith, J. (2002). *Agile software development ecosystems*. Boston: Addison Wesley.
36. Snowden, D. (2015, July). *Cognitive edge, of practical wisdom*. http://cognitive-edge.com/blog/of-practical-wisdom/

Designer Systems of Systems: A Rational Integrated Approach of System Engineering to Tailored Aerodynamics, Aeroelasticity, Aero-viscoelasticity, Stability, Control, Geometry, Materials, Structures, Propulsion, Performance, Sizing, Weight, Cost

Harry H. Hilton, Steven J. D'Urso, and Noe Wiener

1 Introduction

The governing mantra of this chapter is that design/synthesis is not a process but rather a strategy made up of processes and analytical protocols that lead the engineers' and scientists' need to embrace the scientific totality of the system consisting of subsystems, i.e. the rigorous analysis of system of systems (SoS). In other words it should impel professionals to understand the big integrated picture instead of the limited provincial thinking of "I will do my part and just pass it on." Nor should the connotation of the word "design" conjure up visions of practitioners of magic arts. Rather it should be rightfully regarded and understood as system syntheses based on rigorous mathematical and physical principles governing ubiquitous analyses resulting from fundamental research endeavors.

The cartoon of Fig. 1 represents the hypothetical conditions where the various aerospace interest stake holders are left to their own uncoordinated devices and exaggerated views of their possible self-importance without regard to the fact that a flight vehicle is a closed loop system. However, this cartoon provides a catalogue of the

H.H. Hilton (✉)
Aerospace Engineering Department in the College of Engineering and Private Sector Program Division at the National Center for Supercomputing Applications, University of Illinois at Urbana-Champaign (UIUC), 104 South Wright Street, MC-236, Urbana, IL 61801-2935, USA
e-mail: h-hilton@illinois.edu

S.J. D'Urso
Aerospace Engineering Department in the College of Engineering, University of Illinois at Urbana-Champaign (UIUC), 104 South Wright Street, MC-236, Urbana, IL 61801-2935, USA
e-mail: sjdurso@illinois.edu

N. Wiener
Economics Department, The New School for Social Research, 6 East 16th Street, New York, NY 10003, USA
e-mail: wiener@newschool.edu

F.-J. Kahlen et al. (eds.), *Transdisciplinary Perspectives on Complex Systems*,
DOI 10.1007/978-3-319-38756-7_3

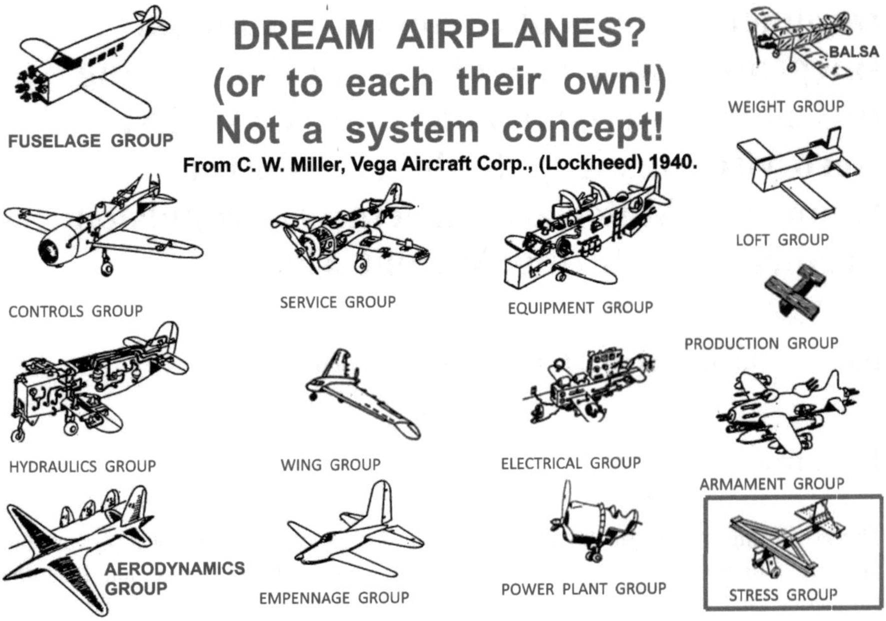

Fig. 1 The worst case scenarios [1]

separate systems contributing to the overall designer SoS concept. The currently developed structures/sizing optimization part, described here as a demonstration means of the SoS protocol, is limited to the block in the lower right hand corner of Fig. 1 and hence represents but a small beginning of the overall SoS optimum tailoring process.

The never ending continuous quest for light weight/high strength materials by the aerospace,[1] automotive and naval industries, as well as a host of other product manufacturers, has brought about the pervasive use of reinforced fiber and high polymer matrix composites [2–11] as primary structural materials of choice. However, the behavior of these composite and sandwich materials, unlike their elastic counterparts, leads to creep strains, stress relaxation, energy dissipation, time dependent failures, structural lifetimes, etc., which in some cases can be helpful while in others, such as closed loop systems for instance, may prove detrimental, unless their properties are selected and controlled appropriately [12, 13].

Emphasis is placed on viscoelastic media because the currently prevalent high polymer composites and elevated temperature metals behave in such a manner. Additional elasticity forms the initial viscoelastic conditions and hence is automatically a part of viscoelastic analyses. In general, viscoelasticity involves behaviors and responses that are fundamentally distinct from their elastic counterparts. Consequently, the nature of solutions to problems with identical geometries, boundary,

[1]Civil and military airplanes, missiles, spacecraft, UAVs, MAVs, wind turbine and helicopter blades, helicopters, etc.

initial and environmental conditions is radically altered due to the ever present time dependence of viscoelastic media. At the very least, quasi-static elastic formulations are replaced by time dependent viscoelastic responses, which may also develop dynamic conditions due to material time dependent properties where none exist in elastic equivalent problems. The viscoelastic dissipative material properties may stabilize or destabilize closed loop configurations depending on the altered phase relations. This latter phenomenon is particularly applicable to self excited systems such as aero-viscoelastic torsional divergence, flutter, buckling, deflections, etc.

The designer material/sizing concept for viscoelastic materials was developed in [14]. It was preceded by elastic formulations as seen in [15–33]. Viscoelastic examples may be found in [34–63]. Other pertinent examples are covered in [64] (polymers), [65] (metals), [66–69] (morphing) and [70] (composites). Similar approaches to the analytical determination of airfoil shapes tailored to deliver optimum aerodynamic properties, such as low drag, high lift, $(L/D)_{max}$, delayed stall, morphing, etc., have been successfully studied [71–88]. On a parallel path, the genesis of entire aeronautical system experiments (tests) are described in [89].

In current and future vehicles, aerodynamics, stability and control, flight performance, propulsion, etc., are and will be all impacted by the highly flexible structure and its materials. Each vehicle after all is a unique closed loop self-excited SoS and should be modeled as such. Rather than accomplishing this goal by the current trial and error "available off the shelf" piece-by-piece selections and then iterating portions of the ensemble, an integrated optimized protocol for designer materials, designer structures, designer aerodynamics, designer cost functions, etc., is described. Each discipline area is a system in its own right and the ensemble of these parts should be considered and analyzed as an *optimized* designer SoS tailored to conform to prescribed constraints, such as weight, size, performance, cost, etc., in order to achieve the best individual configurations.

A literature survey of aerospace systems can be found in [90]. Extensive works on various phases of aerospace design/synthesis are available in works such as [89–119]. Last but by no means least the influence of cost in terms of cost functions [120–124] is included in the SoS analyses.

In the final analysis, the SoS process reduces to the modeling of a host of subsystems and to the models describing their interactions with each other. One must never loose sight of the fact that one is solving the mathematical relations attempting to describe such models and not necessarily actual real world problems. Consequently, one should be constantly mindful of the overriding dictum "*... a map is not the territory it represents ...*",[2] so amply formulated in [125]. Or as Heisenberg has so forcefully reminded us that "*What we observe is not nature itself, but nature exposed to our method of questioning*" [126]. Furthermore, the historical

[2]Entire passage: "Two important characteristics of maps should be noticed. A map is not the territory it represents, but, if correct, it has a similar structure to the territory, which accounts for its usefulness" [125].

data that one uses to construct these models often incompletely and imperfectly chronicles past and present experiences and, therefore, may be partially or even totally unrepresentative of these as well as future events. Weight functions and sparse data protocols [127] may then have to be used in attempts to cover such deficiencies and to move the model and its analysis in the proper physical directions.

2 General Considerations

Because of the delayed availability[3] of the NCSA/NSF/Cray Blue Waters™ advanced sustained petascale (10^{15} flop/s) high performance computing system [128, 129] and its simulator, a series of modest, but novel, first steps of viscoelastic designer materials has been selected and the results have been reported in the literature [34–63]. In its own right, this designer material and structures protocol may be considered as the future of structural analysis and design. The current conventional structural analysis and design protocol consists of selections from *off the shelf* "best" available materials and structural geometries. On the other hand, the new designer/tailored material and sizing protocol calls for analytically engineered geometries and material properties and their distributions in structural components to optimally perform prescribed tasks under specified constraints, such as weight, cost, survivability, observability, vehicle performance, and others.

Elastic structural optimization analyses may be found in [15–33]. A number of elastic and viscoelastic problems have been formulated and investigated using a different approach consisting of designer structural optimization protocols, as seen in [14, 34–63]. Examples are beams, columns, plates and shells—creep buckling and delamination, torsional creep divergence, flutter, aero-acoustic noise attenuation, temperature effects, piezo-electric devices, magneto restrictive and smart materials, aero-servo-viscoelastic controls, wing shape and camber line morphing [14–66, 130–132].

For example, the designer material protocol basically consists of a series of inverse problems, where material properties as well as their spatial distributions—and *not structures* per say—are being designed/tailored/engineered to perform prescribed tasks subject to well identified constraints. These constraints are, but not limited to, geometry, sizing, composite fiber orientation and stacking sequences, weight, cost, lifetimes, failure probabilities, maximum deformations and stresses, etc. Therefore, for specific loading, boundary and initial conditions and constraints:

[3]First online operation in 2013.

1. Elastic and viscoelastic properties, i.e. moduli or compliances, are analytically tailored to provide best performances in terms of failure probabilities, deformations, minimum weight, cost, survival times, etc.
2. Material properties are strategically distributed in one or more structural elements and in uni- or multi- axial directions, which results in what is popularly called elastic or viscoelastic functionally graded materials (FGM and EFGM or VFGM[4]), but effectively are materials with optimized non-homogeneous properties. In short, FGMs are materials where the manufacturing process distributes properties according to a priory predetermined functions. For instance, the elastic Young's modulus instead of being constant becomes a variable function through the blending of multiple materials during the manufacturing of a structural part (see Fig. 8).

In this chapter the protocol is generalized to include the entire vehicle combining designer/tailored/engineered aerodynamics, propulsion, stability and control, cost, etc., into one combined analysis to produce an optimized synthesis subject to imposed constraints.

3 System Engineering Influence on Designer SoS: Flight Vehicle Synthesis and Analysis

A system engineering (holistic) approach to the synthesis (design) and optimization of entire or major portions vehicles is needed to confront this complex systems engineering problem.

All major system component characteristics need to be considered in the context of interdependency and the interacting whole. Advanced engineering methodologies are required to adequately characterize the various components and interfaces of the systems in a multi-disciplined evaluation in a holistic analysis.

Modeling of many of the system components has led significant developments in the disciplines of tailored aerodynamics, aeroelasticity, stability, control, geometry, materials, structures, propulsion, performance, sizing, weight and cost, to mention a few. Unfortunately, except perhaps for aeroelasticity and control problems, they have been considered much too often in isolation from each other as a single system element rather than as part of an integrated closed loop SoS. See the cartoon in Fig. 1 as extreme examples of possible exaggerated self importance.

Economic feasibility of the systems is driven by the cost of fabrication, operation and in a number of instances involving civilian enterprises by the cost of liability insurance.[5] Multidiscipline optimization allows for total system optimization but requires a balance of method fidelity to achieve the appropriately significant results.

[4]FGM = functionally graded materials, EFGM = elastic FGM, VFGM = viscoelastic FGM.

[5]See also Sect. 12 for details on cost functions.

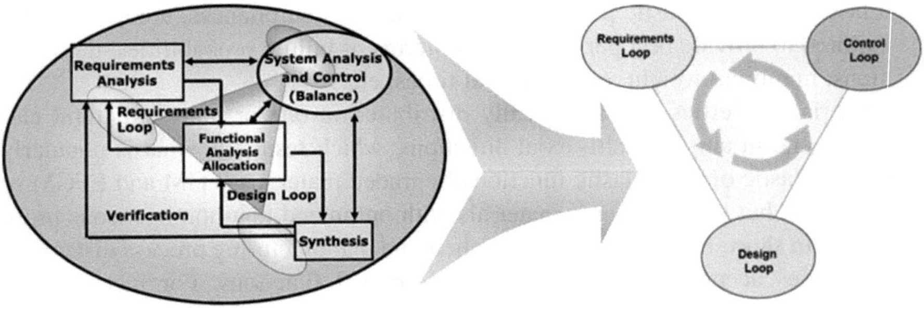

Fig. 2 System engineering flow chart

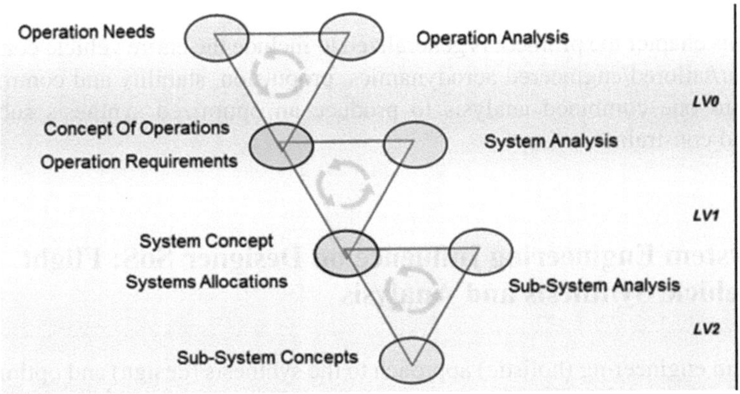

Fig. 3 Schematic of system engineering process

Figure 2 depicts a typical flow chart of a system engineering process [134] on the left hand side and can be simplified into the constituent activities as shown on the right of the figure. The iterative nature of system definition is recognized and expands characteristically in complexity as the system is defined from the top down.

This is that the activities of proposing an objective, then a prospective solution, and the triangle is considered closed when the solution is tested and shown to meet the intent of the objective. Figure 3 depicts a typical flow chart of a system engineering process [134] and shows that as the system is matured the pattern repeats and expands. In the initial stages these iterations are at the SoS and then level then the systems level with an ever increasing number of lower level elements. The nature of the starting point is that of a question or an objective which transforms into a requirement as maturation continues. With each iteration, the solution becomes the requirement (desired condition) for the next level down. In this same way the number of elements that need to be converged on increase at each lower level as is depicted in Fig. 4.

When considering the levels of abstraction of a system it can be said to have a number of levels as shown in Fig. 5. Not all systems have all levels, but generally a

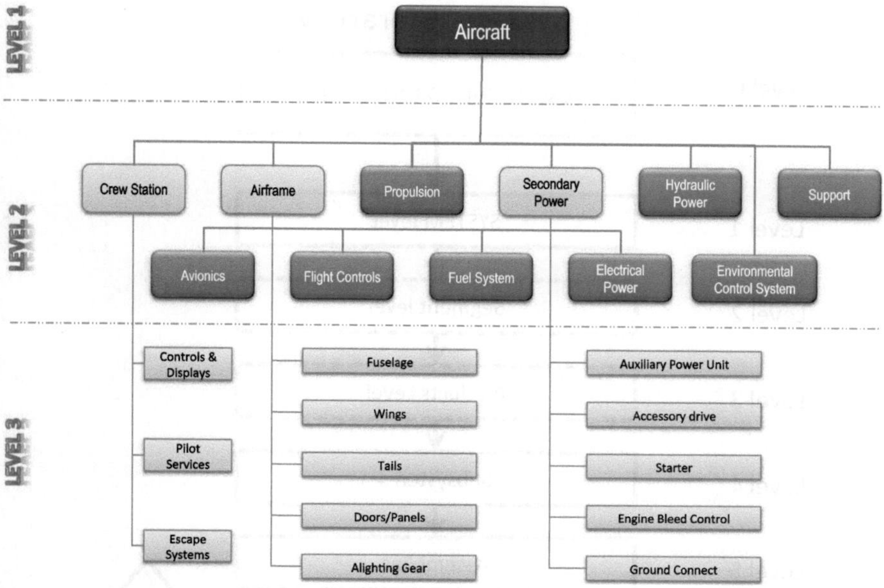

Fig. 4 Systems engineering aircraft hierarchy

systems may have up to eight levels. The lowest level being sub-assembly, parts and the constituent materials, directionality, and geometric features.

4 Identification of Systems, State Variables, Parameters, Constraints, etc.

The chart in Fig. 4 identifies the major components to be considered in an overall system engineering analysis. In each of the sub-systems listed in Fig. 1 sets of parameters and constraints must be identified and catalogued. Although such areas as human factors, environmental issues, etc., are not specifically spelled out they, as well as cost, are implied pervasive components that apply to all areas and can be readily incorporated as needed. Some examples of cost functions are specifically treated in this chapter.

4.1 Materials and Structures

Based on the operational requirements, the flight vehicle structural and material system properties can be synthesized to best perform to the mission situations and constraints over the total life cycle. Loading and fatigue spectrums are determined

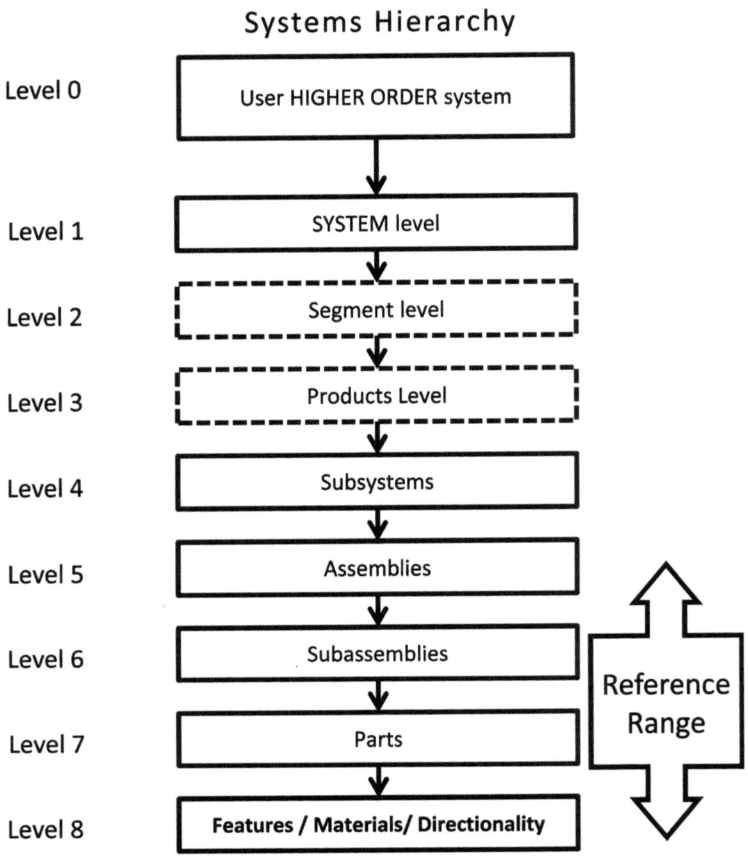

Fig. 5 Generalized system engineering hierarchy—adopted from [135]

from the intended use and in multiple operational modes. These will vary vehicle type and usage situations from military to commercial operations in many varied extreme environments.

The governing relations define conservation of linear and angular momentums for structural elements (beams, columns, plates, shells, etc.) for classes of chosen materials such as elastic, viscoelastic, plastic, viscoplastic, etc.

The optimization parameter subset $\boldsymbol{S} = \{S_n\}$ and their constraints are chosen from a set of effective sizing, minimum structural weights, cost, moduli, compliances, relaxation times, yield stresses, maximum stresses/displacements, lifetimes, weight to strength ratios, flutter and divergence velocities, load factors, failure probabilities, functionally graded material functions, fibers, plies, stacking sequences, etc.

The "design of experiments" method may be used to explore the relationships between flight vehicle physical parameters of span, wing area, thickness-to-chord ratio, and sweep which makes up the constituent elements of lift and drag and maneuvering characteristics.

4.2 Aerodynamics

The governing relations pertaining to aerodynamics and lifting surface shapes are the Navier-Stokes equations and those defining lift and drag generations.

The pertinent optimizing parameters and their constraints are selected from a set containing aerodynamics derivatives, stability derivatives, lift curve slopes, unsteady aerodynamic coefficients, $\left(\frac{L}{D}\right)_{MAX}$ or $\left(\frac{C_L}{C_D}\right)_{MAX}$, airfoil shapes, morphing of airfoil and lifting surface shapes, etc. Aerodynamic properties are strategically distributed and yield span-wise distributed airfoil shapes. Camber lines as well as upper and lower airfoil surfaces can be morphed and/or build into the vehicle to produce optimum shapes that yield desirable aerodynamic performances [66–69].

4.3 Propulsion, Stability and Control, etc.

The same inverse protocols are applicable to other sub-systems as well as to the SoS and similar approaches can be used. In general the overall procedures consider the following:

1. Define problem, goals, constraints, parameters
2. Tailor/engineer optimal parameters for best performance
3. Achieve goals: effective sizing, weights, cost, stresses, lifetimes, flutter and divergence velocities, thrust, $\left(\frac{L}{D}\right)_{max}$, stability and control, performance, etc.

5 Designer SoS Analysis: A Generalized System Engineered Synthesis Case

In Hilton et al., [14] through an analytical formulation of inverse problems [130] it was proven that elastic and viscoelastic material properties and structural sizing can be designed/synthesized/tailored/engineered to render desired performances according to prior specifications and constraints. For instance, these protocols can be used to analytically design/engineer optimum elastic and/or relaxation moduli that guarantee say a maximum strength to weight ratio, maximum dissipation in a given time range or a maximum flutter speed or other constraints.

These formulations will now be generalized to designer SoS in a Cartesian space with co-ordinates $x = \{x_i\}$ with $i = 1, 2, 3$. The Einstein summation convention applies throughout. The number of systems is $\mathcal{P}$ and each system is identified by the superscript p. The state variables in each system are u_m^p and the ensemble of all state variables is designated by $\boldsymbol{u}$ as

$$\boldsymbol{u} = \boldsymbol{u}(x,t) = \left\{u_m^p(x,t)\right\} \qquad m = 1,2,\cdots,\mathcal{M}^p \qquad \text{and} \qquad p = 1,2,\cdots,\mathcal{P} \quad (1)$$

and each system has $\mathcal{M}^p$ of variables.

The entire ensemble of analysis/synthesis parameters to be optimized are defined by

$$\boldsymbol{S} = \{S_n\} \qquad n = 1,2,\cdots,\mathcal{N} \tag{2}$$

where $\mathcal{N}$ is the sum total of the individual sub-system parameters or

$$\mathcal{N} = \sum_{p=1}^{\mathcal{P}} \mathcal{N}_p \tag{3}$$

Some of the variables u_m and parameters S_m will appear in more than one system since each system and the SoS are coupled.

Each system p of the total number of systems $\mathcal{P}$ has $\mathcal{Q}^p$ governing relations and $\hat{L}^p$ constraints

$$\mathcal{L}_q^p(x,t,\boldsymbol{u},\boldsymbol{S}) = 0 \qquad p = 1,2,\cdots,\mathcal{P} \quad \text{and} \quad q = 1,2,\cdots,\mathcal{Q}^p \tag{4}$$

with constraints

$$\mathcal{C}_\ell^p(x,t,\boldsymbol{u},\boldsymbol{S}) = 0 \quad \ell = 1,2,\cdots,\hat{L}^p \qquad \text{and} \qquad p = 1,2,\cdots,\mathcal{P} \tag{5}$$

This set of relations can now be individualized to each of the groups (i.e. system) of Fig. 1. Their total ensemble then forms the SoS simultaneous relations, which can be expressed as

$$\text{governing relations} \quad \Rightarrow \quad \mathcal{L}(x,t,\boldsymbol{u},\boldsymbol{S}) = \left\{\mathcal{L}_q^p(x,t,\boldsymbol{u},\boldsymbol{S})\right\} = 0 \tag{6}$$

and

$$\text{constraints} \quad \Rightarrow \quad \mathcal{C}(x,t,\boldsymbol{u},\boldsymbol{S}) = \left\{\mathcal{C}_\ell^p(x,t,\boldsymbol{u},\boldsymbol{S})\right\} = 0 \tag{7}$$

Equations (6) and (7) are the governing relations for the optimized SoS and their solutions for the parameter set $\boldsymbol{S}$ analytically defines its many details.

A brief outline of the protocol to be followed is shown in Fig. 6. This inverse procedure consists of the following:

1. Derive governing relations for the problem, which in generic form are shown in Eq. (4).
2. For each system formulate desired constraints $\mathcal{C}_\ell^p$ based on the prescribed specifications for the entire vehicle. However, in many cases these specifications

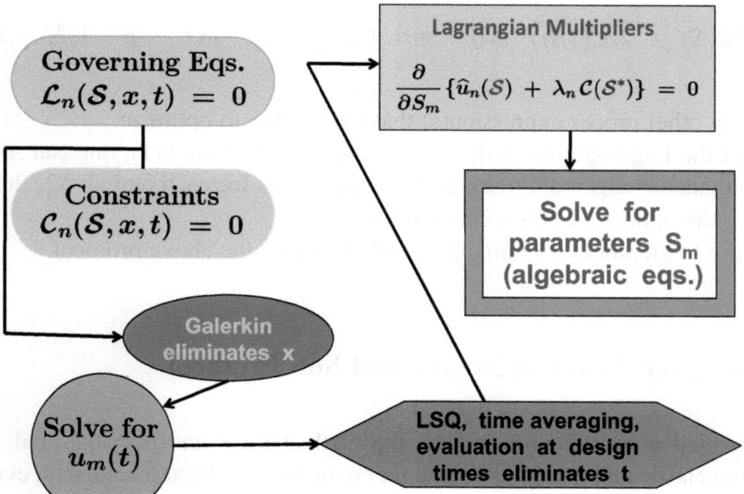

Fig. 6 Designer calculus of variations flow chart where $S = \{S_n\}$ are the parameters to be optimized [14]

may be derived requirements from the overall vehicle specifications in the system engineering sense.

3. Eliminate the spatial dependence of the state variables by applying Galerkin's procedure.
4. Solve the governing relations (4) for the remaining temporal functions

$$\hat{\boldsymbol{u}}(\boldsymbol{S}, t) = \int_a^b \boldsymbol{u}(x, t)\ \boldsymbol{u}_m^p(x, t)\ dx \qquad p = 1, 2, \cdots, \mathcal{P} \quad \text{and} \quad q = 1, 2, \cdots, \mathcal{Q} \tag{8}$$

5. Eliminate the temporal dependence by least square fits or through the collocation method or by evaluation at prescribed life times

$$\widetilde{\boldsymbol{u}}(\boldsymbol{S}) = \hat{\boldsymbol{u}}(\boldsymbol{S}, t_{LF}) \tag{9}$$

or other specified times. Alternately, another specification could involve a time averaging process, such that

$$\widetilde{\boldsymbol{u}}(\boldsymbol{S}) = \frac{1}{t_{LF}} \int_0^{t_{LF}} \hat{\boldsymbol{u}}\left(t'\right)\ dt' \tag{10}$$

6. Formulate $\mathcal{M} \times \mathcal{P}$ simultaneous equations of the $\mathcal{M} \times \mathcal{P}$ unknown parameters through the application of Lagrange multipliers λ_ℓ [136–139], such that

$$\frac{\partial}{\partial S_m^p}\left\{\widetilde{\boldsymbol{u}}(\boldsymbol{S}) + \lambda_\ell C_n^p(\boldsymbol{S})\right\} = 0 \qquad m = 1,2,\cdots,\mathcal{M} \qquad \text{and} \qquad p = 1,2,\cdots,\mathcal{P} \quad (11)$$

or any other proper expression(s) that one wishes to optimize.

7. After the Lagrange multipliers λ_ℓ are eliminated in Eq. (11), one can solve the simultaneous algebraic transcendental equations for each and all S_m^p, thus realizing the optimized SoS configuration.
8. Figure 6: graphically summarizes and illustrates the above protocol.

6 Designer/Tailored/Engineered SoS Protocol

The detailed general foundation for the mathematical and physical analyses for optimum elastic and viscoelastic designer materials has been formulated, evaluated and published in the archival literature [14]. This fundamental formulation and scope can be readily extended to other system and to SoS optimizations. The protocol basically consists of

1. Defining the problem, goals, governing relations, constraints, parameters.
2. Designing/tailoring/engineering optimal parameters for best performance. Additionally some parameters may be optimally spatially distributed.
3. Achieving goals: Effective sizing, weights, cost, maximum stresses, lifetimes, performance, cost, etc.

Measures of effectiveness (MoE) are the usual parameters that vehicles will probably be optimized to. Generally, they are a non-dimensional algorithm relating state variables like speed, payload, g loading, etc. They can take on many forms and levels of complexity and are normally derived from operational simulation analyses. They are also highly dependent of the type of flight vehicle and the overall vehicle operational objectives. An example is

$$MoE \approx aR^\alpha + bP^\beta + cV_{max}^\gamma + \cdots + zN^\omega \qquad (12)$$

where R = range, P = payload, and V_{max} is the maximum velocity at some flight altitude and so on. However, MoE can be much more complex than Eq. (12).

The contributing areas of interest for a flight vehicle SoS study are shown in the Fig. 1 cartoon. The structure and material properties are represented by the rectangle in lower right corner. Each area of interest will have one or more governing relation of the type (Eq. 6) and constraints (Eq. 7) in the state variables $\boldsymbol{u}(x,t,\boldsymbol{S})$ and with sets of parameters $\boldsymbol{S}$ to be optimized. The state variables $\boldsymbol{u}$ represent deformations, stresses, control forces, aerodynamic pressures, lift, drag, velocities, aerodynamic stability derivatives, airfoil shapes, etc. They may be associated with local conditions (wing panel deflection) or they may be global and apply to the entire vehicle. Not all state variables are necessarily coupled with each other.

Each relation has a distinct number of different unknown parameters $\boldsymbol{S} = \{S_m^p\}$ and different sets of unknown state variables $\boldsymbol{u}$. These parameters are related to material properties and their distributions, composite properties, geometries, aerodynamic properties, weight, cost, performance, control and stability functions and others. Associated with each system (area) of interest are also some a priori defined constraints of the type (Eq. 7). Optimizing the MoE here would be equivalent to selecting the globally best value(s) of the MoE algorithm. It must be emphasized that all the equations defining governing relations and constraints must be written symbolically because the parameters are the unknowns for which one seeks optimized values. Consequently, all equations must be solved analytically to formulate the relations for the unknown parameters S_m. The flow chart of Fig. 6 indicates the various steps which need to be executed. First the spatial dependence on the x coordinates must be eliminated by use of for example Galerkin's method or analytical finite element method (FEM) formulations. This reduces the governing partial differential equations to ordinary ones in time. These must now be solved for the state variables $u_m(t, \boldsymbol{S})$. Next the temporal dependence of the $\boldsymbol{u}$ is eliminated by a time averaging approach. Subsequently, Lagrange multipliers [136–139] are employed to formulate the relations for the unknown parameters S_m.

Effectively the optimal designer system analyses can be stated in a simplified illustrative fashion and the schematic of this designer protocol is displayed in Fig. 6.

The analytical operations can be carried out with the aid of such commercially available programs as MATLAB™, MATHEMATICA™, MAPLE™, etc. The last step in the flow chart, i.e. the actual solution for the optimized unknown parameters, is the only instance in the entire protocol when numerical solutions are obtained for the parameters. However, this last ensemble of relations in these unknown parameters $\boldsymbol{S}$ is algebraic, albeit nonlinear and transcendental.

7 Structural/Material/Sizing Examples

The purpose of the designer SoS protocol is determine the important parameters to be optimized. These parameters come from numerous subsystem, such as aerodynamics, structures, materials, propulsion, stability, control, dynamics, etc. Ultimately, their interdependence and simultaneous optimization provide the cohesiveness that holds the designer SoS together as one system.

For example, in the case of structures and materials classes of properties for selected materials that need to be chosen a priori to the optimization. Such candidate materials could come from elastic, viscoelastic, plastic or viscoplastic media. The parameters to be optimized come from the set consisting of moduli, relaxation times, yield stresses, failure stresses or strains. For the selected material class the end product is a hypothetical material with optimized properties and structural sizes, which represent the best configurations. Whether or not such a material exists or can be reasonably manufactured is not within the purview of designer systems. The analysis simply points to what should be the best choice and can be used, if necessary, as a guide to a more precise selection from catalogued existing materials and structural shapes.

For instance, the constitutive relations are for

- *An anisotropic functional graded material (FGM) elastic material*

$$\sigma_{ij}^{E}(x,t) = E_{\underline{ijkl}}^{E}[\mathcal{F}(x)]\left\{\varepsilon_{kl}^{E}(x,t) - \alpha_{\underline{ij}}\vartheta(x,t)\right\} \tag{13}$$

where the parameters to be optimized are the elastic moduli E_{ijkl}^{E}, the FGM function $\mathcal{F}(x)$ and the coefficients of thermal expansion α_{ij}. The FGM function can be expressed in a convenient series, such as

$$\mathcal{F}(x) = \sum_{\ell=0}^{M_\ell^{\mathcal{F}}}\sum_{m=0}^{M_m^{\mathcal{F}}}\sum_{n=0}^{M_n^{\mathcal{F}}} a_{\ell mn}\, x_1^{\ell} x_2^{m} x_3^{n} \qquad x_i \in [0 \;\; a_i] \tag{14}$$

where the coefficients $a_{\ell mn}$ now become the parameters to be optimized. Other examples of the FGM $\mathcal{F}$ function may be found in [140] and the consistency limitations on the function $\mathcal{F}$ for linear viscoelastic materials have been derived in [141].

- *An anisotropic FGM viscoelastic material*

$$\begin{aligned} \sigma_{ij}(x,t) = & \int_{-\infty}^{t} E_{ijkl}\left[t,t',\mathcal{T}\left(x,t'\right),\mathcal{F}(x)\right]\frac{\partial \varepsilon_{kl}(x,t')}{\partial t'}\,dt' \\ & - \int_{-\infty}^{t} E_{ij}^{T}\left[t,t,'\mathcal{T}\left(x,t'\right),\mathcal{F}(x)\right]\frac{\partial\left[\alpha\vartheta(x,t')\right]}{\partial t'}\,dt' \end{aligned} \tag{15}$$

where the relaxation moduli are expressed in as Prony series [142]

$$E_{ijkl}(x,t) = \widetilde{E}_{ijkl}[x,\xi(x,t)] = \sum_{n=1}^{\underline{N_{ijkl}}} E_{\underline{ijkl}}[\mathcal{F}(x)]\exp\left(-\frac{\xi[\mathcal{F}(x),t]}{\tau_{\underline{ijkl}}{}^{0}}\right) \tag{16}$$

with

$$\xi(x,t) = \int_{-\infty}^{t} a_T\left[\mathcal{F}(x),\mathcal{T}\left(x,t'\right)\right]\,dt' \tag{17}$$

with a_T the temperature-time shift function [143–194], which is a material property. This transformation changes (Eq. 15) to

$$\sigma_{ij}(x,t) = \hat{\sigma}_{ij}[x,\xi(x,t)] = \int_{-\infty}^{t} \hat{E}_{ijkl}\left[\mathcal{F}(x),\xi(x,t)-\xi'\right]\frac{\partial \hat{\varepsilon}_{kl}(x,\xi')}{\partial \xi'}\,d\xi'$$
$$-\int_{-\infty}^{t} \hat{E}^{T}_{ij}\left[\mathcal{F}(x),\xi(x,t)-\xi'\right]\frac{\partial\,\widehat{[\alpha\vartheta(x,\xi')]}}{\partial \xi'}\,d\xi' \quad (18)$$

- *Anisotropic plastic or viscoplastic FGM material (media with yield surfaces or stresses)*

Let σ_{yield} represent the chosen yield condition, then

$$\text{plastic} \Rightarrow \sigma_{ij}(x,t) = \begin{cases} f^{el}_{ijkl}[\varepsilon_{kl}(x,t),\mathcal{T}(x,t)] & \sigma_{ij}(x,t) \le \sigma_{yield} \\ f^{pl}_{ijkl}[\varepsilon_{kl}(x,t),\mathcal{T}(x,t)] & \sigma_{ij}(x,t) \ge \sigma_{yield} \end{cases} \quad (19)$$

$$\text{viscoplastic} \Rightarrow \sigma_{ij}(x,t) = \begin{cases} f^{el}_{ijkl}[t,\varepsilon_{kl}(x,t),\mathcal{T}(x,t)] & \sigma_{ij}(x,t) \le \sigma_{yield} \\ f^{vepl}_{ijkl}[t,\varepsilon_{kl}(x,t),\mathcal{T}(x,t)] & \sigma_{ij}(x,t) \ge \sigma_{yield} \end{cases} \quad (20)$$

where the four f functions in Eqs. (19) and (20) symbolically represent distinct function classes.

- *Failure conditions for all materials*

Let $F_{SURF}[\sigma^{max}_{ij}(x,t)]$ represent a failure surface associated with a particular failure theory[6] and $F_{max}[\sigma_{ij}(x,t)]$ be a similarly constructed surface but containing the actual stresses. Failure will occur whenever

$$F_{max}\left[\sigma_{ij}(x,t)\right] \ge F_{SURF}\left[\sigma^{max}_{ij}(x,t)\right] \quad (21)$$

The $F_{SURF}[\sigma^{max}_{ij}(x,t)]$ function is representable by an analytic function or series with their open parameters becoming the quantities to be optimized, cf. Eq. (14) and [132].

- *Non-catastrophic system failures*

[6] Such as von Mises, maximum stress or strain, octahedral shear [195], deterministic Shanley-Ryder [196], probabilistic Shanley-Ryder [21, 132], etc.

Another type of failure is an excessive but non destructive deflections that unfavorably alter the vehicle's electromagnetic signature and/or produce undesirable aerodynamic forces. Such a possible pattern of traveling electromagnetic waves is shown in Fig. 7. Furthermore, these optimized solutions did not contain any restrains to properly line up the peaks and valleys parallel to the flow thus resulting in unacceptable high drag. It is further possible that electromagnetic and aerodynamic constraints are incompatible with each other and a combined optimal compromise is far from individual optimum solutions.

8 A FGM Manufactured to Order

Non-homogeneities are due to multiple dissimilar materials, nano to macro material inclusions, fluid mixtures, composite fibers and their stacking/orientation sequences, temperature gradients, and viscoelastic functionally graded materials (VFGM)[7] among others. VFGMs are produced through specific manufacturing processes as noted for example in [133, 197]. The latter can be readily uniquely and fully described by a continuous or piece-wise continuous FGM $\mathcal{F}(x, t)$.

The term "composites" is generically used to denote any substance made of two or more materials with distinct properties. Notable examples are textiles, reinforced concrete [151], asphalt, polymer or ceramic matrices with polymer or metal fibers, etc., [8–11]. Another example is a plasma sprayed five-layer functionally graded ZrO_2/Ni Co Cr AlY thermal barrier coating of a μ–meter scale thickness described in [133] and shown in Fig. 7. These tailored entities are also referred to as meta-

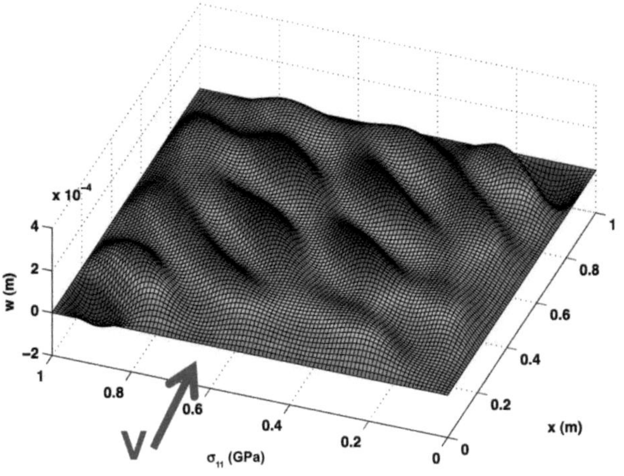

Fig. 7 Deflection waves in a composite panel [52]

[7]From a fundamental mechanics point of view, FGMs are essentially non-homogeneous materials and should be treated as such. Additionally and separately they may also be anisotropic.

materials. Composites have found wide acceptance in the aerospace, automotive and naval structures, to mention a few. However, manufacturing designer composites to order to fill prescriptions of material properties still remains elusive and needs intensive theoretical research undertakings [64, 65].

9 Cause and Effect: A Brief Case Study of a Structural Weight Change

There are numerous significant drivers that permeate throughout the SoS. One of the most pervasive is structural weight, which affects all systems except communication electronics[8] and manifests itself in a closed loop dynamical system of simultaneous integral-partial differential relations. The question then becomes: What happens when the structural weight W_{ST} is altered by an increment $\pm \Delta W_{ST}$? Actual historical statistical non-optimized data indicates that one unit of ΔW_{ST} structural weight change contributes a sevenfold factor of that unit addition/subtraction to the gross weight of the flight vehicle. Similar rules of thumb multipliers can be quoted for changes in L/D, propulsion, electronics, fuel, range, performance, pay load, etc. In other words an SoS is influenced individually and collectively by any and all of its parts as well as by the sum of its parts. Of course, such multipliers depend on the vehicle mission(s).

In the currently prevalent non-optimized heuristic analysis and design modes such structural weight changes are generally compensated by appropriate changes in payload. The SoS protocol derived here allows for the introduction of desired constraints and for simultaneous solutions of general governing relations that will result in an optimized complete vehicle SoS study of pertinent parameter ensembles resulting in the "best" achievable configurations for specifically designated mission performances and their impact on engineering, manufacturing and operational costs.

10 Probability of Failure, Redundancies and Weghing Functions

While ideally one tries to model, analyze and synthesize systems to be devoid of failures, nature presents realities where failures are ever present possibilities. Consequently, eventually one must introduce probabilities of input occurrences, of expected/actual performances and of system properties and failures in order to analytically mimic the real world [16–21]. Of course, component, sub-system and/or SoS failures can be minimized by introducing excessive safety factors,

[8]Excluded: radio, TV, GPS, monitors, autopilot, navigation, etc.; not necessarily excluded: controls.

large numbers of redundancies and by excessively over-designing. Such magnanimous prescriptions are counter productive and lead to additional costs and weight that ultimately will drastically reduce performances and conceivably prevent the vehicle from rising off the ground. Consequently, the above prescriptions must instead be optimized or introduced as judicious constraints as part of any SoS.

Separately, but equally significant, weighing functions that ascribe differential influences to various components and/or systems may be introduced as constraints to effect their relative importance. Additionally, system or component redundancies may be used to improve failure probabilities by prescribing for instance multiple electrical systems. Such practices ultimately increase costs and weight, but may be unavoidable for safety reasons or requirements [90].

In the current analyses described in this chapter, such phenomena are not included and will be subject of future research.

11 Approaches to SoS Cost Estimation

Cost is one of the most significant considerations in the design phase of almost any new product. Since a large share of costs are determined and locked in before production occurs [198], fast and accurate methods of estimating cost are highly desirable. The main approaches to cost modeling are briefly reviewed after the components of cost are defined. Then some possible specifications of cost functions are formulated. For aerospace and other fields SoS cost factors include among others: research and development (R&D); engineering, materials; component transportation to assembly; production; sales; liability, fire, theft insurance; profit; electronic and electrical components; legal advise; overhead; etc.

11.1 Components of Cost

An encompassing notion of costs will consider the entire life cycle of a product, including design and testing, production, operation and disposal. A separate category of costs is insurance, which is offered to both flight vehicle manufacturers and operators to cover various adverse events. Although costs of production and operation are generally borne by different agents, there has been a shift in recent years toward manufacturers leasing out vehicles to operators [199]. This would suggest that manufacturers are becoming more acutely aware of the whole life cycle costs of their product, including possible tradeoffs between costs of production and maintenance. In the following, the focus will nevertheless be on design and production costs as the elements determining producers' costs, directing the interested reader to the literature on operating costs [200].

Within production costs, one may distinguish between recurring and nonrecurring costs, the former composed of labor and material costs incurred for each flight vehicle, and the latter of overheads (such as equipment, tooling, buildings and their maintenance) that are spread over all produced units. Costs of design are non-recurring, but usually quite small. On the other hand, the manufacturing cost of modern aerospace vehicles is substantial and consists of both recurrent and non-recurrent costs. Once the design is chosen, material costs can be estimated with a greater degree of confidence. However, the costs of tooling and labor are more uncertain, being affected by institutional factors, e.g. multiple work shifts [201] that do not affect material costs. Production costs requiring labor input are the product of the "going" rate of pay, the total number of hours worked and further adjustments for learning-by-doing.

11.2 Approaches to Cost Estimation

There are, broadly, two approaches to cost estimation in aerospace production. The parametric method uses historical cost data to estimate a functional relationship between cost and a (small) set of general design parameters (mass, materials, dimensions, etc.). A discussion of some forms that these cost functions may take will be presented. The hope is that the historical cost data from earlier projects could provide an adequate approximation for the purposes of the new project, but this is not guaranteed. This method is most useful at the early design stage when the vehicle design is still not fixed.

Whereas parametric models may be described as "top-down" [198], the generative [202] or analytical [203] approach may be called "bottom-up". Here explicit use is made of the engineering knowledge about the production process, relying on detailed descriptions of individual features and process times. This latter approach has far greater informational requirements and may therefore be impractical at early stages of design. However, the generative approach emphasizes the search for the "true" drivers of cost instead of the proxies often considered in the parametric approach. For instance, in [204] the authors argue in favor of design information as a causal cost driver of flight vehicle parts over mass and material. The latter can be very problematic as shown in [204] because efforts at reducing mass may lead to increases in costs, and materials may only help to forecast manufacturing costs because they proxy for processing costs.

11.3 Cost Functions in Economics

What are the kind of cost functions that may be useful for cost estimation purposes? The guidance offered both by economists and the cost estimation literature proper will be considered. Cost estimation is an engineering discipline with little overlap with economics. Generally, the level of aggregation and time horizon considered by neoclassical economists severely limit their applicability in modeling concrete production decisions [205]. Yet there are some concepts that have proven useful for industry use that will be considered first.

The empirical observation of airframe production led to the important insight that labor productivity usually increases with production volume in a systematic fashion, leading to falling unit costs [120, 206]. The average hours of direct labor "congealed" in the parts and assembly of the first N airframes was found to decline even if non-recurrent tasks were excluded and even if no new net investment took place. The conclusion was drawn in [121] that learning on the part of labour and management was responsible for a good part of the effect, together perhaps with innovations in other sectors. This finding led to the notion of a "learning curve" traced out when plotting direct unit costs against cumulative output (also known as progress ratios). Learning curves differ between industries and tend to be steeper the higher the labor-intensity of production. Thus, to arrive at valid estimates of costs at various levels of production, there is a need for dynamic cost functions that take into account the learning effect [122]. For a recent perspective on incorporating learning into a dynamic process-based cost model in manufacturing, see [123].

The interest in cost functions in contemporary neoclassical economics often results from the use of duality theory to reveal features of the underlying production function from observed costs [124], based on Minkowski's theorem [207] on supporting half-spaces to closed and convex sets. Consequently, a certain number of restrictions are imposed on the cost functions in order to be able to interpret the result as reflecting a neoclassical production function, which will be discussed below. Economists have also estimated production functions directly using very high level aggregate measures of labor and capital (rather than design or production parameters), usually at the industry level since firm data is not readily available. Attempts at testing neoclassical production theory at the macro-level are often vitiated by the fact that such tests merely reflect the accounting identity between the values of inputs and outputs (see [208, 209]). In general, the economics literature emphasizes consistency of the estimated models with the basic neoclassical axioms of subjective expected utility theory, rather than the correspondence with observed behaviors [210].

11.4 Specification of Cost Functions

With this background, one can now move to consider the specification of cost functions. For each production function there is a corresponding cost function and the "classic" production function has the Cobb-Douglas (CD) functional form

$$Y = AL^{\alpha}K^{\beta} \qquad \text{with parameters} \qquad A, \alpha, \beta > 0 \tag{22}$$

where Y is output, L labor input and K "capital" input. This result can be derived as the profit-maximizing production decision assuming perfect competition and returns to scale that are de- or in-creasing depending on $(\alpha+\beta)\gtrless 1$. The average cost function for input prices p_L (the wage rate) and p_K (the rental price of capital)

$$c \;=\; \frac{(\alpha+\beta)\;Y^{[1-\alpha-\beta/(\alpha+\beta)]}\;p_L^{[\alpha/(\alpha+\beta)]}\;\;p_K^{[\beta/(\alpha+\beta)]}}{\alpha^{[\alpha/(\alpha+\beta)]}\;\beta^{[\beta/(\alpha+\beta)]}} \tag{23}$$

Average costs increase with output if $(\alpha+\beta)<1$ and decrease if $(\alpha+\beta)>1$.

The Constant-Elasticity-of-Substitution (CES) production function[9] [211] extends the CD form to one that allows for elasticities of substitution different from unity. The CES is defined as

$$Y \;=\; \gamma[\delta K^{\rho}+(1-\delta)L^{\rho}]^{1/\rho} \tag{24}$$

with associated unit cost function

$$c \;=\; \frac{1}{\rho}\left[\delta^{\sigma}p_K^{(1-\sigma)}+(1-\delta)^{\sigma}p_L^{(1-\sigma)}\right]^{1/(1-\sigma)} \tag{25}$$

where $\sigma \;=\; 1/(1-\rho)$ is the elasticity of substitution. As the name suggests, this function has constant elasticities of substitution. The CES model class implies identical constant elasticities of substitution among all inputs, which calls for a less restrictive functional form in the general case with more than two inputs.

More flexible functional forms that have been proposed include the generalized Leontief [212] and the transcendental logarithmic (translog) function [213]. These functional forms do not impose any a priori restrictions on substitution elasticities and allow for approximation of arbitrary functions to the second order. The generalized Leontief (GL) unit cost function with two inputs has the form

$$c \;=\; \alpha_0\;p_K \;+\; \alpha_1\sqrt{p_K p_L} \;+\; \alpha_2 p_L \tag{26}$$

and the translog (TL) unit cost function is

$$\begin{aligned}\ln[c(p_K,p_L)] \;&=\; \alpha_0 \;+\; \alpha_1\ln(p_K) \;+\; (1-\alpha_1+\beta_{11})\ln(p_K)\\ &\quad-\;\beta_{11}[\ln(p_L p_K) \;-\; \ln(p_L)]\end{aligned} \tag{27}$$

In order to distinguish between the GL and TL functional forms, a comparison in terms of theoretical consistency, domain applicability and flexibility, among other criteria, has been suggested in [214]. Theoretical consistency refers to the implications of the cost-minimization hypothesis, namely that the total cost function be homogeneous of degree one, non-decreasing in both input prices and output quantity, and concave in input prices. One also needs to find the parameter values

[9]In economics, elasticity is defined as the degree to which a demand or supply is sensitive to changes in price or income.

for which the cost function satisfies the requirements of theoretical consistency given any set of positive prices (domain of applicability). For both the GL and TL the parameter restrictions required to ensure global theoretical consistency severely reduce flexibility and so one may have to be satisfied with a smaller domain of theoretical consistency [214].

It should be noted that all of the cost functions considered above were deterministic. In light of the uncertainty with regard to input prices however, a specification in terms of prior probability distributions may be desirable to allow for a propagation of uncertainty to the total cost estimate.

12 Discussion

While no overall protocols or solutions are as yet available for designer SoS, a substantial number of sub-system pertaining to structures [15–63, 131, 132], airfoil shapes [71–88], morphing [66–69] and general synthesis protocols [104–127, 134, 198–219] have been formulated, analyzed and reported previously.

The previously referred to Fig. 7 depicts the optimized wave pattern of a wing panel under a minimum deflection/weight configuration constraint and exposed to aero-acoustic loads [52]. Note that without an additional minimum drag constraint, this wave pattern solution makes no attempt to orient itself parallel to the flow field (Fig. 8).

Figure 9 represents the results showing the optimum fiber orientation of a single ply high polymer fiber-matrix viscoelastic panel [52]. Note the uncommon slowly converging Galerkin protocol with respect to the 625 terms needed to achieve convergence in this particular example.

Figure 10 exhibits analytical aeroelastic results for a tower of an electricity generating wind turbine [42]. The formulation takes into account the loads

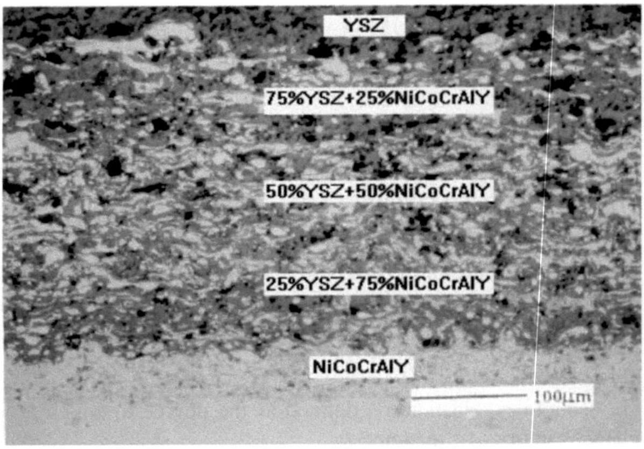

Fig. 8 Five-layer functionally graded ZrO_2/Ni Co Cr AlY coating [133]

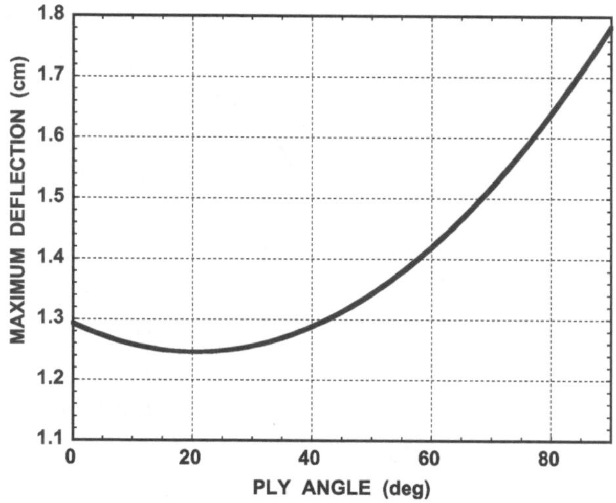

Fig. 9 Optimum fiber orientation [52]

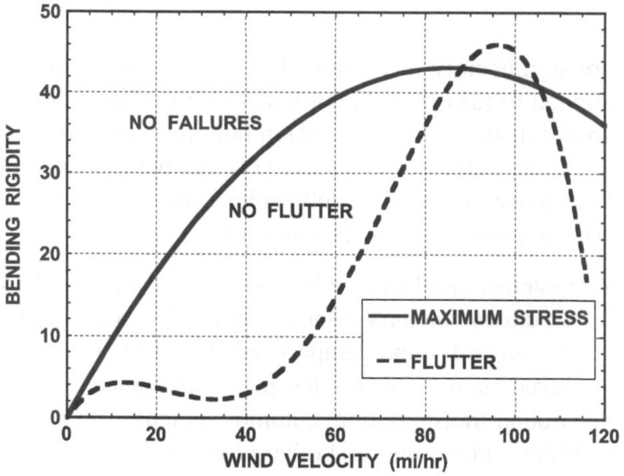

Fig. 10 Evolution of optimum conditions in bending of a wind turbine tower [42]

produced by the weight and thrust of the turbine assembly as well as the weight and drag of the tower itself. Two mutually distinct phenomena are illustrated, i.e. material failures and flutter as functions of incident wind velocity. Considered one by one, the tower has a "safe" design at velocities to the left of each curve. The optimum maximum velocity occurs at the first crossing of the two curves after the trivial solution at $V_{\infty} = 0$.

Figure 11 illustrates some typical results from a designer viscoelastic material study on a small scale for a fiber-matrix viscoelastic composite plate subjected to aerodynamic noise [40]. The flow is in the x_1 direction and material properties are optimized in either and both coordinate directions. Clearly, the probabilities of delamination (structural failure) are the lowest by orders of magnitude and lifetimes

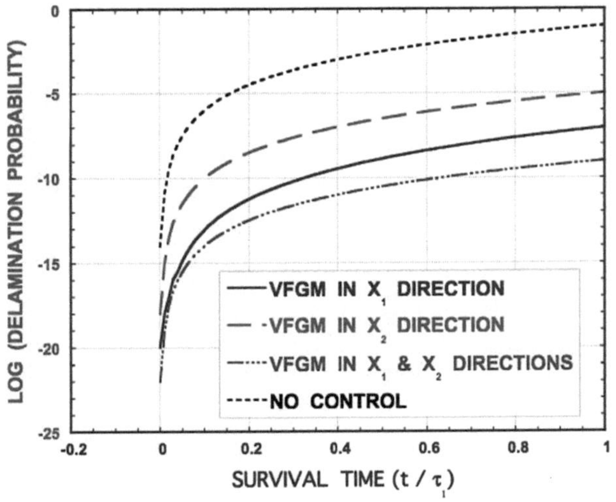

Fig. 11 Viscoelastic composite panel delamination failure probability [40]

are significantly increased when the designer material concept is simultaneously applied to (a) the optimum material property parameters themselves and (b) their spatial distributions (VFGM) in both in-plane directions.

At this stage of the research efforts, the fundamental formulations lead and can be extended to many varied applications of elastic and viscoelastic designer materials and sizing. A few examples are:

- Systems: airplanes, missiles, space vehicles, U/MAVs, satellites, gossamer structures, antenna dishes, space suits, body armor, solar sails, circuit boards, motion and sound dampers, wind turbine and helicopter blades
- Aerodynamics: airfoil design, morphing, etc.
- Ground transportation: automotive, tanks, railroads, cargo containers, etc.
- Ships: submarines and other navy vessels
- Materials: composites, concrete, asphalt
- Medical: artificial limbs, biological materials (tissue, blood), stents, implanted monitoring sensors and actuators
- Food stuffs: cheese, pasta, Jello, etc.
- and more

Specific investigations have been carried out to optimized conditions of column creep buckling, plates under aerodynamic noise, torsional divergence, wing flutter, Boeing's PRSEUS stitched composites (Fig. 12), etc., [216–218].

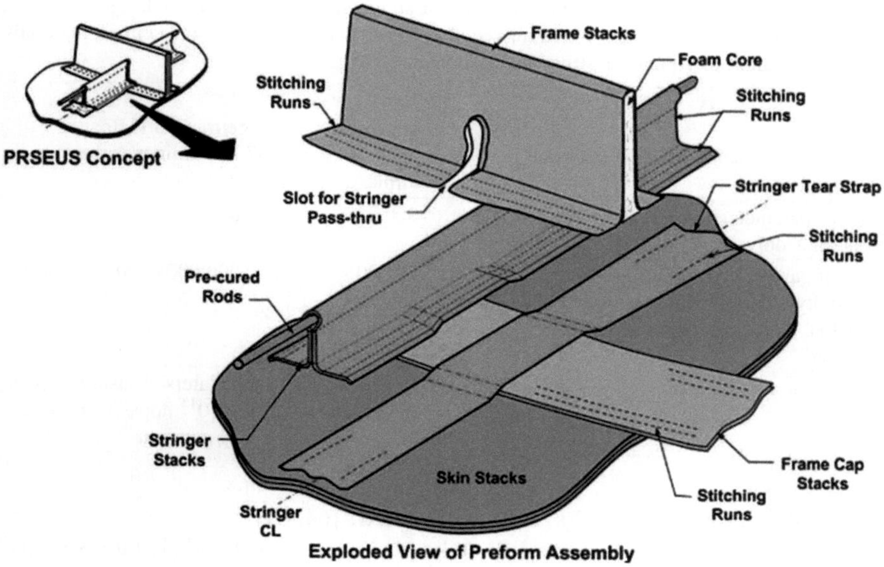

Fig. 12 PRSEUS composite configuration [52, 54, 216–218]

13 Future Expansions to Entire Vehicles: Designer SoS

As demonstrated in above sections, the inverse calculus of variations designer material/sizing concept and protocol can be readily applied in order to establish optimum moduli, sizing, etc. subject to weight, cost, aeroelastic, failure, maximum deflection/stress, etc., constraints.

More system areas beyond material properties and structural sizing will be added in the future in the quest to optimize the entire vehicle and thus eventually achieve the overall integrated desired SoS concept. For instance, the designer airfoil aerodynamics concept has been successfully carried out in [71–85], to mention a few instances. Additionally, a wealth of aerodynamic airfoil data has been collected and catalogued in [88].

Estimates for the number of unknown parameters and simultaneous algebraic transcendental relations to be solved defining various SoS are shown in Table 1. At this point for the larger vehicles, *big data* transitions to *big compute* in defining SoS optimized analyses.

The present analysis concentrated on material, structural and dimensional optimizations - the item in the lower right corner of Fig. 1. Aerodynamic analyses of high L/D airfoils have been formulated in [71–84] among others.

There remain also a few important unanswered questions that need to be discussed here in detail, such as

Table 1 Computational requirement estimates for an integrated SoS optimum designer/tailored/engineered parameter approach for entire vehicles: aerodynamics, stability and control, materials, propulsion, flight performance, structures, sizing, cost, etc.

Vehicle ⇒ Parameters/Computational Demands ⇓	UAV/ MAV	Small general aviation airplane	Large transport (Boeing 787 Dreamliner Airbus A350–800)
No. of parameters for structure and materials only	90	700	300,000
No. of parameters for entire vehicle (stability, control, aerodynamics, propulsion, structures, materials, sizing, cost, etc.	500	70,000	800,000,000–1,200,000,000
Computational hardware	Laptop	HPC mainframe	Blue Waters™ sustained peta scale (10^{15} flops/s) Big data ⇒ Big compute

- Can the designer material(s) be manufactured? [64, 67]
- Can the optimization meaningfully take place within cost and sizing restraints?
- Can the National Center of Supercomputing Applications at the University of Illinois at Urbana-Champaign Blue Waters™ [128, 129] sustained peta scale computing system handle a problem of such magnitude efficiently and effectively?

This last computational issue can be resolved, if necessary, by separately considering optimization of only major components (wings, fuselages, tail assemblies, power plants, etc.) rather than an entire vehicle. In the final analysis what matters is the understanding of the SoS concepts and their applications and as Richard Hamming [220] phrased it "The purpose of computing is insight, not numbers."

Of course, the designer SoS concept as outlined here is not limited to airplanes but can be applied to any large or small scale entity, such as

- Airplanes, missiles, space vehicles, U/MAVs, satellites, gossamer structures, antenna dishes, space suits, body armor, solar sails, circuit boards, wind turbine and helicopter blades, helicopters, etc.
- Ground transportation: automotive, tanks, railroads, cargo containers, etc.
- Submarines and other navy, commercial and recreational ships

14 Conclusions

The SoS essentially consists of the coupling of parameters and governing relations between subsystems that make up in the aggregate the actual SoS.

The original structural material/sizing optimal analysis points to a path through the herein formulated generalized analysis encompassing subsystems, systems and/or SoS optimizations as novel protocols to be pursued with definite advantages. Among others, the SoS protocol offers significant realistic bonuses of considering the synthesized object as an entire closed loop entity.

The recent fully operational advent of the UIUC NCSA/NSF/Cray Blue Waters™—the advanced sustained petascale high performance computing system—makes the symbolic developments and the needed computation of hundreds of millions of simultaneous relations a realistic possibility. Even if limited to only subsystems, such as wings, fuselages, etc., use of the designer protocols will produce more efficient, light weight, cost effective, high performance aerospace vehicles.

The economic sub-system examines in detail the introduction of cost functions into SoS engineering analyses. An active coupling is thus created between the physical synthesis and the cost factors associated with the real world production of such systems of systems.

Acknowledgement Support by grants from the Private Sector Program Division (PSP) of the National Center for Supercomputing Applications (NCSA) and from the Department of Aerospace Engineering both at the University of Illinois at Urbana-Champaign (UIUC), and from the Economics Department at the New School for Social Research is gratefully acknowledged.

References

1. Bruhn, E. F., Bollard, R. J. H., Hackman, L. E., Lianis, G., William, M., Schmitt, A. F., et al. (1973). *Analysis and design of flight vehicle structures*. Indianapolis: S. R. Jacob.
2. Altus, E. (1989). *Mechanics of composite materials: An introductory course*. Toronto: Ontario Centre for Materials Research.
3. Gibson, R. F. (1994). *Principles of composite material mechanics*. New York: McGraw-Hill.
4. Hyer, M. W. (Ed.). (1993). *Mechanics of composite materials—Nonlinear effects, AMD-159*. New York: ASME.
5. Hyer, M. W. (2008). *Stress analysis of fiber-reinforced composite materials*. Lancaster, PA: DEStech Publications.
6. Mallick, P. K. (2008). *Fiber-reinforced composites, materials, manufacturing, and design* (3rd ed.). New York: CRC Press.
7. Jones, R. M. (1999). *Mechanics of composite materials* (2nd ed.). New York: CRC Press.
8. Kuhn, W. (1939). Elastizität und Viscosität hochpolymerer Verbindungen. *Angewandte Chemie, 52*, 289–301. doi:10.1002/ange.19390521602.
9. Sierakowski, R. L., & Chaturvedi, S. K. (1997). *Dynamic loading and characterization of fiber-reinforced composites*. New York: John Wiley & Sons Inc.
10. Vinson, J. R. (1999). *The behavior of sandwich structures of isotropic and composite materials*. Lancaster, PA: Technimic Publishing Co.
11. Vinson, J. R., & Sierakowski, R. L. (2011). *The behavior of structures composed of composite materials* (2nd ed.). Amsterdam: Kluwer.
12. Hilton, H. H. (2010). Aeroelasticity and aero-viscoelasticity: A critical appreciation of similarities and differences. In *Proceedings 51st AIAA/ASME/ASCE/AHS/ASC Structures, Structural Dynamics and Materials (SDM) Conference, AIAA Paper 2010-2702*. Orlando, FL.

13. Hilton, H. H. (2011). Equivalences and contrasts between thermo-elasticity and thermo-viscoelasticity: A comprehensive critique. *Journal of Thermal Stresses, 34*, 488–535. doi:10.1080/01495739.2011.564010.
14. Hilton, H. H., Lee, D. H., & El Fouly, A. R. A. (2008). General analysis of viscoelastic designer functionally graded auxetic materials engineered/tailored for specific task performances. *Mechanics of Time-Dependent Materials, 12*, 151–178.
15. Weaver, P. M., & Ashby, M. F. (1996). The optimal selection of material and section shape. *Journal of Engineering Design, 7*, 129–150.
16. Shanley, F. R. (1948). *Principles of optimum structural design as applied to aircraft weight analysis*. Santa Monica, CA: Douglas Aircraft C.
17. Lin, Y.-K. (1967). *Probabilistic theory of structural dynamics*. Huntington, NY: Krieger.
18. Lin, Y.-K., & Cai, G.-Q. (1995). *Probabilistic structural dynamics: Advanced theory and applications*. New York: McGraw-Hill.
19. Elishakoff, I. (2010). *Optimization and anti-optimization of structures under uncertainty*. London: Imperial College Press.
20. Gallagher, R. H., & Zienkiewicz, O. C. (Eds.). (1973). *Optimum structural design—Theory and applications*. New York: John Wiley & Sons.
21. Hilton, H. H., & Feigen, M. (1960). Minimum weight analysis based on structural reliability. *Journal of the Aero/Space Sciences, 27*, 641–652.
22. Lemanski, S. L., & Weaver, P. M. (2003). Analytical optimization of composite cylindrical shells to meet given cross-sectional stiffness properties. In *Proceedings 44th AIAA/ASCE/ASME/AHS SDM Conference*. Reston, VA.
23. Cecchini, L., & Weaver, P. M. (2003). The optimization of foam-filled cylindrical shells subject to flexural loading. In *Proceedings 44th AIAA/ASCE/ASME/AHS SDM Conference*. Reston, VA.
24. Weaver, P. M. (2004). On optimization of long anisotropic flat plates subject to shear buckling loads. In *Proceedings 45th AIAA/ASCE/ASME/AHS SDM Conference*. Reston, VA.
25. Lemanski, S. L., & Weaver, P. M. (2006). Optimization of a 4-layer laminated cylindrical shell to meet given cross-sectional stiffness properties. *Composite Structures, 72*, 163–176.
26. Herencia, J. E., Weaver, P. M., & Friswell, M. I. (2006). Local optimization of long anisotropic laminated fibre composite panels with T-shape stiffeners. In *Proceedings 47th AIAA/ASCE/ASME/AHS SDM Conference*. Newport, RI.
27. Herencia, J. E., Weaver, P. M., & Friswell, M. I. (2007). Optimization of long anisotropic laminated fiber composite panels with T-shaped stiffeners. *AIAA Journal, 45*, 2497–2509.
28. Herencia, J. E., Weaver, P. M., & Friswell, M. I. (2007). Optimization of anisotropic plates that vary in thicknesses and properties. In *Proceedings 16th International Conference on Composite Materials*. Kyoto.
29. Herencia, J. E., Weaver, P. M., & Friswell, M. I. (2007). Local optimization of anisotropic composite panels with T-shape stiffeners. In *Proceedings 48th AIAA/ASCE/ASME/AHS SDM Conference*. Waikiki, HI.
30. Bloomfield, M. W., Herencia, J. E., & Weaver, P. M. (2008). Optimization of anisotropic composite plates using an increased design envelope of ply orientations. In *Proceedings 49th AIAA/ASCE/ASME/AHS SDM Conference*. Schaumburg, IL.
31. Herencia, J. E., Hatfka, R. T., Weaver, P. M., & Friswell, M. I. (2008). Lay-up optimization of composite stiffened panels using linear approximations in lamination space. *AIAA Journal, 46*, 2387–2391.
32. Blanchard, I. (2014). Composite design optimization for automated fiber placement. *SAE Aerospace Engineering, 4*, 14–19 (Also *SAE Paper 2014-01-2261*).
33. Herencia, J. E., Weaver, P. M., & Friswell, M. I. (2008). Initial sizing optimization of anisotropic composite panels with T-shaped stiffeners. *Thin-Walled Structures, 46*, 399–412.
34. Hilton, H. H., & Yi, S. (1992). Analytical formulation of optimum material properties for viscoelastic damping. *Journal of Smart Materials and Structures, 1*, 113–122.

35. Beldica, C. E., & Hilton, H. H. (1999). Analytical simulations of optimum anisotropic linear viscoelastic damping properties. *Journal of Reinforced Plastics and Composites, 18*, 1658–1676.
36. Hilton, H. H., & Sossou, G. (2012). Viscoelastic and structural damping analysis with designer materials. In *Proceedings 50th AIAA Aerospace Sciences Meeting Multidisciplinary Design Optimization (MDO), AIAA Paper 2012-1256.*
37. Hilton, H. H. (2003). Optimum viscoelastic designer materials for minimizing failure probabilities during composite cure. *Journal of Thermal Stresses, 26*, 547–557.
38. Hilton, H. H. (2005). Optimum linear and nonlinear viscoelastic designer functionally graded materials—characterization and analysis. *Composites Part A: Applied Science and Manufacturing, 36*, 1329–1334.
39. Hilton, H. H. (2006). Designer linear viscoelastic material properties tailored to minimize probabilistic failures or thermal stress induced dynamic column creep buckling. *Journal of Thermal Stresses, 29*, 403–421.
40. Hilton, H. H., & Lee, D. H. (2006). Designer functionally graded viscoelastic materials performance tailored to minimize probabilistic failures in viscoelastic panels subjected to aerodynamic noise. In *Proceedings 9th International Conference on Recent Advances in Structural Dynamics*, 18–39. Southampton.
41. Hilton, H. H., & El Fouly, A. R. A. (2007). Designer auxetic viscoelastic sandwich column materials tailored to minimize creep buckling failure probabilities and prolong survival times. In *Proceedings 48th AIAA/ASME/ASCE/AHS/ASC SDM Conference, AIAA Paper 2007-2400.*
42. Hilton, H. H., & D'Urso, S. J. (2013). Designer Euler and elastica columns subjected to aerodynamic loads—System engineering of the aeroelasticity of wind turbine towers. In *Proceedings 54th AIAA/ASME/ASCE/AHS/ASC Structures, Structural Dynamics and Materials (SDM) Conference, AIAA Paper 2013-1821*. Boston, MA.
43. Naraghi, M., Chasiotis, I., & Hilton, H. H. (2009). Theory of designer nano – viscoelastic composites. In *Proceedings World Scientific and Engineering Academy International Conference on Continuum Mechanics 09*, 225–233. Cambridge.
44. Hilton, H. H. (2009). Analytical formulation of optimal viscoelastic designer material properties for sandwich/composites. In *Proceedings 16th Journées Nationales sur les Composites, Paper No. 119*. Toulouse.
45. Hilton, H. H., & Lee, D. H. (2006). Designer functionally graded viscoelastic materials performance tailored to minimize probabilistic failures in viscoelastic panels subjected to aerodynamic noise. In *Proceedings 9th International Conference on Recent Advances in Structural Dynamics, CD-ROM*:1-16. University of Southampton.
46. Hilton, H. H. (2006). Tailored designer functionally graded materials for minimizing probabilistic creep buckling failures in linear viscoelastic columns with large deformations and follower loads. In *Proceedings 47th AIAA/ASME/ASCE/AHS Structures, Structural Dynamics and Materials Conference, AIAA Paper AIAA-2006-1629*. Newport, RI.
47. Hilton, H. H. (2006). Tailored designer functionally graded materials for minimizing probabilistic creep buckling failures in linear viscoelastic columns with large deformations and follower loads. In *Proceedings 47th AIAA/ASME/ASCE/AHS Structures, Structural Dynamics and Materials Conference, AIAA Paper AIAA-2006-1629*. Reston, VA.
48. Hilton, H. H. (2007). Designer auxetic viscoelastic materials for sandwich plates tailored to minimize failure probabilities and prolong survival times. In *Proceedings of the 2007 International Mechanical Engineering Congress and Exposition. ASME Paper IMECE 2007-41079*. Seattle, WA.
49. Hilton, H. H. (2008). Functionally graded designer viscoelastic materials tailored to perform prescribed tasks with failure probabilities and survival times. In G. H. Paulino, M. J. Pindera, R. H. Dodds, F. A. Rochinha, E. V. Dave, & L. Chen (Eds.), *Proceedings Multiscale and Functionally Graded Materials Conference (FGM-IX)*, AIP-973:410–415. American Institute of Physics, Melville, NY.

50. Hilton, H. H., Lee,D. H., & Merrett,C. G. (2009). Wing torsional divergence avoidance through designer viscoelastic material properties and tailored aero-servo-controls. In *Proceedings International Forum on Aeroelasticity and Structural Dynamics, Paper IFASD-2009-146*. Reston, VA.
51. Hilton, H. H. (2009). A novel approach to structural analysis: Designer/engineered viscoelastic materials vs. 'off the shelf' property selections. *Journal of Vacuum Technology and Coating, 10*, 23–29.
52. Lee, D. H., Hilton, H. H., & Velicki, A. (2010). Optimum designer stitched composites. In *Proceedings 51st AIAA/ASME/ASCE/AHS/ASC Structures, Structural Dynamics and Materials (SDM) Conference, AIAA Paper 2010-2942*. Reston, VA.
53. Hilton, H. H., Lee, D. H., & Velicki, A. (2010). Designer composite materials to alleviate aeroelastic and aero-viscoelastic wing and panel problems (torsional divergence, flutter, aero-acoustics. In *Aerospace Flutter and Dynamics Council Conference*. NASA Langley, VA.
54. Lee, D. H., Hilton, H. H., & Velicki, A. (2011). Optimum designer materials and geometries for stitched composites. *Proceedings 52nd AIAA/ASME/ASCE/AHS/ASC Structures, Structural Dynamics and Materials (SDM) Conference, AIAA Paper 2011-1910*. Reston, VA.
55. Hilton, H. H. (2014). Designer viscoelastic materials for minimizing vibration effects. In *Proceedings 19th Symposium on Vibrations, Shock and Noise (VISHNO), Paper 135*. Aix-en-Provence.
56. Hilton, H. H., & D'Urso, S. (2014). Astro-elastic and astro-viscoelastic system engineering: Optimum solar sail configurations based on astrodynamics, designer materials, sizing and geometry. In *Proceedings AIAA SPACE 2014 Conference, AIAA Paper 2014-4203*. San Diego, CA. doi:10.2514/6.2014-4203.
57. Pedersen, P. (Ed.). (1992). *Optimal design with advanced materials: The Frithiof Niordson Volume: Proceedings of the IUTAM Symposium on Optimal Design with Advanced Materials*. Lyngby.
58. Vasiliev, V. V., & Gürdal, Z. (Eds.). (1999). *Optimal design: Theory and applications to materials and structures*. Lancaster, PA: Technomic Publishing Co.
59. Ricker, K. (2006). The shape of things to come. http://www.trecc.org/features/USSAshville/
60. Strebel, E. (2009). A new approach to structural analysis. *NCSA ACCESS, 22*(1), 23. http://www.ncsa.uiuc.edu/News/09/0129NCSAresearcher.html.
61. Hilton, H. H., Lee, D. H., & Merrett, C. G. (2009). Novel protocols of matching optimized designer aero-servo-controls with engineered viscoelastic materials. In *Proceedings IV ECCOMAS SMART-09, Paper ID038*. Porto.
62. Merrett, C. G., & Hilton, H. H. (2009). Panel flutter and aerodynamic noise attenuation through aero-servo-viscoelastic controls. In *Proceedings 50th AIAA/ASME/ASCE/AHS SDM Conference, AIAA Paper 2009-2512*. Reston, VA.
63. Lee, D. H., Hilton, H. H., & Velicki, A. (2010). Optimum designer/tailored stitched composites. In *Proceedings 51st AIAA/ASME/ASCE/AHS SDM Conference, AIAA Paper 2010-2942*. Reston, VA.
64. Van Krevelen, D. W. (1990). *Properties of polymers—Their correlation with chemical structure; their numerical estimation and prediction from additive group contributions* (3rd ed.). Amsterdam: Elsevier.
65. Emspak, J. (2012). Scientists use mathematical shortcut to create new metals. www.innovationnewsdaily.com/1546-mathematical-shortcut-create-new-metals.html
66. Cagle, C. M., & Schlecht, R. W. (2007). Composite elastic skins for shape changing structures. *NASA Tech Briefs LAR–16599–1*. http://www.techbriefs.com/content/view/1113/34/
67. Spencer, N. D. (2012). *Tailoring surfaces: Modifying surface composition and structure for applications in tribology, biology and catalysis*. Singapore: World Scientific.
68. Valasek, J. (Ed.). (2012). *Morphing aerospace vehicles and structures*. John Wiley & Sons, Hoboken. doi:10.1002/9781119964032.

69. Brinkmeyer, A. W., Santer, M., Pirrera, A., & Weaver, P. M. (2012). Morphing composite panel with pseudo-bistable viscoelastic behavior. In *SEM XII International Congress & Exposition on Experimental and Applied Mechanics, SEM Paper 404*.
70. Boisse, P. (Ed.). (2011). *Composite reinforcements for optimum performance*. Cambridge, UK: Woodhead Publishing.
71. Liebeck, R. H. (1973). A class of airfoils designed for high lift in incompressible flow. *Journal of Aircraft, 10*, 610–617.
72. Liebeck, R. H., & Ormsbee, A. L. (1970). Optimization of airfoils for maximum lift. *Journal of Aircraft, 7*, 409–415.
73. Liebeck, R. H. (1978). Design of subsonic airfoils for high lift. *Journal of Aircraft, 15*, 547–561.
74. Adkins, C. N., & Liebeck, R. H. (1983). Design of optimum propellers. In *AIAA Paper AIAA-1983-190*.
75. Weber, J. (1955). The calculation of the pressure distribution on the surface of thick cambered wings and the design of wings with given pressure distribution. In *R & M No. 3026, R.A.E. Report Aero. 2548*.
76. Weisshaar, T. A. (2006). Induced drag reduction using aeroelastic tailoring with adaptive control surfaces. *Journal of Aircraft, 43*, 157–164.
77. Volpe, G. (1990). Inverse airfoil design: A classical approach updated for transonic flow. *Progress in Astronautics and Aeronautics, 125*, 191–215 (AIAA, Reston, VA).
78. Volpe, G. (1983). The inverse design of closed airfoils in transonic flow. In *AIAA Paper 1983-504*.
79. Adkins, C. N., & Liebeck, R. H. (1994). Design of optimum propellers. *Journal of Propulsion and Power, 10*, 676–682.
80. Giguère, P., & Selig, M. S. (1998). New airfoils for small horizontal axis wind turbines. *ASME Journal of Solar Energy Engineering, 120*, 108–114.
81. Selig, M. S., & Guglielmo, J. J. (1997). High-lift low Reynolds number airfoil design. *Journal of Aircraft, 34*, 72–79.
82. Gopalarathnam, A., Broughton, B. A., McGranahan, B. D., & Selig, M. S. (2003). Design of low Reynolds number airfoils with trips. *Journal of Aircraft, 40*, 768–775.
83. Selig, M. S. (2011). PROFOIL-WWW: Airfoil design software for the web. http://www.profoil.org
84. Eppler, R. (1900). *Airfoil design and data*. Berlin: Springer.
85. Anonymous. (2000). Eppler airfoil design and analysis code. http://www.airfoils.com/eppler.htm
86. Selig, M. S., & Maughmer, M. D. (1992). Generalized multi-point inverse airfoil design. *AIAA Journal, 30*, 2618–2625.
87. Maughmer, M. D., & Somers, D. M. (1989). Design and experimental results for a high-altitude, long-endurance airfoil. *Journal of Aircraft, 26*, 148–153.
88. Selig, M. S. (2012). The UIUC airfoil data site. http://www.ae.illinois.edu/m-selig/ads.html
89. Eden, P. E. (2012). Time for CIASTA. *Aerospace Testing International, 12*, 14–18. http://viewer.zmags.com/publication/91faafef#/91faafef/2.
90. Marzat, J., Piet-Lahander, H., Damongeot, F., & Walters, E. (2012). Model-based fault analysis for aerospace systems: A survey. *Journal of Aerospace Engineering, 226*, 1329–1360.
91. Benavides, E. M. (2012). *Advanced engineering design—An integrated approach*. Philadelphia: Woodland Publishing.
92. Ceruti, A., & Marzocca, P. (2013). A conceptual approach to unconventional airship design and synthesis. *Journal of Aerospace Engineering*. doi:10.1061/(ASCE)AS.1943-5525.0000344.
93. Raymer, D. P. (2006). *Aircraft design: A conceptual approach* (4th ed.). Reston, VA: AIAA Educational Series.

94. Torenbeek, E. (2013). *Advanced aircraft design: Conceptual design, technology and optimization of subsonic civil airplanes*. New York: Wiley.
95. Sadraey, M. H. (2012). *Aircraft design: A systems engineering approach*. New York: Wiley.
96. Carichner, G. E., & Nicholai, L. M. (1984 & 2013). *Fundamentals of aircraft and airship design—Airship design and case studies* (Vol. 1 & 2). AIAA Education Series, Renton, VA.
97. Kundu, A. (2010). *Aircraft design*. West Nyack, NY: Cambridge University Press.
98. Fielding, J. P. (1999). *Introduction to aircraft design*. West Nyack, NY: Cambridge University Press.
99. Teichmann, F. K. (1944). *Airplane design manual*. New York: Pitman.
100. Volodin, V. V., Lyseitsev, N. K., & Maximovich, V. Z. (1985). *Idiosyncrasies of designing IC powered airplanes for vertical takeoff and landing*. Mashinosfroenie, Moscow (in Russian).
101. Austin, R. (2010). *Unmanned aircraft systems—UAVs design, development and deployment* (AIAA educational series). New York: John Wiley & Sons.
102. Moir, I., & Seabridge, A. (2013). *Design and development of aircraft systems* (2nd ed.). New York: John Wiley & Sons.
103. Nicolai, L. M. (1984). *Fundamentals of aircraft design*. Dayton, OH: METS (rev ed.).
104. Braha, D., Minai, A. A., & Ben-Yam, Y. (Eds.). (2006). *Complex engineered systems*. New York: Springer.
105. Ferman, M. A. (2011). *A wing design method for aerospace students and home builders—Strength, weight, flutter, divergence, buckling, deflection and twist*. Bloomington, IN: Trafford Publishing.
106. Shanley, F. R. (1952). *Weight-strength analysis of aircraft structures*. New York: Dover.
107. Dahan, E., Herman, Z. A. C., Procaccino, C. T., Wang, T., Bandyopadhyay, S., Ahern, D., et al. (2013). Integration into system functionality and decomposition as an extension to previous Mars exploration studies. In *AIAA SPACE 2013 Conference & Exposition, AIAA Paper ID 1661909*. San Diego, CA.
108. Hinrichsen, D., & Pritchard, A. J. (2005). *Mathematical systems theory I—Modelling, state space analysis, stability and robustness*. New York: Springer Verlag. ISBN 9783540441250.
109. Klein, V., & Morelli, E. A. (2006). *Aircraft system identification: Theory and practice*. AIAA Education Series, Reston, VA.
110. Martins, J. R. R. A., & Lambe, A. B. (2013). Multidisciplinary design optimization: A survey of architectures. *AIAA Journal, 51*, 2049–2075.
111. D'Urso, S. J., & Sivier, K. R. (1991). An example of industrial/interaction with undergraduate aircraft design program. *American Institute of Aeronautics and Astronautics, AIAA Paper 91-3116*.
112. Sivier, K. R., & D'Urso, S. J. (1994). Tauchi sizing experiments in the aircraft conceptual design process. *American Institute of Aeronautics and Astronautics, AIAA Paper ICAS – 94.1.8.5*.
113. Park, H.-U., Chung, J., & Neufeld, D. (2016). Uncertainty based aircraft derivative design for requirement changes. *The Aeronautical Journal, 120*, 375–389.
114. Teichmann, F. K. (1939). *Airplane Design Manual*. New York: Pitman.
115. Cassidy, P. F., Gatzke, T. D., & Vaporean, C. N. (2008). Integrating synthesis and simulation for conceptual design. *AIAA Paper 2008-1443*.
116. Long, D., & Scott, Z. (2011). *A primer for model-based systems engineering* (2nd ed.). Blacksburg, VA: Vitech Corp.
117. D'Urso, S. J. (1990). Configuring tactical aircraft. *AIAA Paper 90-3305*.
118. Howe, D. (2004). *Aircraft loading and structural layout*. AIAA Education Series, AIAA, Reston, VA.
119. Maier, M. W., & Rechtin, E. (2000). *The Art of Systems Architecting* (2nd ed.). Boca Raton: CRC Press.
120. Wright, T. P. (1936). Factors affecting the cost of airplanes. *Journal of the Aeronautical Sciences, 3*(4), 122–128.

121. Hirsch, W. Z. (1956). Firm progress ratios. *Econometrica, 24*, 136–143.
122. Raymer, D. P. (1992). *Aircraft design: A conceptual approach*. Reston, VA: American Institute of Aeronautics and Astronautics.
123. Nadeau, M. C., Kar, A., Roth, R., & Kirchain, R. (2010). A dynamic process-based cost modeling approach to understand learning effects in manufacturing. *International Journal of Production Economics, 128*, 223–234.
124. Diewert, W. E. (1974). *Applications of duality theory", Frontiers of Quantitative Economics 2:106–176*. Amsterdam: North-Holland Publishing Company.
125. Korzybski, A. (1958). *Science and sanity: An introduction to non-aristotelian systems and general semantics* (4th ed.). Lakeville, CT: Institute of General Semantics.
126. Heisenberg, W. (1962). *Physics and philosophy: The revolution in modern science*. New York: Harper.
127. Rish, I., Cecchi, G. A., Lozano, A., & Niculescu-Mizil, A. (Eds.). (2014). *Practical applications of sparse modeling*. Cambridge, MA: MIT Press.
128. Anonymous. (2011). http://www.ncsa.uiuc.edu/BlueWaters/
129. Anonymous. (2009). http://www.ncsa.illinois.edu/News/Stories/Kramer/
130. Tarantola, A. (2005). *Inverse problem theory and methods for model parameters estimation*. Philadelphia, PA: SIAM.
131. Jones, R. M. (2015). *Design of Composite Structures*. Blacksburg, VA: Bull Ridge Publishing.
132. Hilton, H. H., & Ariaratnam, S. T. (1993). Invariant anisotropic large deformation deterministic and stochastic combined load failure criteria. *International Journal of Solids and Structures, 31*, 3285–3293.
133. Aik, K. K. (2003). Plasma sprayed functionally graded ZrO_2/NiCoCrAlY thermal barrier coating. http://www.ntu.edu.sg/mae/research/programmes/adv_materials/FGM.htm
134. Anonymous. (2001). *Systems engineering fundamentals*. Fort Belvoir, VA: Defense Acquisition University Press.
135. Wasson, C. S. (2005). *System analysis, design, and development: Concepts, principles, and practices*. Hoboken, NJ: Wiley-Interscience.
136. Lagrange, J.-L. (1788). *Mécanique analytique*. Paris: Gauthier-Villars et fils.
137. Lagrange, J.-L. (1811). *Mécanique analytique*. Courcier, Paris. (2009) Reissued by Cambridge University Press, New York. ISBN: 978-1-108-00174-8.
138. Lagrange, J.-L. (1762). Essai d'une nouvelle methode pour déterminer les maxima et les minima des formules integrales indéfinies. *Mélanges de philosophie et de mathématique de la Société Royale de Turin, 1*, 173–195.
139. Komzsik, L. (2014). *Applied calculus of variations for engineers* (2nd ed.). Boca Raton, FL: CRC Press.
140. Elishakoff, I., & Ohsaki, M. (2005). *Eigenvalues of inhomogeneous structures*. CRC Press, Boca Rotan. doi:10.1201/9781420038019.
141. Hilton, H. H. (2012). Generalized fractional derivative anisotropic viscoelastic characterization. *Materials, 5*, 169–191. doi:10.3390/ma5010169.
142. Prony, G. C. F. M. R., & Baron de. (1795). Essai experimental et analytique. *Journal de l'École Polytechnique de Paris, 1*, 24–76.
143. Alfrey, T., Jr. (1948). *Mechanical behavior of high polymers*. New York: Interscience Publishers, Inc.
144. Read, W. T. (1950). Stress analysis for compressible viscoelastic materials. *Journal of Applied Physics, 21*, 671–674.
145. Freudenthal, A. M. (1950). *The inelastic behavior of engineering materials and structures*. New York: John Wiley & Sons.
146. Aklonis, J. J., MacKnight, W. J., & Shen, M. C. (1972). *Introduction to polymer viscoelasticity*. New York: Wiley.
147. Aklonis, J. J., & MacKnight, W. J. (1983). *Introduction to polymer viscoelasticity*. New York: Wiley.

148. Barnes, H. A., Hutton, J. F., & Walters, K. (1989). *An introduction to rheology*. Amsterdam: Elsevier.
149. Bartenev, G. M., & Zuyev, Y. S. (1968). *Strength and failure of viscoelastic materials*. Oxford: Pergamon Press.
150. Bažant, Z. P. (Ed.). (1988). *Mathematical modeling of creep and shrinkage of concrete*. New York: John Wiley and Sons.
151. Bažant, Z. P., & Cedolin, L. (1991). *Stability of structures—Elastic, inelastic, fracture and damage theories*. New York: Oxford University Press.
152. Bergen, J. T. (1960). *Viscoelasticity; phenomenological aspects*. New York: Academic.
153. Betten, J. (2005). *Creep mechanics* (2nd ed.). New York: Springer.
154. Biot, M. A. (1965). *Mechanics of incremental deformations*. New York: John Wiley & Sons.
155. Bland, D. R. (1960). *The theory of linear viscoelasticity*. New York: Pergamon Press.
156. Brinson, H. F., & Catherine Brinson, L. (2008). *Polymer engineering science and viscoelasticity: An introduction*. New York: Springer.
157. Christensen, R. M. (1982). *Theory of viscoelasticity—An introduction* (2nd ed.). New York: Academic.
158. Drozdov, A. D., & Kolmanovski, V. B. (1994). *Stability in viscoelasticity*. Amsterdam: North-Holland.
159. Drozdov, A. D. (1996). *Finite elasticity and viscoelasticity—A course in nonlinear mechanics of solids*. Singapore: World Scientific.
160. Drozdov, A. D. (1998). *Mechanics of viscoelastic solids*. New York: John Wiley & Sons.
161. Drozdov, A. D. (1998). *Viscoelastic structures mechanics of growth and aging*. San Diego: Academic.
162. Eringen, A. C. (1962). *Nonlinear theory of continuous media*. New York: McGraw-Hill.
163. Fabrizio, M., & Morro, A. (1992). *Mathematical problems in linear viscoelasticity*. Philadelphia, PA: SIAM.
164. Ferry, J. D. (1980). *Viscoelastic properties of polymers*. New York: John Wiley & Sons.
165. Findley, W. N., Lai, J. S., & Onaran, K. (1976). *Creep and relaxation of nonlinear materials*. Amsterdam: North-Holland Publ. Co.
166. Flügge, W. (1967). *Viscoelasticity*. Waltham, MA: Blaisdell Pub. Co.
167. Fung, Y. C. (1965). *Foundations of solid mechanics*. Englewood Cliffs, NJ: Prentice-Hall.
168. Golden, J. M., & Graham, C. A. C. (1988). *Boundary value problems in linear viscoelasticity*. Berlin: Springer Verlag.
169. Gross, B. (1953). *Mathematical structure of the theories of viscoelasticity*. Paris: Hermann & Cie.
170. Gunasekaran, S., & Mehmet Ak, M. (2003). *Cheese rheology and texture*. Boca Raton, FL: CRC Press.
171. Gurtin, M. E., & Sternberg, E. (1962). On the linear theory of viscoelasticity. *Archive for Rational Mechanics and Analysis, 11*, 291–356.
172. Hilton, H. H. (1964). An introduction to viscoelastic analysis. In E. Baer (Ed.), *Engineering design for plastics* (pp. 199–276). New York: Reinhold Publishing Corp.
173. Hilton, H. H., & Dong, S. B. (1964). *An analogy for anisotropic, nonhomogeneous, linear viscoelasticity including thermal stresses* (Development in mechanics, pp. 58–73). New York: Pergamon Press.
174. Hunter, S. C. (1960). Viscoelastic waves. *Progress in Solid Mechanics, 1*, 3–57.
175. Lakes, R. S. (1998). *Viscoelastic solids*. Boca Rotan: CRC Press.
176. Lakes, R. S. (2009). *Viscoelastic materials*. New York: Cambridge University Press.
177. Lazan, B. J. (1968). *Damping of materials and members in structural mechanics*. Oxford: Pergamon Press.
178. Le Tallec, P. (1990). *Numerical analysis of viscoelastic problems*. Berlin: Springer.
179. Levi, F., & Pizzetti, G. (1951). *Fluage, Plasticité, Précontrainte*. Paris: Dunod.
180. Lockett, F. J. (1972). *Nonlinear viscoelastic solids*. London: Academic.

181. Lodge, A. S., Renardy, M., & Nohel, J. A. (Eds.). (1985). *Viscoelasticity and rheology*. New York: Academic.
182. Mainardi, F. (2010). *Fractional calculus and waves in linear viscoelasticity: An introduction to mathematical models*. London: Imperial College Press.
183. Marques, S. P. C., & Creus, G. J. (2012). *Computational viscoelasticity*. Berlin: Springer.
184. Nashif, A. D., Jones, D. I. G., & Henderson, J. P. (1985). *Vibration damping*. New York: John Wiley & Sons.
185. Pipkin, A. C. (1972). *Lectures on viscoelasticity theory*. Berlin: Springer.
186. Renardy, M., Hrusa, W. J., & Nohel, J. A. (1987). *Mathematical problems in viscoelasticity*. Burnt Mill: Longmans Scientific and Technical Press.
187. Roland, C. M. (2011). *Viscoelastic behavior of rubbery materials*. New York: Oxford University Press.
188. Salençon, J. (1983). *Viscoélasticité*. Paris: Presses ENPC.
189. Schowalter, W. R. (1978). *Mechanics of non-Newtonian fluids*. Oxford: Pergamon.
190. Scott-Blair, G. W. (1949). *Survey of general and applied rheology*. London: I. Pitman and Son.
191. Shaw, M. T., & MacKnight, W. J. (2005). *Introduction to polymer viscoelasticity* (3rd ed.). Hoboken, NJ: Wiley-Interscience.
192. Tschoegl, N. W. (1989). *The phenomenological theory of linear viscoelastic behavior: An introduction*. New York: Springer.
193. Wineman, A. S., & Rajakopal, K. R. (2000). *Mechanical response of polymers—An introduction*. Cambridge: New York.
194. Zener, C. (1948). *Elasticity and anelasticity of metals*. Chicago: University of Chicago Press.
195. Allen, D. H., & Haisler, W. E. (1985). *Introduction to aerospace analysis*. New York: John Wiley & Sons.
196. Shanley, F. R., & Ryder, E. I. (1937). Stress ratios: The answer to the combined loading problem. *Aviation*, 36, 28–29, 43, 66, 69–70.
197. Lambros, J., Santare, M. H., Li, H., & Sapna, G. H. (1999). A novel technique for the fabrication of laboratory scale model functionally graded materials. *Experimental Mechanics, 39*, 184–190.
198. Asiedu, Y., & Gu, P. (1998). State of the art review. *International Journal of Production Research, 36*, 37–41.
199. Xu, Y., Erkoyuncu, J. A., Bankole, O., Goh, Y. M., Cheung, W. M., Baguley, P., et al. (2012). Cost engineering for manufacturing: Current and future research. *International Journal of Computer Integrated Manufacturing, 25*, 300–314.
200. Wei, W., & Hansen, M. (2005). Impact of aircraft size and seat availability on airlines' demand and market share in duopoly markets. *Transportation Research Part E: Logistics and Transportation Review, 41*, 315–327.
201. Inman, R. R. (1995). Shape characteristics of cost curves involving multiple shifts in automotive assembly plants. *The Engineering Economist, 41*(1), 53–67.
202. Scanlan, J., Hill, T., & Marsh, R. (2002). Cost modeling for aircraft design optimization. *Journal of Engineering Design, 13*, 37–41.
203. Layer, A., Brinke, E. T., Houten, F. V., Kals, H., & Haasis, S. (2002). Recent and future trends in cost estimation. *International Journal of Computer Integrated Manufacturing, 15* (6), 499–510.
204. Collopy, P. D., & Eames, D. J. H. (2001). *Aerospace manufacturing cost prediction form a measure of part definition information* (SAE Technical Paper 001-01-3004).
205. Hackman, S. T. (2008). *Production Economics Integrating the Microeconomic and Engineering Perspectives*. Berlin: Springer.
206. Alchian, A. (1949) "An airframe production function," *Project RAND Paper P-108*.
207. Minkowski, H. (1912). *Geometrie der Zahlen*. Leibzig: Teubner.

208. Shaikh, A. (1987). Humbug production function. In J. Eatwell, M. Milgate, & P. Newman (Eds.), *The new Palgrave: A dictionary of economics* (pp. 690–692). London: MacMillan Press.
209. Simon, H. A. (1979). On parsimonious explanations of production relations. *The Scandinavian Journal of Economics, 81*(4), 459–474.
210. Eiteman, W. J., & Guthrie, G. E. (1952). The shape of the average cost curve. *The American Economic Review, 42*, 832–838.
211. Arrow, K. J., Chenery, H. B., Minhas, B. S., & Solow, R. M. (1961). Capital-labor substitution and economic efficiency. *The Review of Economics and Statistics, 43*, 225–250.
212. Diewert, W. E. (1971). An application of the Shephard duality theorem: a generalized Leontief production function. *Journal of Political Economy, 79*, 481–507.
213. Christensen, L. R., Jorgenson, D. W., & Lau, L. J. (1973). Transcendental logarithmic production frontiers. *The Review of Economics and Statistics, 45*, 28–45.
214. Lau, L. J. (1986). Functional forms in econometric model building. *Handbook of econometric, 3*, 1515–1566.
215. Curran, R., Kundu, A. K., Wright, J. M., Crosby, S., Price, M., Raghunathan, S., et al. (2006). Modeling of aircraft manufacturing cost at the concept stage. *The International Journal of Advanced Manufacturing Technology, 31*(3-4), 407–420.
216. Velicki, A., & Thrash, P. (2008). Advanced structural concept development using stitched composites. In *49th AIAA/ASME/ASCE/AHS/ASC Structures, Structural Dynamics, and Materials Conference, Paper AAIAA 2008-2329*. Reston, VA.
217. Yovanof, N. P., Velicki, A., & Li, V. (2009). Advanced structural stability analysis of a noncircular, BWB–shaped vehicle. In *50th AIAA/ASME/ASCE/AHS/ASC Structures, Structural Dynamics, and Materials Conference, Paper AIAA 2009-2452*. Reston, VA.
218. Velicki, A., Thrash, P., & Jegley, D. (2009). Airframe development for the hybrid wing body aircraft. In *47th AIAA Aerospace Sciences Meeting Including The New Horizons Forum and Aerospace Exposition, Paper AIAA 2009-932*. Reston, VA.
219. Hilton, H. H. (2012). Designer materials for controlling thermally excited viscoelastic lifting surface flutter. In *Proceedings AIAA AFM Conference, AIAA Paper 2012-4952*. Minneapolis, MN.
220. Hamming, R. W. (1962). *Numerical methods for scientists and engineers*. New York: Dover.

Digital Twin: Mitigating Unpredictable, Undesirable Emergent Behavior in Complex Systems

Michael Grieves and John Vickers

1 Introduction

Up until fairly recently, the only way to have extensive knowledge of the physical object was to be in close proximity to that object. The information about any physical object was relatively inseparable from the physical object itself. We could have superficial descriptions of that object, but at best they were limited in both extensiveness and fidelity.

Such basic information as dimensions height, width, and length, only became available in the mid-1800s with the creation of the standard inch and a way to consistently measure that inch. Prior to that period of time, everyone had his or her own version of measurement definitions that meant that interchangeable manufacturing was impossible.

It was then only in the last half of the twentieth century, that we could strip the information from a physical object and create what we are calling a Digital Twin. This Digital Twin started off relatively sparse as a CAD description and is becoming more and more rich and robust over the years. While at first, this Digital Twin was merely descriptive, in recent years it is becoming actionable.

What actionable means is that the CAD object is no longer simply a three dimensional object hanging in empty space, time independent. We can now simulate physical forces on this object over time in order to determine its behavior. Where CAD models were static representations of form, simulations are dynamic representations, not only of form but also of behavior.

M. Grieves (✉)
Florida Institute of Technology, Melbourne, FL, USA
e-mail: mgrieves@fit.edu

J. Vickers
NASA MSFC, Huntsville, AL, USA

F.-J. Kahlen et al. (eds.), *Transdisciplinary Perspectives on Complex Systems*,
DOI 10.1007/978-3-319-38756-7_4

One of the previous issues with only being able to work with a physical object is that the range of investigation into its behavior was both expensive and time-consuming. We first had to physically create the object, a one-off proposition. We then had to create a physical environment in which the object was impacted by actual forces. This meant that we were limited to investigating forces and their associated levels that we thought were of concern. Often, the forces would result in destruction of the object, dramatically increasing the expense.

This meant that the first time we actually saw a condition not covered by a physical test would be when the physical object was in actual use. This meant that there were going to be many unforeseen conditions or emergent behaviors that resulted in failures that could result in harm and even death to its users.

Aggregating these objects into systems compounded the problem. Systems are much more sophisticated than simple objects. We need to have better ways to understand these increasingly sophisticated systems.

The idea of the Digital Twin is to be able to design, test, manufacture, and use the virtual version of the systems. We need to understand whether our designs are actually manufacturable. We need to determine the modes of failure when the system is in use. We need all of this information before the physical system is actually produced. This will reduce failures of the physical system when it is deployed and in use, reducing expenses, time, and most importantly harm to its users.

2 Conventional Approaches and Current Issues

The issue with complex systems is not that they are complex. A complex system that performed its tasks flawlessly, always produced the desired results, and gave forewarning that potential failures were likely in the future so that these potential failures could be corrected before they occurred would be a desirable system to have.

But, unfortunately, that is not usually the case with complex systems. The issue is that the systems do not always either perform flawlessly or do not produce the desired results. More importantly, they often fail without warning and can fail dramatically and catastrophically, with a minor issue cascading into a major failure.

It is this aspect of complex systems that is the thorny problem that needs major mitigation and/or resolution.

2.1 *Defining Complex Systems*

The first task is to define what a system and a complex system are. The definition of a system is taken to be the following [1][1]:

[1]Modified to add that the results could not be obtained from the components individually. While there are much more detailed descriptions of systems and their characteristics (see Ackoff, R. L. [2]) this definition suffices for the points we wish to make here.

A system is two or more components that combine together to produce from one or more inputs one or more results that could not be obtained from the components individually

Systems can be categorized into three types: simple, complicated, and complex.

Simple systems are just that. The outside observer has no problem in discerning the operation of the system. The system is completely predictable. The inputs are highly visible. The actions performed on those inputs are obvious and transparent. The outputs are easily predictable.

Complicated systems are also completely predictable. The system follows well-defined patterns [3]. The difference between simple systems and complicated systems is the component count. Complicated systems have many more components. Complicated systems are often described as intricate. However, the inputs are well known, as are the resulting outputs. The connection between components is linear and straightforward. A mechanical watch would be a representative example of a complicated system.

Complex systems are in a different class of systems entirely. There is not very good agreement as to how to even describe a complex system. In fact, there is little agreement on the term "complexity" [4].

Complex systems have been characterized as being a large network of components, many-to-many communication channels, and sophisticated information processing that makes prediction of system states difficult [5].

We would contend that complex systems have a major element of surprise, as in "I didn't see that coming." That surprise is generally, although not always, an unwelcome one.

2.2 Complex Systems and Associated Problems

While man-made complex systems can be considered relatively new phenomena, the issue of complex systems as it relates to serious and catastrophic problems goes back decades. The first major discussion of this issue is often considered to be Perrow's seminal work on the inherent dangers of complex systems, *Normal Accidents* [6]. Perrow defines the difference between complex and complicated systems.

Perrow defines complexity in terms of interactions (p. 78). He defines linear interactions as "those in expected and familiar production or maintenance sequence, and those that are quite visible even if unplanned". He defines complex interactions as, "those of unfamiliar sequences, or unplanned and unexpected sequences, and are either not visible or not immediately comprehensible." Perrow also uses "tightly-coupled" to describe complex systems and "loosely-coupled" to describe complicated systems.

In other terminology, Perrow's linear interactions and loosely coupled connections would be characteristic of a complicated system, whereas his complex interactions and tightly coupled connections would be the characteristic of a complex system. Perrow's claim, which he supports with numerous examples, is that complex systems lead quite naturally to the idea of "normal accidents".

Perrow's laundry list of normal accidents, better described as disasters, span the land, air, and sea. Examples include the Three-Mile Island nuclear reactor meltdown, the 1979 DC-10 crash in Chicago, and maritime disasters.

The common thread to Perrow's examples is the human element in interacting with complex systems, which makes these systems sociotechnical systems. It takes primarily two forms: human inconsistency, both deliberate and accidental, in following rules, processes, and procedures and a lack of sensemaking, i.e., the ability to make sense out of the inputs and stimuli that are being presented to the human.

The ubiquitous presence of computers that was not present in Perrow's day can prevent human inconsistency that is accidental or even intentional. Computers forget nothing and perform processes over and over again without any deviation. Computers can even go a long way in preventing the deliberate, error-causing human acts, by sensing what the system's state should be and comparing it against what it is, and raising an alarm if the two do not match.

One of the examples that Perrow uses is a fire door being propped open when it should not have been. In today's environment, a sensor would have been placed on that door and triggered an alarm in the computer that the door was open when it should be closed. If nothing else, humans are incredibly creative in trying to bypass such systems. However, the use of computers would and does dramatically decrease the numbers of even deliberate human intent to not follow procedures and processes.

The other source of human interaction problems, sensemaking, has played a role in major system disasters. This is an area that has been explored quite well by Karl Weick. The core issue here is that humans often do not do a good job in making sense of the information streaming at them, especially in stressful situations. As a result, they sometimes do not make the right decisions and not infrequently make exactly the wrong decision in the heat of a crisis.

Weick has his own list of disasters where this is happened, including the NASA disasters, Challenger and Columbia [7], and, what can be classified as a System-of-Systems failure, the air accident involving two 747s colliding on Tennerife in the Canary Islands [8].

This is a much more difficult issue to address, as computers do not do any sensemaking. They simply execute what they have been programmed to do. However, as will be discussed below, what computers can do is perform simulations before the system is deployed to both determine how the system reacts in a wide variety of conditions and train users of the system under abnormal conditions that they might face. We also will propose that simulated systems might "front-run" in real time developing conditions in order to assist humans in making the correct sense of developing situations and overcoming the biases that can negatively affect human decision making [9].

2.3 Defining Emergence and Emergent Behavior

"Emergence" is too general a term to be of much use. In the dynamic universe in which we live, emergence is a keen sense of the obvious. Emergence of the universe itself has existed since the Big Bang.

Discussion about emergence goes back to at least the times of Aristotle who in Metaphysics introduced the concept that the whole is more than the sum of its parts [10]. The idea of emergence spans a wide, wide spectrum. It goes from the idea that a pile of bricks has the potential emergence behavior of a house [11] to a system that learns and adapts to its environment. Emergence covers the idea of both emergence of form and function or behavior.

"Emergent behavior" is a little better in that it narrows the discussion down to function, as opposed to emergent form, which refers to the physical manifestation of a system. However, as we shall discuss, it still is both too general and too ambiguous to pin down exactly what we are referring to. Is emergent behavior a result of new and unique behavior that has arisen or is it behavior that has been a possibility from inception, but simply has not been recognized [12]?

Much more recently, Holland [13] proposes a taxonomy of emergent behavior. His taxonomy is as follows:

Type 0: Constituent (Non-Emergence)
Type 1: Nominal Emergence
Type 2: Moderated Emergence
Type 3: Multiple Emergence
Type 4: Evolutionary Emergence

Of the five types, only Type 4 would qualify as behavior that results from modifications to the system as it reacts with its environment. It is dynamic emergent behavior. Evolutionary emergence involves a deliberate learning/modification behavior loop that humans exhibit. Evolutionary emergence can also be randomly generated, such as nature uses in evolutionary genetics and humans, often to their detriment, exhibit randomness in times of crisis. The Type 4 systems are actually different systems over time.

In the other four types, the emergent behavior exists from the point that the system is completed and deployed. It is static emergent behavior. It is not that something new creeps into the system. It is that we have not foreseen this behavior. Even where the behavior is a result of human interaction, the system at its inception would have performed in exactly this fashion when presented with these inputs.

The system did not change. We just "didn't see that coming!" It is these first four types of static emergent behavior that we will deal with, static emergent behavior that is built into the system and is unforeseen rather than arises from dynamic system modification or randomness.

While this taxonomy is useful in describing the type of emergent behavior as it increases in sophistication of actions and communications, it does not describe what emergent behavior is desirable and predictable. We cannot rely on the name to give us this indication. Holland's Type 2b, Moderated Unstable, which results in rapid growth changes and general instability, e.g., explosions, sounds undesirable and probably is in a number of situations. However, this behavior is quite useful in weapon systems.

2.4 *Four Categories of System Behavior*

Figure 1 characterizes behavior along the lines of desirable and undesirable behaviors. It divides static emergent behavior into two categories: predicted and unpredicted. The predicted and unpredicted categories are also divided into two categories: desirable and undesirable. This gives us four final categories: Predicted Desirable (PD), Predicted Undesirable (PU), Unpredicted Desirable (UD), and Unpredicted Undesirable (UU).

The PD category is obviously the desired behavior of our system. This is the intentional design and realization of our system. In systems engineering terms, it is the requirements our system is designed to meet.

The PU category contains problems we know about and will do something about eliminating. PU problems are management issues. If we know about the PU problems and do nothing about them, that is engineering malpractice. This is the category that expensive lawsuits are made of.

The unpredicted category is our "surprise" category. The UD category contains pleasant surprises. While ignorance is never bliss, this category will only hurt our pride of not having understood our system well enough to predict these beneficial behaviors.

The UU category holds the potential for serious, catastrophic problems. It is this category of emergent behavior that we need to focus on. We need capabilities and methodologies to mitigate and/or eliminate any serious problems and even reduce unforeseen problems that are merely annoying. This is the purpose of the Digital Twin model and methodology.

Finally, we will not abandon emergent form, because that is of great importance in the manufacturing phase and operational phase. However, physical forms have a permanency that behavior does not.

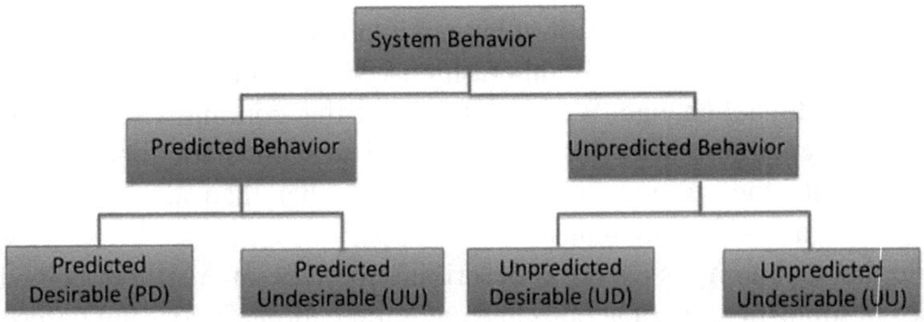

Fig. 1 Categories of system behavior

2.5 *Emergence, Emergent Behavior, and the Product Lifecycle*

Systems do not pop into existence fully formed, exhibiting behavior, emergent or otherwise. Systems get the way they are by a progression through their product lifecycle. This is shown in Fig. 2 [14]. The system's product lifecycle starts at system creation, takes on its physical attributes during the production phase, and then exist as an entity during its operational phase, when its behavior is a result of how the system was created and produced.

During the creation phase, designers create the characteristics and behaviors of the desired system. In terms of systems engineering, this means defining the requirements of the system. It is at this stage that the desirable attributes of the system are defined. Any undesirable behaviors are also identified and counter strategies are created to prevent them from occurring. The issue is that it is easier to enumerate the behaviors that are desired than it is to identify all the possible ways that the system can go wrong.

It is this phase that we should be considering the "ilities" of the system: manufacturability, sustainability, supportability, and disposability. Historically, we have known this. Realistically, we have done a very poor job in this regard. We have operated in a siloed fashion, with system design and engineering performing their tasks of defining the system requirements and verifying and validating them. Design and engineering then throw the plans over the wall to manufacturing. However, this siloing exists even within the design and engineering

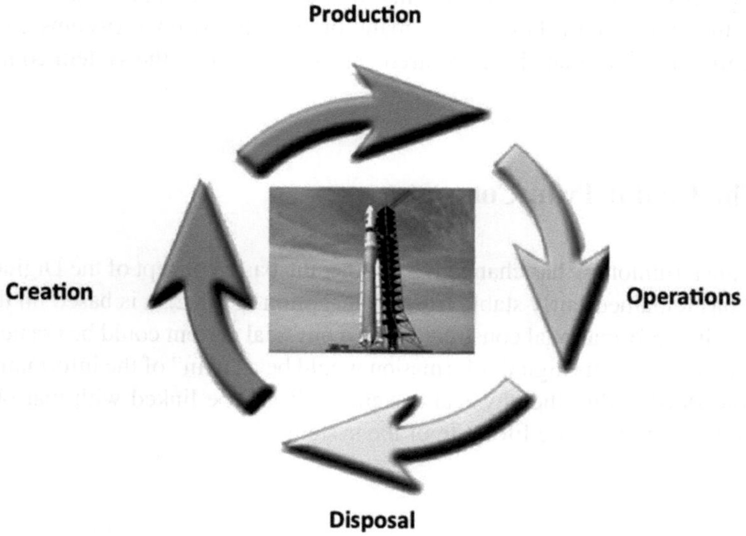

Fig. 2 Product lifecycle—4 phases

function, leading to undesirable behaviors arising because of lack of understanding at interface points between various sub-functions and assemblies.

In the production or manufacturing phase, the system emerges as a physical artifact. It is at this phase that we find whether or not our system design is realizable. We may find out that the manufacturing techniques that we have at our disposal are insufficient to realize the design that we have envisioned. It is at this phase that undesirable behaviors can also start to creep into the system as we change the design to meet the realities of manufacturing.

It is in the operational phase where we find out all our categories of behavior. We obviously would not put into operation those systems that do not meet the requirements that we set out in the creation phase (PD). We also would have eliminated all the known undesirable behaviors (PU). However, given the traditional approach to system development, we would still continue to be bedeviled by unpredicted undesirable behaviors (UU).

The lifecycle ends with the retirement and disposal of the system. This is generally not considered to have relevance to system complexity or emergent behavior. That may indeed be why there is little discussion of it. However, for some systems we might want to rethink that position. Two examples that come to mind are spent nuclear fuel and decommissioned spacecraft in orbit. While we will not dwell on the disposal issue, we will make mention of it when relevant.

We have proposed that Systems Engineering and Product Lifecycle Management highly overlap at a minimum and may in fact be the one and the same [1]. However, as a current practice, they are not generally considered as such.

In theory, Systems Engineering should concern itself with the behavior of the system throughout the entire lifecycle. Product Lifecycle Management (PLM) claims to do just that. As practiced, Systems Engineering has often degenerated into "systems accounting," with systems engineers considering their job done at the end of the create phase [14, 15]. In many organizations, once systems engineers have verified and validated the requirements, they consider the system completed.

3 The Digital Twin Concept

While the terminology has changed over time, the basic concept of the Digital Twin model has remained fairly stable from its inception in 2002. It is based on the idea that a digital informational construct about a physical system could be created as an entity on its own. This digital information would be a "twin" of the information that was embedded within the physical system itself and be linked with that physical system through the entire lifecycle of the system.

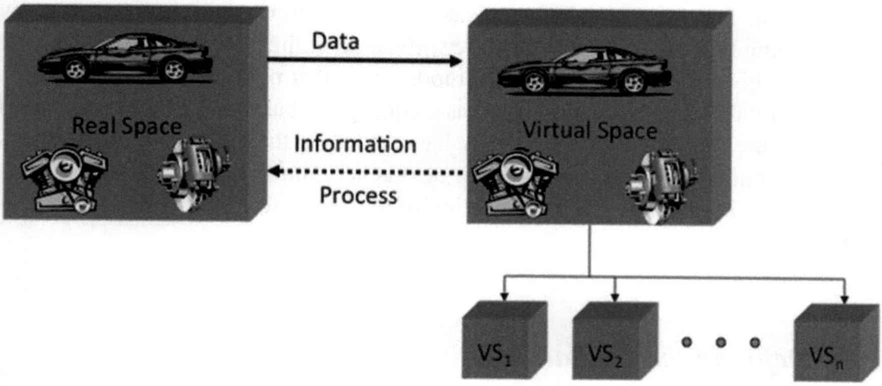

Fig. 3 Conceptual ideal for PLM. Dr. Michael Grieves, University of Michigan, Lurie Engineering Center, Dec 3, 2002

3.1 Origins of the Digital Twin Concept

The concept of the Digital Twin dates back to a University of Michigan presentation to industry in 2002 for the formation of a Product Lifecycle Management (PLM) center. The presentation slide, as shown in Fig. 3, was simply called "Conceptual Ideal for PLM." However, it did have all the elements of the Digital Twin: real space, virtual space, the link for data flow from real space to virtual space, the link for information flow from virtual space to real space and virtual sub-spaces.

The premise driving the model was that each system consisted of two systems, the physical system that has always existed and a new virtual system that contained all of the information about the physical system. This meant that there was a mirroring or twinning of systems between what existed in real space to what existed in virtual space and vice versa.

The PLM or Product Lifecycle Management in the title meant that this was not a static representation, but that the two systems would be linked throughout the entire lifecycle of the system. The virtual and real systems would be connected as the system went through the four phases of creation, production (manufacture), operation (sustainment/support), and disposal.

This conceptual model was used in the first PLM courses at the University of Michigan in early 2003, where it was referred to as the Mirrored Spaces Model. It was referenced that way in a 2005 journal article [16]. In the seminal PLM book, *Product Lifecycle Management: Driving the Next Generation of Lean Thinking*, the conceptual model was referred to as the Information Mirroring Model [17].

The concept was greatly expanded in *Virtually Perfect: Driving Innovative and Lean Products through Product Lifecycle Management* [14], where the concept was still referred to as the Information Mirroring Model. However, it is here that the term, Digital Twin, was attached to this concept by reference to the co-author's way of describing this model. Given the descriptiveness of the phrase, Digital Twin, we have used this term for the conceptual model from that point on.

The Digital Twin has been adopted as a conceptual basis in the astronautics and aerospace area in recent years. NASA has used it in their technology roadmaps [18]. The concept has been proposed for next generation fighter aircraft and NASA vehicles [19, 20], along with a description of the challenges [20] and implementation of as-builts [21].

3.2 Defining the Digital Twin

What would be helpful are some definitions to rely on when referring to the Digital Twin and its different manifestations. We would propose the following:

Digital Twin (DT)—the Digital Twin is a set of virtual information constructs that fully describes a potential or actual physical manufactured product from the micro atomic level to the macro geometrical level. At its optimum, any information that could be obtained from inspecting a physical manufactured product can be obtained from its Digital Twin. Digital Twins are of two types: Digital Twin Prototype (DTP) and Digital Twin Instance (DTI). DT's are operated on in a Digital Twin Environment (DTE).

Digital Twin Prototype (DTP)—this type of Digital Twin describes the prototypical physical artifact. It contains the informational sets necessary to describe and produce a physical version that duplicates or twins the virtual version. These informational sets include, but are not limited to, Requirements, Fully annotated 3D model, Bill of Materials (with material specifications), Bill of Processes, Bill of Services, and Bill of Disposal.

Digital Twin Instance (DTI)—this type of Digital Twin describes a specific corresponding physical product that an individual Digital Twin remains linked to throughout the life of that physical product. Depending on the use cases required for it, this type of Digital Twin may contain, but again is not limited to, the following information sets: A fully annotated 3D model with Geometric Dimensioning and Tolerancing (GD&T) that describes the geometry of the physical instance and its components, a Bill of Materials that lists current components and all past components, a Bill of Process that lists the operations that were performed in creating this physical instance, along with the results of any measurements and tests on the instance, a Service Record that describes past services performed and components replaced, and Operational States captured from actual sensor data, current, past actual, and future predicted.

Digital Twin Environment (DTE)—this is an integrated, multi-domain physics application space for operating on Digital Twins for a variety of purposes. These purposes would include:

Predictive—the Digital Twin would be used for predicting future behavior and performance of the physical product. At the Prototype stage, the prediction would be of the behavior of the designed product with components that vary between its high and low tolerances in order to ascertain that the as-designed product met the proposed requirements. In the Instance stage, the prediction would be a specific instance of a specific physical product that incorporated actual components and component history. The predictive performance would be based from current point in the product's lifecycle at its current state and move forward. Multiple instances of the product could be aggregated to provide a range of possible future states.

Interrogative—this would apply to DTI's. Digital Twin Instances could be interrogated for the current and past histories. Irrespective of where their physical counterpart resided in the world, individual instances could be interrogated for their current system state: fuel amount, throttle settings, geographical location, structure stress, or any other characteristic that was instrumented. Multiple instances of products would provide data that would be correlated for predicting future states. For example, correlating component sensor readings with subsequent failures of that component would result in an alert of possible component failure being generated when that sensor pattern was reported. The aggregate of actual failures could provide Bayesian probabilities for predictive uses.

3.3 The Digital Twin Model Throughout the Lifecycle

As indicated by the 2002 slide in Fig. 3, the reference to PLM indicated that this conceptual model was and is intended to be a dynamic model that changes over the lifecycle of the system. The system emerges virtually at the beginning of its lifecycle, takes physical form in the production phase, continues through its operational life, and is eventually retired and disposed of.

In the create phase, the physical system does not yet exist. The system starts to take shape in virtual space as a Digital Twin Prototype (DTP). This is not a new phenomenon. For most of human history, the virtual space where this system was created existed only in people's minds. It is only in the last quarter of the twentieth century that this virtual space could exist within the digital space of computers.

This opened up an entire new way of system creation. Prior to this leap in technology, the system would have to have been implemented in physical form, initially in sketches and blueprints but shortly thereafter made into costly prototypes, because simply existing in people's minds meant very limited group sharing and understanding of both form and behavior.

In addition, while human minds are a marvel, they have severe limitations for tasks like these. The fidelity and permanence of our human memory leaves a great deal to be desired. Our ability to create and maintain detailed information in our memories over a long period of time is not very good. Even for simple objects, asking us to accurately visualize its shape is a task that most of us would be hard-pressed to do with any precision. Ask most of us to spatially manipulate complex shapes, and the results would be hopelessly inadequate.

However, the exponential advances in digital technologies means that the form of the system can be fully and richly modeled in three dimensions. In the past, emergent form in complex and even complicated system was a problem because it was very difficult to insure that all the 2D diagrams fit together when translated into 3D objects.

In addition, where parts of the system move, understanding conflicts and clashes ranged from difficult to impossible. There was substantial wasted time and costs in translating 2D blueprints to 3D physical models, uncovering form problems, and going back to the 2D blueprints to resolve the problems and beginning the cycle anew.

With 3D models, the entire system can be brought together in virtual space, and the conflicts and clashes discovered cheaply and quickly. It is only once that these issues of form have been resolved that the translation to physical models need to occur.

While uncovering emergent form issues is a tremendous improvement over the iterative and costly two-dimensional blueprints to physical models, the ability to simulate behavior of the system in digital form is a quantum leap in discovering and understanding emergent behavior. System creators can now test and understand how their systems will behave under a wide variety of environments, using virtual space and simulation.

Also as shown in Fig. 3, the ability to have multiple virtual spaces as indicated by the blocks labeled $VS_1 \ldots VS_n$ meant that that the system could be put through destructive tests inexpensively. When physical prototypes were the only means of testing, a destructive test meant the end of that costly prototype and potentially its environment. A physical rocket that blows up on the launch pad destroys the rocket and launch pad, the cost of which is enormous. The virtual rocket only blows up the virtual rocket and virtual launch pad, which can be recreated in a new virtual space at close to zero cost.

The create phase is the phase in which we do the bulk of the work in filling in the system's four emergent areas: PD, PU, UD, and UU. While the traditional emphasis has been on verifying and validating the requirements or predicted desirable (PD) and eliminating the problems and failures or the predicted undesirable (PU), the DTP model is also an opportunity to identify and eliminate the unpredicted undesirable (UU). By varying simulation parameters across the possible range they can take, we can investigate the non-linear behavior that may have combinations or discontinuities that lead to catastrophic problems.

Once the virtual system is completed and validated, the information is used in real space to create a physical twin. If we have done our modeling and simulation correctly, meaning we have accurately modeled and simulated the real world in virtual space over a range of possibilities, we should have dramatically reduced the number of UUs.

This is not to say we can model and simulate all possibilities. Because of all the possible permutations and combinations in a complex system, exploring all possibilities may not be feasible in the time allowed. However, the exponential advances in computing capability mean that we can keep expanding the possibilities that we can examine.

It is in this create phase that we can attempt to mitigate or eradicate the major source of UUs—ones caused by human interaction. We can test the virtual system under a wide variety of conditions with a wide variety of human actors. System designers often do not allow for conditions that they cannot conceive of occurring. No one would think of interacting with system in such a way—until people actually do just that in moments of panic in a crisis.

Before this ability to simulate our systems, we often tested systems using the most competent and experienced personnel because we could not afford expensive failures of physical prototypes. But most systems are operated by a relatively wide range of personnel. There is an old joke that goes, "What do they call the medical student who graduates at the bottom of his or her class?" Answer, "Doctor." We can now afford to virtually test systems with a diversity of personnel, including the least qualified personnel, because virtual failures are not only inexpensive, but they point out UUs that we have not considered.

We next move into the following phase of the lifecycle, the production phase. Here we start to build physical systems with specific and potentially unique configurations. We need to reflect these configurations, the as-builts, as a DTI in virtual space so that we can have knowledge of the exact specifications and makeup of these systems without having to be in possession of the physical systems.

So in terms of the Digital Twin, the flow goes in the opposite direction from the create phase. The physical system is built. The data about that physical build is sent to virtual space. A virtual representation of that exact physical system is created in digital space.

In the support/sustain phase, we find out whether our predictions about the system behavior were accurate. The real and virtual systems maintain their linkage. Changes to the real system occur in both form, i.e., replacement parts, and behavior, i.e., state changes. It is during this phase that we find out whether our predicted desirable performance actually occurs and whether we eliminated the predicted undesirable behaviors.

This is the phase when we see those nasty unpredicted undesirable behaviors. If we have done a good job in ferreting out UUs in the create phase with modeling and simulation, then these UUs will be annoyances but will cause only minor problems. However, as has often been the case in complex systems in the past, these UUs can be major and costly problems to resolve. In the extreme cases, these UUs can be catastrophic failures with loss of life and property.

In this phase the linkage between the real system and virtual system goes both ways. As the physical system undergoes changes we capture those changes in the virtual system so that we know the exact configuration of each system in use. On the other side, we can use the information from our virtual systems to predict performance and failures of the physical systems. We can aggregate information over a range of systems to correlate specific state changes with the high probability of future failures.

As mentioned before, the final phase, disposal/decommissioning, is often ignored as an actual phase. There are two reasons in the context of this topic why the disposal phase should receive closer attention. The first is that knowledge about a system's behavior is often lost when the system is retired. The next generation of the system often has similar problems that could have been avoided by using knowledge about the predecessor system. While the physical system may need to be retired, the information about it can be retained at little cost.

Second, while the topic at hand is emergent behavior of the system as it is in use, there is the issue of emergent impact of the system on the environment upon disposal. Without maintaining the design information about what material is in the system and how it is to be disposed of properly, the system may be disposed of in a haphazard and improper way.

3.4 System Engineering Models and the Digital Twin

Systems Engineering is commonly represented by a few different models: the Waterfall Model, the Spiral Model, and the Vee Model (Fig. 4) [14, 15, 17]. What these models have in common is a sequential perspective. The Waterfall model is clearly sequential as the sequence flows from design to operation. The Spiral model reflects the same, although there is an iterative aspect to it. The Vee model implies a deconstruction and push down of requirements to the component level and a building back up from components to the complete system.

While these are conceptual models, the messy reality of systems development is that the ideal forward flow from inception to system is simply that—an ideal. What actually happens is that there is a great deal of determining that the system as designed does not really deliver the desired behavior, cannot be manufactured, and is not supportable or sustainable at the desired cost levels. Even when the design goes according to plan using the Vee model, the claim is that it leads to highly fragile systems [22].

The Digital Twin implementation model, as shown in Fig. 5, attempts to convey a sense of being iterative and simultaneous in the development process. Unlike the Waterfall or even Spiral Models, the downstream functional areas as conventionally thought of are brought upstream into the create phase. The "ilities" are part of the considerations of system design.

In fact, these areas can and should influence the design. For example being able to manufacture a honeycombed part through additive manufacturing would result in a part meeting its performance requirement at a significant savings in weight. Without that consideration, designers would specify a more expensive material in order to meet the weight and performance requirements.

What makes this new model feasible is the ability to work in the virtual space of the Digital Twin. The classic sequential models of Systems Engineering were necessitated by the need to work with physical objects. Designs had to be translated into expensive physical prototypes in order to do the downstream work of say,

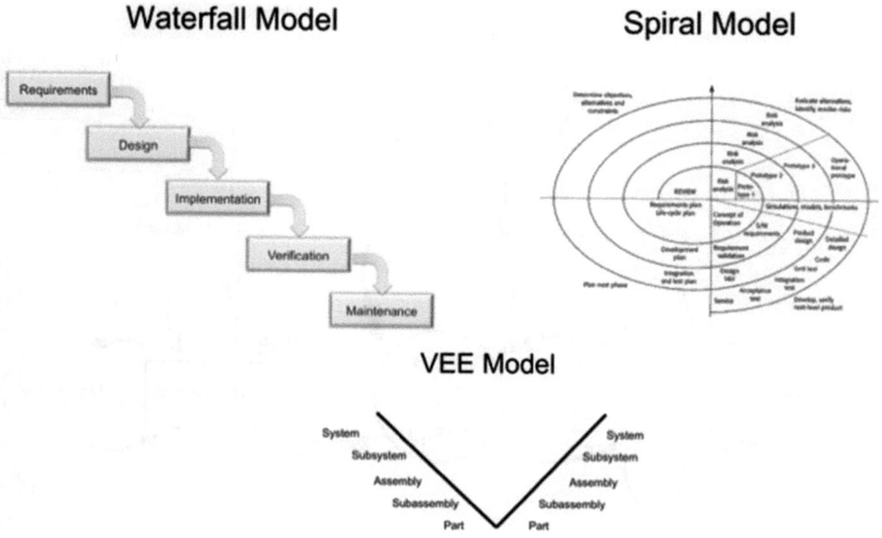

Fig. 4 System engineering model

manufacturing. This meant that only a subset of designs could be considered, because the cost of getting it wrong and having to go back and redesign was expensive and time consuming.

The Digital Twin changes that with its ability to model and simulate digitally. As indicated in Fig. 5, downstream functional areas can influence design because working with digital models in the create phase is much cheaper and faster and will continue to move in that direction.

While this new model is advantageous with traditional subtractive manufacturing methods, it will be required as additive manufacturing technologies advance and become mainstream production capabilities. As described by Witherell et al. [23] in another chapter ("Additive Manufacturing: A Trans-disciplinary Experience") here, the design-to-production process for additive manufacturing is much more integrative than subtractive manufacturing. Integration is a major hallmark of the Digital Twin Implementation Model. Additionally, the authors describe Digital Twin like modeling and simulation as having high potential for additive manufacturing part qualification.

While Digital Twin Implementation Model needs more detail and maturation, the concepts behind it, addressing the "ilities" as early as possible, integration across the lifecycle, and fast iterations, address the shortcomings of the current Systems Engineering models. The maturation and adoption of additive manufacturing will only serve to highlight this need.

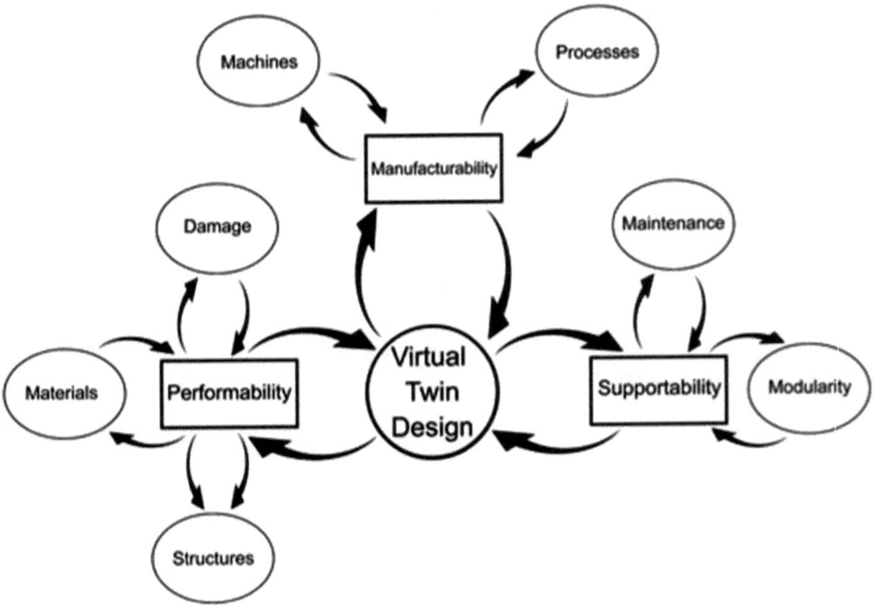

Fig. 5 Digital twin implementation model

3.5 The Digital Twin and Big Data

The Digital Twin model requires massive amounts of information and computing capability. Its progression will rely on advances in computer and communications technology. As things currently stand, this should not be a barrier to the advancement of this model. Moore's law is still alive and well, with not only computing technology growing exponentially, but also storage and bandwidth.

The advantages of using the virtual model rather than relying on expensive prototypes can be illustrated in Fig. 6. This is the cost of working with physical material versus virtual. If we assume that the costs are equal today, and project those costs into the future, we can see how they diverge with physical costs increasing at the rate of inflation and virtual cost decreasing on an exponential basis.

The other advantage is that with atom-based models, they have to exist in a certain geographical location. The parts of the system cannot be distributed if we want to execute the system itself. With virtual systems, the components of the system can be widely distributed, without the user even knowing where the components of the system reside. All this requires coordination among the virtual components. While this is a nontrivial requirement, the trajectory of computing technology is enabling this capability.

Fig. 6 Real vs. virtual costs

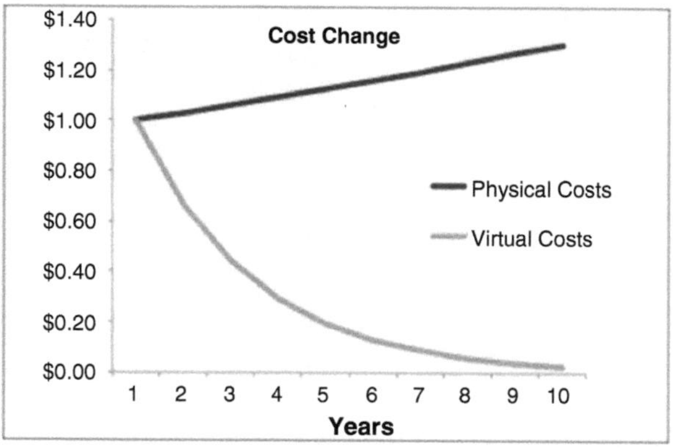

We are also driving deeper and deeper into the make-up of physical structures. The expectation is that we will eventually model at the atom level, incorporating the physical rules of behavior that are embedded with physical components and systems. Boeing's "Atoms to Airplanes" is an example of the initiatives in this area.

While there are many challenges in the area of big data, all indications are that the necessary technical capabilities exist or will exist to enable the Digital Twin.[2]

4 Value of the Digital Twin Model

So what is the value of the Digital Twin model? The Digital Twin is about information. While much has been written about information, it still is a relatively fuzzy concept. Many focus on the "inform" part of information and deal with it as a transmission issue.

However, the core premise for the Digital Twin model is that information is a replacement for wasted physical resources, i.e., time, energy, and material. Take any task, designing an airplane, manufacturing a fuel tank, or diagnosing and repairing an automobile. As shown in Fig. 7 [14, 15, 17], we can represent that task quantitatively as the sum of the cost of all resources required to complete the task. Since we live in a monetized society, we can cost the labor time, the material costs, and the energy costs over the life of the task.

Furthermore, as shown on the left side of the figure, we can divide the tasks into two parts. The bottom part is the minimum amount of physical resources that it would take to perform the task. If we were omniscient and omnipotent, we would

[2]The technical issues will most likely not be what gate this advancement. The lack of interoperability between vendor software systems that implement different aspects of this Digital Twin functionality will be the major issue. The user community will need to be proactive in encouraging the various vendors to play nice in the digital sandbox.

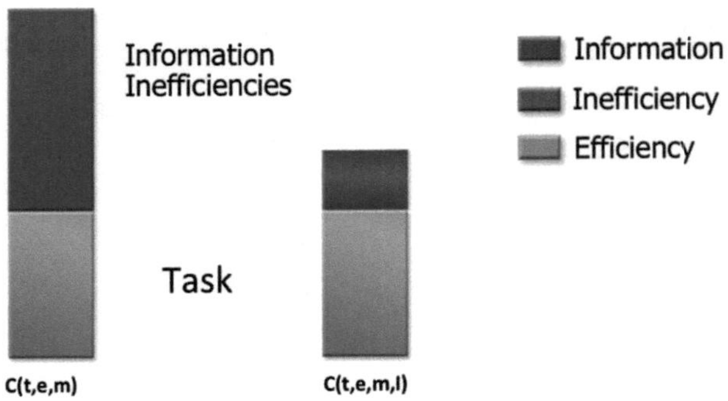

Fig. 7 Information as time, energy, material trade-off

know what the minimum amount of resources it took to perform the task and then perform the task in the exact way that minimizes those resources.

The upper part of the bar represents the waste in resources that we incur in actually performing the task. Because we are not omniscient, we waste resources in a wide variety of ways. We design parts that do not fit with each other. We design entire systems that do not produce the behavior we think that they will produce. We design components that cannot be manufactured. We produce system designs that require far more upkeep than we predicted.

On the right side of the Fig. 7, we have the same task. The amount of resources needed to perform the task without waste is exactly the same. But on the upper part of the bar we have information that allows us to replace some, but not usually all, of the wasted resources necessary to perform this task.

Information is never free to acquire or use, so there is always some use of resources in developing and using information. The key assumption here always is that the cost of these resources of information will be less than the cost of the wasted resources. For repetitive tasks and tasks of complexity, this is usually a reasonable assumption.

Obviously this figure is meant to be illustrative and ideal and not definitive. Humans are neither omniscient nor omnipotent. There will always be information that we do not have or task execution that is not perfect. There will always be wasted physical resources that information does not replace. However, this does not invalidate the point that the use of information can greatly reduce the waste of physical resources.

If we look at the Digital Twin through its lifecycle, we can see where we can obtain this value in each of the phases.

In the create phase, being able to model and simulate both form and behavior via a DTP allow us to save wasted physical resources and the time-consuming and costly production of physical prototypes. Determining predicted undesirable

(PU) behavior and discovering unpredicted undesirable (UU) behavior with simulations save resources both in the create phase and downstream.

In the production phase, simulating manufacture before it occurs allows us to reduce trial and error manufacturing. Collecting the as-built information allows us to understand what components with what specifications are in each system that we produce.

In the operations or support/sustainment phase, the Digital Twin allows us to understand how to more efficiently and effectively maintain the system. In addition, if we can prevent undesirable behaviors, both predicted and unpredicted, we can potentially save the cost of unforeseen "normal accidents". While the claim that human life is priceless, we do put a cost on life as we limit the tests for the states that systems can take. That is we only test for a subset of states the systems can take. However, with the use of the Digital Twin we can dramatically lower the cost that we incur in preventing loss of life by testing more states.

In the disposal phase, having the information about how the system was designed to be safely retired and/or decommissioned will greatly reduce the impact cost on the environment.

The Digital Twin has the potential to have a major impact on reduction of wasted resources in the lifecycle of our systems. Preventing a catastrophe caused by undesirable emergent behavior that results in loss of life is, as they said in the MasterCard ad, priceless.

4.1 Evaluating Virtual Systems

If we are going to create virtual systems, then we will need to evaluate how well these virtual systems mirror their physical counterparts. Grieves has proposed a series of tests, called the Grieves' Tests of Virtuality (GTV), to evaluate how well virtual systems mirror their physical counterparts ([17], pp. 18–19).

These test are based on the Turing tests of intelligence, proposed by Alan Turing in 1950 [24]. Turing called this the "Imitation Game".[3] The premise of the test was that an observer, isolated from a human and a computer could ask questions to both. If the observer could not tell the computer from the human, the computer passes the test. To date, in spite of some claims to the contrary, no computer has passed this test.

However, with all deference to Turing, he had the right idea, but was focusing on the wrong subject. The imitation that computers may be capable of is not imitating human intelligence, but imitating the physical world, *except* for human intelligence.

The premise of the GTV tests is very similar to Turing's Imitation Game. In the test, a human observer is positioned in a room that has two screens. One screen is a

[3]We suspect that Turing would be astounded that this would be the name of a movie featuring his life and work.

video feed from an actual room where a physical system is present. The other screen is connected to a computer.

There are three tests that the observer can perform. These three tests are: a Sensory Visual Test, a Performance Test, and a Reflectivity Test.

When first proposed, the second two tests were the same. However, the first test was called the Visual Test. It was pointed out that we have other senses than simply our eyes. This indeed is true. However, our visual sense is our highest bandwidth sense.

The other senses are important, and we do not want to give them short shrift. In fact, with respect to the sense of hearing, the throaty roar of a Chevrolet Corvette and the rumbling of a Harley-Davidson motorcycle are clearly distinctive. In fact, with respect to the Corvette, engineers actually tune the exhaust system in order to give it the right tone and pitch. So a Sensory Audio Test certainly is not only feasible but is actually being done.

In addition, the work in haptics technology continues to evolve. We have the ability to put on special gloves and actually feel virtual surfaces, with the gloves providing the necessary feedback to convince us that we have touched a solid object. We can grasp a virtual steering wheel. We can turn a virtual wrench and feel the resistance of the virtual bolt.

However, because of the importance of the visual sense, we are going to only focus on a Sensory Visual Test. We will leave it up to the reader to fashion tests for the other senses. It will be along the lines of the sensory visual test.

For the Sensory Visual Test, the observer can ask that any spatial manipulation can be performed on both the physical and virtual system. These systems can be rotated, lifted, and even disassembled. If the observer cannot tell which is the physical system and which is the virtual system, then the Sensory Visual Test is said to be passed.

For the Performance Test, the observer can ask to have any action performed on the physical and virtual system. He or she can ask to have power applied to it; to have stresses placed on it; to place it in a wind tunnel; or any other action that could be applied to the system. Again, if the observer cannot tell the difference between the behavior of the physical system and the behavior of the virtual system, then the virtual system is said to have passed the Performance Test.

The last test is the Reflectivity Test. This test requires a little imagination, because the observer is going to know that the physical product is out and about in the physical world. Again, we have the same observer. This observer looks at the same two screens, one of which holds the physical system and the other of which holds the virtual system.

The observer can ask to see the status of any aspect of the product. For example, he or she could ask for the odometer reading of an automobile or the fuel gauge reading in a helicopter. Or, he or she could ask to see the model and serial number of the fuel pump on an airplane or the nozzle pressure of a jet engine at takeoff. If this observer can get the same information from the virtual system as he or she could get from inspecting the physical product, the virtual system is said to pass the Reflectivity Test.

4.2 *Virtuality Test Progress*

So as we come to the middle of the second decade of the twenty-first century, how are we doing with our Tests of Virtuality?

4.2.1 Visual Tests

If we look at the Sensory Visual Test, the answer is that we have made a great deal of progress since the year 2000. While at the beginning of this millennium, the visual representation of systems was getting pretty good, it was fairly easy to tell the virtual system from the physical system. As the years moved along, the visual representations kept getting better and better.

Today, in many situations, virtual systems pass the Sensory Visual Test. Many companies now have life-size power walls upon which they project their digital designs. By rotating them on the screen, these virtual systems can be looked at from any angle, and, in the case of automobiles, doors and hoods can be opened and closed. The view from the interior of the car is as if we were sitting in it.

Since the technological improvements will continue, for all practical purposes we can say that we are passing the Sensory Visual Test. No doubt we will continue to expand our requirements in this area.

The requirement for the 3-D version is to no longer confine the observer to looking through the windows of a physical and virtual room. In this test, we can require that the observer be allowed to walk around the system. In keeping with this being a visual test, we can confine the observer to simply visually inspecting the systems, still allowing him or her to perform any spatial operations. We have some of this capability in 3D Caves and Virtual Reality (VR) goggles today. Holographic displays of tomorrow are within technological reach.

What this means for the Digital Twin is that we should have no UUs of emergent form. We can fit together the entire system virtually, even allowing for tolerances. As components are being manufactured, we can scan and replace the designed parts with actual as-built parts. Our Digital Twin can both predict what the actual form should be and then reflect what the actual form is. There should be no surprises in form anymore.

4.2.2 Performance Tests

The Performance Test is moving along more slowly. The requirements for this test are much more expansive. We not only have to sense the virtual system as if it was physical, but this virtual system must also behave exactly like the physical system. The requirement of this test is that we can have any operation performed on this system that we could have performed on the corresponding physical system.

This means that our virtual environment would need to have all the physical laws of the universe built into it—a daunting task. But what we can hope for in the near future is to build in to our virtual environment selected laws of the universe.

This is where we are today. The issue is that we can abstract certain features of systems today and subject them to specific kind of tests. These tests are generally referred to as "solvers", and there are different solvers for different kind of problems.

We have solvers for finite element analysis, stress testing, heat analysis, material deformation, aerodynamic airflow, and other tests. The issue is that we generally have to prepare the system for the test by abstracting the relevant information that we are interested in studying. This abstracted information is then used in the solver. This information is changed until the solver produces the results that we are looking for. The resulting information then needs to be incorporated back into the virtual system.

Because this information has been abstracted and subject to manipulation in the solver, integrating it back into the virtual system may require interpretation and trial and error. In the example of finite element analysis, the trend is to integrate these solvers with a virtual object so as to eliminate abstraction and reinterpretation of the solver results back into the virtual product.

We would expect this trend to continue. Although unlike the Sensory Visual Test, there may not be a specific point in time where we can say that we have "passed" the Performance Test. It may be that we simply integrate more and more solvers and make them multi-domain with their virtual systems.

One of the main issues of the Performance Tests is how do we evaluate performance? What we have done with physical systems is to create instruments that measure the characteristics that we are interested in. We then correlate those instrument readings with performance characteristics so as to make decisions as to whether we are getting the required performance. Some of the instruments that we have created are temperature gauges, force meters, flow meters, and impact gauges.

While we do use our visual senses to evaluate performance, they are generally not broad or precise enough and need to be augmented to really assess performance. For example, we insert smoke into a wind tunnel test in order to "see" aerodynamic performance. We use high-speed photography in order to capture the results of a crash tests, because our natural vision is not fast enough to either see the details of the crash as it occurs nor to accurately capture the sequence of events.

The virtual environment cannot only emulate the creation of our instruments, but it can translate those results so that our senses can be brought to bear in evaluating virtual product performance. So for example we can see the heat of a virtual system by using a color-coded gradient to indicate the temperature of each element of that system. We can see the aerodynamic properties of the vehicle in motion by making visible the virtual particles of air.

It is with these Performance Tests that we can seek to validate the Predicted Desirable behaviors of our system, validate that we have eliminated the Predictable Undesirable behaviors, and attempt to uncover Unpredicted Undesirable Behaviors. We still have much more to do in this area.

4.2.3 Reflectivity Tests

With respect to the Reflectivity Test, it is clearly in its infancy. The idea of a company maintaining a link with its product after it has left the factory door is very much a twenty-first century concept. The links in the past were only maintained by taking possession of the physical product on a periodic basis, such as bringing it into a repair hub or bringing it back to the factory for an overhaul.

Today, it is difficult to find a product that does not have a microprocessor integrated into it. There is more computing value in today's cars than there is steel value. But it is not just large ticket items such as automobiles and airplanes. Almost any product that we can think of has a microprocessor in it. This includes washers, dryers, refrigerators, thermostats, pacemakers, and even running shoes.

With these microprocessors and their ability to communicate, companies can continue to stay in touch with their product long after the product has left their factory door. This will allow for these products to communicate back the current state of their condition at all times.

This embedding of connections is becoming much more ubiquitous with the rise of the Internet of Things (IoT) [25]. Components within a system can interact with each other, exchanging statuses and making request for services. This can and most likely will result in emerging behaviors that are currently unforeseen and unpredicted.

It is going to be important to have a Digital Twin that reflects this activity for a number of reasons. First, capturing these interactions will be critical in understanding what emerging behaviors occurred and having an audit trail of activities that led up to these behaviors. Second, only by seeing what is occurring can humans have a hope of stepping in when UU behaviors start occurring.

We are on the path of having autonomy in vehicles such as automobiles, farm equipment, and drones. We have not given as much thought to reflectivity as we should. The concepts of the Digital Twin are something we should strongly consider as we move in this direction.

5 Digital Twin Obstacles and Possibilities

As with all new concepts, there are both obstacles and possibilities that present themselves. In this section, we will discuss a few of them.

5.1 Obstacles

As we see it, there are three main obstacles that will need to be addressed. These obstacles are organizational siloing, knowledge of the physical world, and the number of possible states that systems can take.

Organizational siloing probably presents the biggest obstacle to the Digital Twin. Organizations are divided into functions such as design, engineering, manufacturing, and support. There is a natural siloing of information within these functional areas. Each of these informational silos has information about the systems. However there may be very little sharing across functions.

To use but one example, most organizations still have the issue of engineering and manufacturing having different bill of materials for the same components. Even though the component that engineering deals with and the component that manufacturing deals with are one and the same, each of these areas approaches the parts that make up these components in a different way. The Digital Twin concept requires a homogeneous perspective of this information that persists across functional boundaries.

Even within functions there is siloing and fragmentation. The domains of mechanical engineering, electrical engineering, programming, and systems engineering exists separate and apart from each other. In many if not most organizations, simulation exists specific to a domain and does not have a multi-domain focus.

Even with areas that one might think should be the same such as manufacturing and materials, the work in one domain does not carry over to another. Mechanical engineers may be developing a structure that requires a certain weight limit. The material people may have different perspectives on how material affects not only weight but also structure. It is not until later in the development cycle that these two issues, which should be related, are reconciled, adding additional costs and delays.

These are cultural issues that will need to be addressed. Cultural issues are much more difficult to address than technical issues. We expect that the technology needed to address these issues will be available much earlier than the cultural changes necessary to fully adopt and make use of the technology.

The next obstacle is simply our understanding of the physical world. Technology in general requires us to capture and understand physical phenomena [26]. The Digital Twin concept is built on understanding and being able to simulate natural phenomena. While this area is rapidly increasing, the Digital Twin requires that we need to know how our system will react to the forces that it will encounter. Being able to understand, model, and simulate structures, materials, and force phenomena will be critical in doing meaningful digital analysis of how our system will actually perform.

Currently we are hard at work at understanding how structures respond to forces and how materials respond, fracture, and deteriorate in the face of the substantial forces that we encounter in airplane and rocket flight.

The third obstacle is simply the sheer number of states that the system can take over time. If we are to tease out undesirable emergent behaviors, we will need to simulate the conditions that systems face under a range of parameter changes, where we may to be dealing with thousands of parameters. We simply may not have the amount of computing capability required to perform all the computations that we require. This is a problem that diminishes in time as computing and its associated technologies continue to advance at exponential rates.

5.2 Possibilities

We are already seeing the possibility of trading costly and time-consuming physical prototypes for modeling and simulation. We are seeing great strides in these areas. As an example, Boeing recently patented an "Atoms to Airplanes" perspective that deals with the modeling and simulation of composite materials.

There are two interesting possibilities to deal with complexity that we will discuss. The first is capturing in-use information to feed back into the creation phase. The second possibility is the idea of "front-running" a simulation in real time. The first possibility could allow us to uncover complexity issues, i.e. UUs, and deal with them before the system is deployed. The second could help mitigate UUs that arise as the system operates.

With the ability to capture data on how the system is produced and used in the Production and Operational phases, we can collect system states that would collectively be profiles that we can use in future development activities to simulate the new system's behavior. Running these states through the simulation of a new system could help point out particular profiles that would give UUs. Since these profiles would reflect human interactions, designers would give particular attention to those profiles where humans interacted in unexpected ways.

As Fig. 5 shows, these profiles could be run via simulations in a simultaneous and iterative fashion, well before a physical system is produced. While there might be many more possible system states, this would cover many states that might not be considered by the system designer. It would be helpful in identifying complexity because uncovering UUs would point out areas that need simplification.

The second possible opportunity is to help mitigate complexity. In this possibility we would have our simulation "front-run" the system in actual use in real-time. This means that we would run the simulation with real-time feeds for what is happening to the system in actual use. The simulation could present a window into the future of the possible system states that might occur.

This might be helpful in the case of system crises, where it is unclear as to what is happening with the system. As was pointed out earlier, humans often have a problem with sensemaking. They jump to a final conclusion almost immediately, even though working through the issue with a systematic methodology would present other possibilities they should consider [27]. Once there, they lock onto a paradigm that they cannot easily discard. System front running would show them other possibilities of what is occurring and possibly help in making better sense of the situation.

This obviously would work best when the crisis is developing over time. The BP Gulf Disaster would be such an example [28]. There were a number of indications that something was going awry. However, the operators were so locked into a frame of reference that they could not see other alternatives.

It would even be useful in events that are fast moving and looking seconds ahead could be the difference between life and death. The example of this is the Air France 447 flight that crashed off of the coast of Brazil. A simulation front running in advance of what was occurring could have alerted the pilots that they were doing exactly the wrong thing in pulling the nose of the airplane up and that continuing their actions would result in a fatal stall.

Obviously the front running capability requires computing capability that can run faster than the physical activity it is simulating. As computing power continues to increase, there will be more and more instances where front running will be possible.

With the introduction of new, conceptual models, there will always be both unforeseen obstacles and possibilities. The key will be to remain alert to both address the obstacles and exploit the possibilities.

6 A NASA Approach to the Digital Twin

NASA has three major issues that most organizations do not have. These issues are: (a) the systems that they create are very expensive; (b) they make very few of these systems; and (c) the systems that they make have not been made before.

NASA's systems certainly qualify as complex systems. Components in its launch systems range from the nanometer scale to components that are tens of meters in size. NASA also has system of systems, because launch systems include, rockets, solid-state boosters, capsules, launch pads, and flight operations. NASA's launch systems are susceptible to "normal accidents" where minor problems cascade into major catastrophes. An O-Ring that failed caused the space shuttle Challenger's explosion. A strike of insulating foam falling off at launch critically damaged the space shuttle Columbia.

In the past, NASA has addressed these issues with additional resources. They had the resources in order to do expensive, time-consuming testing with physical prototypes. However, this is no longer a viable alternative. NASA has added affordability as a major criterion for their space launch capabilities. They are under extreme pressure to create their systems, especially their new Space Launch System, using their resources much more efficiently and effectively than may have been done in the past.

It is in this new climate that NASA is investigating the usage of the Digital Twin model. The proponents of this model within NASA believe that it can take cost out of the budget and time off the schedule. This has the possibility in resulting in major resource savings for NASA.

NASA recently completed a project in developing a composite tank. While the project was successful, the project pointed out issues that need addressing. A major issue was that there were major discrepancies between the performance predicted by the Advanced Concepts Office (ACO), the performance predicted by the in-depth analysis of experts, and the actual physical tests. In addition, the in-depth analysis was not completed until after the production of the composite tank was well underway.

Both the gaps between the two predicted results and the actual results and the timing of the in-depth analysis are problems that need to be resolved and done better. We need the predicted behavior and the actual behavior to be much closer. We also need the in-depth analysis to be completed before we begin actual production.

The proponents of the Digital Twin concept within NASA think that this concept has the opportunity to help in both these areas. By focusing on creating better and more complete information about the virtual system, NASA can do a better job in predicting the actual performance of its systems and reducing the possibility of UU problems.

In addition, unlike physical systems, which either exists or don't exist, the Digital Twin allows for maturity of information so that analysis can begin much earlier and not wait until the design is complete. This should allow for in-depth analysis to start much earlier than it has in the past and finish before the manufacture of the actual system.

The Digital Twin is not the main thrust of this project. However, success in proving the viability of its concepts has the possibility to have a major impact on affordability within NASA.

7 Conclusion

The premise driving the Digital Twin concept was that each system consisted of two systems, the physical system that has always existed and a new virtual system that contained all of the information about the physical system. Because information is a replacement for wasted physical resources, we can use much less costly resources in creating, producing, and operating systems. Through modeling and simulation in virtual space we can better understand the emergent form and behaviors of systems and diminish the "I didn't see that coming" factor.

Virtual systems can be a great assist in dealing with the four categories of system behaviors. We can ensure we will obtain the Predicted Desirable (PD), eliminate the Predictable Undesirable (PU), and decrease the Unpredicted Undesirable (UU) [If we have Unpredicted Desirable (UD), that will not hurt us, although it does point out that we do not full understand our system].

Complex systems are susceptible to "normal accidents". While the term "emergent behavior" is used, we would contend that it is behavior that potentially existed from the inception of the system, although one of its main causes is the system's interaction with humans who behave in unexpected ways. This is often caused by a failure in sensemaking. By the Digital Twin's simulation, we may be able to reduce the UUs caused by unexpected human interaction. We specifically are not dealing with evolving systems: systems that have randomness built in or change the rules that govern their behaviors in operation.

Systems do not burst forth fully formed. They progress through a lifecycle of creation, production, operation, and disposal. With "physical-only" systems, this was a linear progression. The Digital Twin allows for a more iterative, simultaneous development to consider the "ilities". While there are obstacles ahead, especially cultural ones, we have made significant progress in the last decade and a half as we have shown via the Grieves Virtual Tests. We should expect future advances, as computing technology shows no sign of slowing its rate of progress.

Finally the Digital Twin and its reflection of the physical system mean that we can potentially use the virtual system even when the physical system is in operation. Two potential uses are capturing and using in-use information and system front running.

The Digital Twin concept has the opportunity to change how we view system design, manufacturing and operation, reduce the UUs of complex systems, and augment Systems Engineering.

References

1. Grieves, M. (2012). Virtually indistinguishable: Systems engineering and PLM. In L. Rivest, A. Bouras, & B. Louhichi (Eds.), *Product lifecycle management: Towards knowledge-rich enterprises* (pp. 226–242). Montreal: Springer.
2. Ackoff, R. L. (1971). Towards a system of systems concepts. *Management Science (pre-1986) 17*(11), 661.
3. Sargut, G., & McGrath, R. G. (2011). Learning to live with complexity. *Harvard Business Review, 89*(9), 68–76.
4. Nature. (2008). Language: Disputed definitions. *Nature, 455*, 1023–1028.
5. Mitchell, M. (2009). *Complexity: a guided tour*. Oxford: Oxford University Press.
6. Perrow, C. (1984). *Normal accidents: Living with high-risk technologies*. New York: Basic Books.
7. Weick, K. E. (2005). Making sense of blurred images: Mindful organizing in mission STS-107. In M. Farjoun & W. Starbuck (Eds.), *Organization at the limit: Lessons from the Columbia disaster* (pp. 159–177). Malden, MA: Blackwell.
8. Weick, K. E. (1990). The vulnerable system: An analysis of the Tenerife air disaster. *Journal of Management, 16*(3), 571–593.
9. Troyer, L. (2015). Expanding sociotechnical systems theory through the trans-disciplinary lens of complexity theory. In F.-J. Kahlen, S. Flumerfelt, & A. Alves (Eds.), *Trans-disciplinary perspectives on system complexity*. New York: Springer.
10. Aristotle, H. T., & Armstrong, G. C. (1933). *The metaphysics*. London, New York: W. Heinemann, G. P. Putnam's Sons.

11. Ablowitz, R. (1939). The theory of emergence. *Philosophy of Science, 6*(1), 16.
12. Felder, W. N. (2013). *Interactions among components in complex systems, AIAA CASE 2013 Framing Paper*, Complex Aerospace Systems Exchange, San Diego, CA, 10-12 Sept 2013, http://www.aiaa.org/uploadedFiles/Events/Conferences/2013_Conferences/2013__Aviation/Detailed_Program/CASE%202013%20Framing%20Paper.pdf, Accessed 22 June16.
13. Holland, O. T. (2007). Taxonomy for the modeling and simulation of emergent behavior systems. In *Proceedings of the 2007 spring simulation multiconference-Volume 2, Society for Computer Simulation International.*
14. Grieves, M. (2011). *Virtually perfect: Driving innovative and lean products through product lifecycle management*. Cocoa Beach, FL: Space Coast Press.
15. Grieves, M. (2009). *Virtually perfect: Driving innovative and lean products through product lifecycle management*. Cocoa Beach, FL: Space Coast Press.
16. Grieves, M. (2005). Product lifecycle management: The new paradigm for enterprises. *International Journal Product Development, 2*(1/2), 71–84.
17. Grieves, M. (2006). *Product lifecycle management: Driving the next generation of lean thinking*. New York: McGraw-Hill.
18. Piascik, R., Vickers, J., Lowry, D., Scotti, S., Stewart, J., & Calomino, A. (2010). *Technology area 12: Materials, structures, mechanical systems, and manufacturing road map*. NASA Office of Chief Technologist.
19. Glaessgen, E. H., & Stargel, D. (2012). *The digital twin paradigm for future NASA and US air force vehicles*. AAIA 53rd Structures, Structural Dynamics, and Materials Conference, Honolulu, Hawaii.
20. Eric, J. T., Anthony, R. I., Thomas, G. E., & Michael Spottswood, S. (2011). Reengineering aircraft structural life prediction using a digital twin. *International Journal of Aerospace Engineering, 2011*(154798), 14. doi:10.1155/2011/154798.
21. Cerrone, A., Hochhalter, J., Heber, G., & Ingraffea, A. (2014). On the effects of modeling as-manufactured geometry: Toward digital twin. *International Journal of Aerospace Engineering* 2014.
22. Eremenko, P., & Wiedenman, N. L. (2010). *Adaptive Vehicle Make (AVM)*. Driving Innovation for Superior Defense Manufacturing, Washington, DC, NDIA – DARPA.
23. Witherell, P., Jones, A., & Lu, Y. (2015). Additive manufacturing: A trans-disciplinary experience. In F.-J. Kahlen, S. Flumerfelt, & A. Alves (Eds.), *Trans-disciplinary perspectives on system complexity*. New York: Springer.
24. Turing, A. (1950). Computing machinery and intelligence. *Mind, LIX*(236), 433–460.
25. Porter, M. E., & Heppelmann, J. E. (2014). How smart, connected products are transforming competition. *Harvard Business Review, 92*(11), 64–88.
26. Arthur, W. B. (2009). *The nature of technology: What it is and how it evolves*. New York: Free Press.
27. Grieves, M. (2014). Process, practice, and innovation. In P. Gupta & B. Trusko (Eds.), *Global innovation science handbook* (pp. 159–172). New York: McGraw Hill Professional.
28. Graham, B., Reilly, W., Beinecke, F., Boesch, D., Garcia, T., Murray, C., et al. (2011). *Deep water: The gulf oil disaster and the future of offshore drilling*. National Commission on the BP Deepwater Horizon Oil Spill.

Managing Systems Complexity Through Congruence

Shannon Flumerfelt, Anabela Alves, Javier Calvo-Amodio, Chris Hoyle, and Franz-Josef Kahlen

1 Introduction

The chapter supports the thesis that the body of knowledge and practice of systems management is now encumbering the non-linear attributes of systems, complexity, interrelatedness and complexity through objective analysis. Attending to this, the chapter presents a tool, called the CX Tool, from which it is possible to participate in objective analysis to improve system impact and outcomes.

All organizations suffer from a missing link between what they think they do and what they really do. Tending to an equal relationship between thinking and doing is described here as *creating congruence* between what is known and what is done.

S. Flumerfelt (✉)
Department of Organizational Leadership, School of Education and Human Services Oakland University, 475C Pawley Hall, Rochester, MI 48309-4494, USA
e-mail: flumerfe@oakland.edu

A. Alves
Department of Production and Systems, School of Engineering University of Minho, Campus of Azurém, Guimarães 4800-058, Portugal

J. Calvo-Amodio
School of Mechanical, Industrial, and Manufacturing Engineering, Oregon State University, 410 Rogers Hall, Corvallis, OR 97331, USA
e-mail: Javier.Calvo@oregonstate.edu

C. Hoyle
School of Mechanical, Industrial, and Manufacturing Engineering, Oregon State University, 204 Rogers Hall, Corvallis, OR 97331-6001, USA
e-mail: chris.hoyle@oregonstate.edu

F.-J. Kahlen
Kahlen Global Professional Solutions, Fabrikstr. 3, Gronau 48599, Germany

Department of Mechanical Engineering, University of Cape Town, Cape Town, South Africa

F.-J. Kahlen et al. (eds.), *Transdisciplinary Perspectives on Complex Systems*,
DOI 10.1007/978-3-319-38756-7_5

The term, the CX Tool, stands for this exact meaning, with "C" referring to "Congruence" and "X" referring to "Multiple Ways." The CX Tool, therefore, translates as a method for demonstrating and creating congruence in various ways between thinking, or Organizational Intelligence, and doing, or Performance Management. Understanding this state of congruence between Organizational Intelligence and Performance Management is critical to managing system performance. To account for the reality of system attributes and dynamics, there are four tiers or four views of system stakeholders' perspectives, complexity and complicatedness made possible through this tool. In this chapter, the Tier 1 view is largely described with some examples of Tier 2 analysis as well.

In order to understand how to maximize system management through this tool, this chapter is divided into five sections total. After this introduction, the second section describes the theoretical case and extant literature for systems-based approaches. The third section presents attributes of the CX Tool. The fourth section briefly highlights the applications of Tiers 1 and 2 of the CX Tool in various empirical applications and case studies. Finally, the last section provides conclusions on the state of the tool's development and future applications.

2 Background and Motivation for Better Systems Management

This section presents the theoretical foundations of the CX Tool. The basics of the relationship between systems and people, systems types and attributes, organizational learning, systems management theories, the congruence concept and system metrics are described as important tenets of the CX Tool.

2.1 Systems and People

Everyone interacts reciprocally with systems everyday. The relationship between systems and people is continually present. For instance, the impact of the weather conditions, the laws of physics, the technology of a smart phone and the policies and culture of the workplace, provide a broad range of examples of systems that are both needed by people and are affected by people. Systems derive from various sources, such as natural (i.e., the weather), technological (i.e., the laws of physics), machine (i.e., smart phones) and human/information (i.e., organizational policies and culture). System competency held by humans, therefore, is identified as a key attribute in influencing system consequences. In other words, the relationship between systems and people places the need for people to have acumen to know, value and work effectively using systems-based management approaches to ensure better results.

Some systems are self-initiated, such as in nature. People are the recipients of the processes and outcomes of self-initiated systems. So this means that people will be influenced by and will influence the subsequent state of nature-based systems. However, it is difficult for people to manage self-initiated systems and typically people serve as passive actors in self-initiated systems and do not serve as creators or designers of these systems. People can take certain actions to impact the outcomes of these self-initiated systems, much like Skinner's (1904–1990) stimulus-response model of behaviorism [1]. For example, the ecosystem will respond to decreasing environmental pollutants. The stimulus is the human choice to manufacture within sustainability standards and the environmental response belongs to the ecosystem in the form of cleaner natural resources.

Other systems are human-initiated systems, such as with technical, information or human systems. In turn, people are the designers and facilitators of technical, information and human-based systems and will determine, either intentionally or unintentionally the processes and outcomes of these systems. Managing the outcomes of human initiated systems, therefore, is important and is a preferred approach to non-management of systems. Further, due to the control that people can retain over human-based systems, there is a need for optimal management of those systems. It is possible for people to be pro-active in the management of human-initiated systems, an idea support by Glasser [2] as lead-management.

There is a substantial interrelated dynamic between systems and people, whether these systems are self-initiated or human-initiated or hybrids. This dynamic interrelationship, mainly the one that relates technical systems and people (social system) is broadly known as Sociotechnical systems. These systems are largely discussed in the chapter by Troyer in this volume.

The problem of how to develop systems competency for better management of human-initiated systems holds great potential for increasing organizational leadership outcomes. Systems competency is isomorphic with Flood and Jackson's [3] system of systems methodology (SOSM). Systems competency arises from the need for a systems intervention, when the iterations of actions, processes and tools produce problems. In other words, systems competency is viewed as the application of the right tool in the right manner to solve the right system dysfunction. Therefore, concerns for systems competency encumbers both functionalist and interpretive sociological paradigms and methods.

Systems-based approaches are critical, for they clarify the sources of system dysfunction. Felder and Collopy [4] described one source of dysfunction as a lack of systems competency, or system ignorance, or lack of systems management of the system itself. Miller and Page [5] explained that when systems involve human operators, they can seem even more difficult to manage. However, from a systems competency perspective, it is possible to draw on the commonalities of various system types, such as shared attributes and common sources of dysfunction. For example, human systems hold the same properties that the machine, information and technical systems do to an extent. They are complex, interdependent, emergent, organic and adaptive. This issue has been examined for engineers by Jones et al. [6], citing the American Institute of Aeronautics and Astronautics as:

> Technical accomplishments can overshadow significant schedule delays, cost overruns, and many other shortcomings associated with environmental issues, societal impacts, and political difficulties. The bottom line is that current processes and methods may not ensure overall program/project success, particularly if we consider meeting the budget, schedule, and operational effectiveness requirements.

In other words, it is not enough to possess domain and/or subject expertise and mastery, one has to be able to actualize through system competency mastery successful management of the system at hand, tending to results, but also to the processes used to obtain the results and the various systems-based dynamics in play that both hinder and enable the system.

2.2 Systems Defined

Systems hold certain attributes, such as emergence, interrelatedness and complex adaptivity. These characteristics mean that systems will inevitably produce either more or less than is intended because a different state is generated with each iterative interaction or operation. Senge [7] described this dynamic with systems as resulting in the entire system consisting of more than the addition of system elements individually. In other words, it is understood that systems cannot be fully represented in linear formulas, based on the paradigm that one plus one equals two. In systems, the formulas are generative, whereby one plus one may equal ten or negative five. The consequences of systems attributes, therefore, may result in a variety of consequences. The ability to mitigate the variation of system consequences, therefore, is a highly desirable capability. And this is why system management is an exciting area of organizational development and leadership. The beauty and horror of system attributes can increase or destroy the potential of the system itself.

The consequences of systems can produce three types of outcomes in three general realms. The outcomes of consequences may be either (1) positive, (2) negative, or (3) some combination of positive and negative or neutral. The three realms of consequences may be either (1) succinct and directional, (2) broad and non-directional, or (3) some combination of succinct/directional and broad/non-directional or neutral. Drucker et al. [8] informed leaders that it is critical to ascertain valued outcomes and consequences through answers to five questions, such as, "What is Our Mission?," "Who Is Our Customer?," "What Does The Customer Value?," "What Are Our Results?," and "What Is Our Plan?" (p. 11–76). The answers to these questions are context-bound. Therefore, both desired outcomes and consequences of systems will vary widely and are not prescriptive or standardized. In any case, the basis for systems-based management is to create conditions for desirable outcomes and consequences, whatever they may be. As stated by Wheatley [9], "The complexity of modern systems cannot be understood by our old ways of separating problems, or scapegoating individuals, or rearranging boxes on an org chart" (p. 13). Wheatley's work paralleled the complexity of living systems with organizational systems in order to better understand the task of management of human-initiated systems. Wheatley's metaphors provide excellent mental models for the reality of workplace systems.

Systems can be perplexing because in practice they elude the straightforward conceptualization of three sequential elements of inputs feeding processes resulting in outputs, encumbered by feedback loops. This three-step mental model of systems is not problematic, but it is operationally simplistic. Therefore, the input-process-output feedback loop model is misaligned with complexity, adaptivity and emergent attributes of systems in practice. Scholars have urged enriched views of systems to inform both knowledge and practice of systems management, including consideration of the context of systems [10], the evolution of systems [11], and leveraging system outcomes [12]. Felder [13] expressed the need for management of systems in regard to obtaining proactive outcomes as a complexity on top of a complexity, in the context of influence by rapid technological change. And because of this fluid context, Felder advocated for proactive systems management where positive system outcomes can be realized. For example, this approach is seen when the advantages of rapid deployment of innovations are realized. Under this scenario, systems management allows for adjusting to market demands, changing economic conditions, or designing for new compliance regulations. In this regard, organizations such as the Mayo Clinic have taken on systems management approaches through teamwork, shared responsibility and collaboration in an era of healthcare mandates. In doing so, they are recognized for top-notch patient care [14].

Additionally, in the chapter in this volume by Hester et al. the authors advocated that a transdisciplinary view and a systemic approach can help researchers and health professionals to identify and recognize different perspectives. And further it creates better understand of emerging and re-emerging infectious diseases in order to decrease the gap in infectious disease knowledge around the world.

The evolution of the body of knowledge of systems management is ongoing. This situation is quite similar to the developmental path of other fields and topics. For example, Wittgenstein (1889–1951), an analytic and linguistic philosopher, after penning his ultimate response to all problems of philosophy (*Tractaus Logico-philosphicus*), changed his mind. His initial work focused on the finality of words as streamlining to ultimate objectivity. However, he found that his intended terminal work was, in fact, a flawed approach. Wittgenstein subsequently wrote an open-ended work, *Philosophical Investigations*. This book reversed the former reductionist approach to thought and word development. In reverse, he called for the recognition of the subjectivity in language and the use of language as a tool, not as an end in itself [15]. In this regard, Jones et al. [6] advocated to not to take a reductionist view in system analysis, but to let an ever-expanding view come forth about system attributes and performance. These authors explained this stance as a continuous learning loop, where each cycle provided a better sense of the system, even if the probe is about system dysfunction. For example, NASA recognizes this as Deutero Learning (DL), whereby the complexities of the operational system, such as a bureaucracy, are reviewed and reviewed again through organizational learning loops [16].

The point is that the body of knowledge and practice of systems and systems management will predictably continue to evolve and develop from subjectivity to objectivity. This was described by Axelrod and Cohen [17] as, "With the Advent of the Information Age, the bottom-up style of thinking associated with Complex Adaptive Systems may well become a greater part of people's understanding of how economic, social and political processes actually function and change" (p. 61). The transition to objectivity is a critical issue for systems management as this condition allows for the objective analysis of the nonlinear attributes of systems, complexity, interrelatedness and complexity. Further, it is posited since it is possible to participate in objective analysis, it is, therefore, possible to improve system impact and outcomes.

Systems conceptualizations are now more broadly conceived of as representative of systems attributes in degrees of emergence, complexity and adaptivity, rather than as only a straight line of end run interactions. According to Sterman [18] and Checkland [19], system interactions are now considered to either be isolated/terminal or iterative/circular depending on the degree of interrelatedness, complexity and adaptivity that the system's elements possess. Ackhoff [20] advocated for the deliberate examination of the comprehensive set of static and emergent attributes, rather than maintaining a passive stance toward system complexity and reacting to isolated outcomes. For example, Amen et al. [21, 22] supported systems perspectives in mechanical engineering, while Felder et al. [23] and Felder and Collopy [4] supported active systems management in chemical engineering and aerospace, and Calvo-Amodio et al. [24–26] and Calvo-Amodio [27] supported systems management for healthcare systems.

Systems can be visualized using mental models, depicting their attributes and dynamics. Virtual models such as the Digital Twin concept described by Grieves in his chapter in this volume are also being used to link the physical system with its virtual equivalent and mitigate undesirable and unpredicted emergent behavior of systems, especially complex and complicated systems.

Vitiello and Kalawksy [28] advocated for visual management as the method for making sense of systems. Shewhart [29] developed a frontline mental model of system process management and decision-making. The Plan-Do-Check-Adjust (PDCA) cycle, also known as the continuous improvement cycle or Shewhart cycle, separates four distinct but connected reciprocal phases of decision analytics. As Deming [30] stated, the PDCA cycle presented a theoretically and empirically tested approach to system-based decision analytics. Bahill and Gissing [31] conducted a literature review and analysis and built a more detailed decision analytics process than the PDCA cycle. Their work is based on the Shewhart cycle and uses the acronym SIMILAR (State the problem, Investigate alternatives, Model the system, Integrate, Launch the system, Assess performance, and Re-evaluate). Lean Six Sigma, with its Define-Measure-Analyze-Improve-Control (DMAIC) model is another example of systems-based approaches. Further,

Snowden's [32] award-winning Cynefin model is another example of a systems-based approach that has informed managers and leaders about the selection of strategies based on emerging system attributes ranging from states of chaos, obviousness, complicatedness to complexity through sensing and probing. Bolman and Deal [33] advocated for a systems-based view of the organization through a four-frame model of leadership. Their four-frame model was inclusive of structural, human resource, political and symbolic considerations.

Holt, Collopy and DeTurris in their chapter in this volume reviewed other techniques and methods to cope with complex systems such as Cynefin framework referred to above in this chapter and Agile Software Development methodology.

2.3 *Organizational Learning and Systems Management*

Senge [7] also provided insights into systems-based views of organizational practice which included the notion that organizations can become better at systems management, as if solving a puzzle or learning, based on strategies of personal mastery, mental models, shared vision and team learning. Senge also advocated for organizational learning as the gateway behavior to understanding the system at hand. Drucker enlightened organizational learning theory with his recognition of the methods of organizational learning. He extended knowledge management well beyond the cold cognitive approaches of training to also encompass, reflections and observations as valid methods of learning. Drucker [34] arranged the systems of organizational sense making into arenas of explicit learning and tacit learning. These arenas afforded learning opportunities whereas an individual can learn about organizational culture, through explicit to explicit venues of conditioning, through tacit to tacit venues of socialization, through explicit to tacit venues of internalization and/or through tacit to explicit interfaces of articulation. Drucker's work brought recognition of organizational learning systems in all aspects of organizational life. His views added urgency to what can be done through better knowledge management as a systems-based approach.

Organizational learning theory is paralleled with individual learning theory. For example, Burke [10] highlighted how waves of thinking have described understanding of the power of the human brain, ranging from reductionist states to innovative states over time. He described systems conceptualization of the brain, with trillions of connectors that lace information together as, "its [the brain's] connective potential makes it easy to run scenarios in real time" (p. 189). It is now believed that the brain learns best through selection and survival, represented by systems-based schema of abilities. This information challenges the traditional notions of learning through direct instruction in training and education programs, as well as through directive management strategies in the workplace. Brain theorists now contend that problem-based, experiential, active learning is needed. This is a substantial finding because it means that individuals and organizations have a dominant role in developing their own knowledge schemata based on experience through constructivism. It also means something more that they also have to figure out what the problem is. Therefore, it can be drawn out that the ideal situation for organizational learning is for problem-based learning challenges to intersect with sense making exercises of constructivism. This position represents a substantial connection between 'theory and practice' based on biology and brain science and organizational practice and learning. As described by Jensen [35], this occurs as, "Nature's biological imperative is simple: No intelligence or ability will unfold until or unless given the appropriate model environment" (p. 6). In other words, the brain is a multi-processing system and it provides a solid metaphor for our conceptions of organizational learning and improvement. For example, Gardner's [36] work in the expression of intelligence extends recognition to the multiplicity of intelligence, again confirming the complexity of the learning process and the variety in learning outcomes. This enforces for us the need for systems based management that encumbers the adaptiveness, complexity and interrelatedness of human existence.

Also, organizations must improve organizational systems, making them faster, better and at lower costs, continuously searching for the systems' ideality, this means to increase the number of beneficial functions and decrease harmful functions. This could be achieved by some techniques and analytical tools of the Theory of Creative Problem Solving (TRIZ) methodology reviewed by Navas in her chapter in this volume. This methodology contributes to accelerate and improve the process of problem solving and to increase an ideality of complex systems in almost all human activities.

Further, this theoretical foundation is important because Argyis and Schön [37] drew the process of organizational learning as connected to organizational effectiveness. They concluded that individual and corporate learning were mutually beneficial co-development initiatives. In other words, there should be a way for individual mastery to transfer into organizational benefit. Described as characterization by Jones et al. [6], it is the dynamic of co-authored learning between individuals and systems and vice versa that enables the leveraging of a system. So in the end, it becomes more difficult to separate the essence of human systems from organizational life.

2.4 System Congruence of Organizational Thinking and Doing

In the sense that the brain works to relate theory to practice and practice to theory through its connections, there is a similar need to interlace the spheres of organizational thinking and organizational doing in leadership practice. For the last two decades, leaders have been urged to move away from singular inspirational campaigns of improvement toward a method of studied actions of systems-based management. Moss-Kanter [38] described this as, "I have sensed growing weariness with hearing one more heroic story that simply does not match the mundane issues managers struggle with everyday. . .—a backlash against legends of still one more Superwoman or Supermanager or Supercompany" (p. 21). The call to innovation, for instance, is challenged by a lack of systemization according to Moss-Kanter. Her concerns were identified largely within the traditional corporate ethos of the Great Man/Woman theory that questioned whether these organizations have the capacity to transition to agility, continuous improvement and quality paradigms and operations. In essence, she reflected on the gap between "the talk versus the walk" of organizations. Moss-Kanter [38] offered implications for a focus on enlightened paradigms and informed process improvement to replace silos of plans, often protected by spin over substance. She outlined the need to reconcile the differences in what is thought about against what is actually done organizationally.

Therefore, sources of system dysfunction can be similarly categorized broadly for machine, information, technical and human systems. When a lack of systems management is present, dysfunction can be addressed through systems-based tools. Frank [39] stated that these tools should build systems competency, including acumen in systems thinking, applications of circular reasoning, extending solutions beyond design and production, integrating perspectives into problems and solutions, managing through visual techniques to "see" the system at hand, and placing solutions into a continuous improvement cycle.

Because system behavior ranges broadly, systems have been a stumbling block for organizations. To overcome this problem, managing to the constancy of and the presence of systems elements, each related to the realms of thinking and doing, serves as a means for managing systems effectively. For example, Pfeffer [40] aligned with Moss-Kanter's perspectives and described organizational learning as systems-based sense making between two realms, organizational thinking and organizational doing. When these two realms align, then system capacity increases as organizational learning heightens. And when these two realms do not align, the system suffers from dysfunction due to barriers to organizational learning. In other words, by perceiving that a system exists and that the system draws meaning from the alignment of two spheres of organizational work, thinking and doing, then the framework for systems management is formed.

In earlier works, Argyis [41] and Arygis and Schon [42] also provided confirmation into the potential that can occur from perceiving organizational dynamics in this regard. They identified how powerful the connections between organizational thinking and organizational doing can be when these spheres of work are properly aligned. For example, when an organization launches a new campaign stemming from organizational thinking, such as Ward's [43] description of NASA's Faster, Cheaper, Better (FCB) campaign (a new/improved idea), and corresponding change occurs in performance management strategies, such as managers working to standards of FBC (a new/improved standard), inevitably the system's function will increase and impact performance. This interrelationship between thinking and doing, therefore, is a source of organizational capacity through systems management. In contrast, when the relationship between organizational thinking and doing is adrift, the system will produce multiple sources of confusion, ambiguity and waste [33]. This is, in fact, what occurred at NASA when the Essential Idea within Organizational Intelligence was not congruent with the Essential Standards within Performance Management. Dissonance was present and the organization found it difficult to manage the system to congruence and thereby abandoned FBC.

In adding to the need to create a correlation between organizational thinking and doing, Bennis [44] provided insights into the need for systems-based management as he urged leaders to, "learn more about the complex roots of our various predicaments . . . learn how to identify the real questions and there begin to come up with some workable answers" (p. 114). Bennis [45] also described the reconciliation of thinking and doing as a leadership need, "I've felt that there must be a better way than that of Prospero, who threw away his book when he had to leave the island and take his place in society as a leader" ([44, 45], p. ix). Bennis [45] highlighted that leaders cannot only dwell on what they know, but that they need to also do something. In turn, leaders cannot only act, but need to do so based on theories with fidelity.

Therefore, there is a need to explore the equality of the relationship between thinking and doing, also thought of as *creating congruence* between what is known, organizational intelligence, and what is done, performance management. The term, the CX Tool, stands for this exact meaning, with "C" referring to "Congruence" and "X" referring to "Multiple Ways" to create congruence or state of equality between thinking and doing in a system. The CX Tool, therefore, translates as a method for demonstrating and creating congruence in various ways between thinking, Organizational Intelligence, and doing, Performance Management. Understanding this state of congruence between Organizational Intelligence and Performance Management is critical. This was cited by Glasser [2] quoting Deming as, ". . . knowledge is theory. Experience teaches nothing without theory… Unless you understand the theory behind it [success], trying to copy it can lead to complete chaos" (p. 50). So, managing the critical relationship between the realms of Organizational Intelligence and Performance Management as measured through congruence represents great potential for systems management.

These views on organizational thinking and doing inform the need for better systems-based approaches to organizational work. Pfeffer and Sutton [46] provided insights into the leadership tendency to dwell on organizational thinking over organizational doing. They described the need to create bridges between what a business knows and what a business does as an organizational value proposition. They have provided an excellent conceptualization of organizational sense making and what is needed to promote its outcomes.

We find examples of this in the corporate world. For instance, as Hewlett-Packard struggled with market share in 2012, many pegged the root of this problem as a Performance Management problem [47], where the company lost its "HP Way," described as the "values and behaviors and principles and commitments" ([48], online) of the organization. While HP was talking about its revitalization as an Essential Idea, an Organizational Intelligence element, the deployment of its growth plan stumbled. There was a disconnection between organizational thinking and organizational doing by focusing too much on organizational thinking and not deploying congruent organizational doing. In contrast, Arthur Anderson suffered from focusing too much on organizational doing and letting ethical accounting practice lapse, departing from one of its critical tenets of its organizational thinking. As it was, the firm struggled to stay competitive in the organizational doing sphere, its employees were signing off to generate revenue. This impetus diverted focus from the firm's thinking that ethics mattered. Arthur Anderson was subsequently the first accounting firm to be convicted in 2002. And while the conviction was overturned in 2005 [49], the firm as it had stood since 1913 folded. What is interesting is that while organizational doing failed in this system, the organizational thinking was solid enough to resurrect the firm as Andersen Tax [50]. For this new firm to survive, it will be essential to redo its organizational doing to align with organizational thinking. Therefore, guarding against the tendency to focus too much on either Organizational Intelligence or Performance Management is assisted by "seeing" that equal value is given to both realms of organizational activity through congruence measures.

2.5 *Testing Congruence Through System Metrics*

It must noted that congruence between Organizational Intelligence and Performance Management as a desired state for systems is only useful when valued metrics for the system are present and often when valued perspectives are in play. In other words, congruence may be present in a system between thinking and doing with metrics that are questionable. An example would be the deployment of a technical system whereby organizational knowledge is actualized as organizational doing, but where the system is built on flawed metrics. Organized crime is an example of congruent system of aligned thinking and doing that lacks metrics of value to civil society, namely transparency, honesty and ethics. So, even within a congruent system of thinking and doing, metrics of the system need attention. Since congruence can be managed as an emergent condition of a system, metrics can also be managed. Axelrod and Cohen [17] described the management of system metrics as, "... harnessing complexity does not take any performance as 'given'" and "performance measures are defined within the system" (pp. 194–195). This means that the best condition for a system is for congruence to be present between organizational thinking and organizational doing and for metrics of value to either be in place or be managed to the value proposition—and for the value proposition to be deeply understood. In addition, attaining a state of congruence in a system with valued metrics, and understanding how that congruence and benchmarks were attained, means that scaling up and/or replicating of success is possible.

3 The CX Tool

This section introduces the CX tool, the need for this tool and the continuous improvement related to it. The Tier 1 and Tier 2 of the CX Tool design and method is also presented.

3.1 The Need for the CX Tool

The CX Tool is presented next as one systems-based tool to provide insight into sources of dysfunction for systems. The need for this tool was described by the American Institute of Aeronautics and Astronautics as Jones et al. [6] stated,

> CASE organizers have recognized that current engineering or mathematical tools are not necessarily deficient, however we now face an increasingly complex mix of challenges that include examining our internal processes, organizations, policies, contracting, and other facets that stretch beyond traditional engineering competencies (p. 1).

There are long-term perspectives on the CX Tool that can be found in the work of Frederic Winslow Taylor (1856–1915) the Father of Scientific Management (1911). He overtly recognized and the realms of organizational intelligence and organizational performance. The recognition of these two spheres of required organizational activity is an instrumental tenet in the CX Tool. Unfortunately, Taylor chose to separate, not reconcile, these realms via hierarchal structures, disassembling the system into silos. As Markman [51] advocated, the approach enforced in the CX Tool contrasts the separation of planning and production by ensuring that thinking and doing are connected as an organizational system. So, the basis for the CX Tool is rooted in urgency around the need for systems competency as a sense making engagement as described by Vitiello and Kalawsky [28].

The CX Tool offers these two spheres of engaging in socio-technical learning about a system, a thinking side, called Organizational Intelligence, and a doing side, called Performance Management. The thinking side, Organizational Intelligence, is destined to be reconciled with the doing side, Performance Management, based on benchmark metrics. And the same improvement thinking applies in reverse, where congruence between thinking and doing and doing and thinking is bi-directional and multi-directional in a highly functional system.

3.2 The CX Tool and Continuous Improvement

The CX tool, therefore, is based on the desire to provide equal attention, or to create congruence between and among Organizational Intelligence and Performance Management distributed over six system elements. There is a reason that the state of congruence is an important outcome from the CX Tool. Organizational

Intelligence and Performance Management represent the same flow of activity that is found in the continuous improvement cycle, Plan-Do-Check-Adjust (PDCA). Plan and Check are fundamentally Organizational Intelligence activities. Do and Adjust are likewise Performance Management activities. Therefore, the PDCA cycle guarantees congruence between organizational thinking and doing. The commitment to PDCA as a cultural affair has been identified as a benchmark practice [52]. And, therefore, maintaining system congruence is an extension of this. In turn, Deming [30] stated that organizations will suffer when there is a break in the continuous improvement cycle of PDCA. This will occur when dissonance between organizational thinking and doing is present. For example, if "doing" or Performance Management is dominant, then work will focus on meeting production quotas or changing job duties to hit metrics. Over time, the culture gets sets toward "doing" as the frequent and valued activity of employees. This phenomenon is common in high volume production settings, where pushing out work is valued. On the other hand, if "thinking" or Organizational Intelligence is dominant, then work will focus on process design or assessing performance and how it occurs. Over time, the culture gets set toward "thinking" as the frequent and valued activity of employees. This phenomenon is common in bureaucratic low to moderate volume settings, where checking already checked work is valued. The CX Tool offers a visual management system for congruent foci on the spheres of Performance Management and Organizational Intelligence based on metrics of value. This method in the CX Tool resides in the essence of PDCA.

3.3 The CX Tool Design and Method

Within the CX Tool, there are three elements that define Organizational Intelligence. These three elements are: Essential Ideas; Essential Processes/Protocols/Structures; and Essential Assessments/Audits. There are three elements that define Performance Management. These three elements are: Essential Actions; Essential Standards; and Essential Deliverables. Because these six elements are an organizational system held together by the shoelace of congruence and measured by metrics of value, these six elements do influence each other, creating paired and tripled interactions between and among them. For each of the six elements, therefore, it is possible to outline the current state of the system and identify its weaknesses. And the future state of the system can then be readily set out on the CX Tool as well (see Fig. 1).

The ability to see the current state of congruence between Organizational Intelligence and Performance Management, the presence of any and all of the six system elements, and the evidence of the system's metrics are important for systems management. The current state's points of improvement are then depicted in the future state of the CX Tool. This dynamic examination, embedded in tenets of continuous improvement, informs both the problems and promise of the system and highlights actionable points of improvement (See Fig. 2). Further this examination

CX TOOL				
Metrics?			Metrics?	
Future State	Current State		Current State	Future State
ORGANIZATIONAL INTELLIGENCE	ORGANIZATIONAL INTELLIGENCE		PERFORMANCE MANAGEMENT	PERFORMANCE MANAGEMENT
Essential Ideas	Essential Ideas		Essential Actions	Essential Actions
Essential Processes, Protocols, Structures	Essential Processes, Protocols, Structures		Essential Standards	Essential Standards
Essential Assessments	Essential Assessments		Essential Deliverables	Essential Deliverables

Fig. 1 CX Tool model with selected system metrics

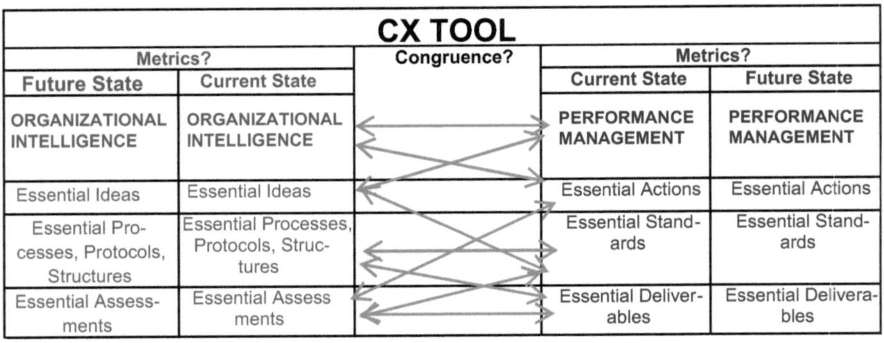

Fig. 2 CX Tool model with selected system metrics and congruence measures

can take place from a Tier 1 to a Tier 4 level, enabling more for complexity and complicatedness at each tier.

The Organizational Intelligence and Performance Management realms and their six systems elements create multiple opportunities to examine system congruence, system metrics, and possible points of improvement. The CX Tool, therefore, has several possible decision points. But as any improvement is introduced, those changes will be fully mapped to depict changes in the elements, the metrics and the overall state of congruence. The status of the system and the points of improvement for the system are discovered through data-driven inquiry, probing and sensing at Tier 1 to Baysein network analysis at Tier 4. In any case, these questions are posed to ascertain the Current State:

1. Are both Organizational Intelligence and Performance Management represented?
2. Are all six elements in the system represented?
3. What are the metrics on the system?
4. What is the status of the metrics on each element in the system?
5. Where is congruence present-horizontally, diagonally, vertically with the six elements?
6. What is the state of congruence in the system?

Once the Current State of the system is understood, these questions should be posed to develop the Future State:

7. Where is improvement needed regarding element metrics?
8. Where is improvement needed regarding system congruence?

In iterative rounds of continuous improvement work using the CX Tool, decisions are made as to whether metrics on system elements are excellent or poor and/or system congruence or dissonance is present. Strategies for disruption of poor metrics and/or system dissonance are based on the Future State plan. Strategies for homeostasis of excellent metrics and/or system congruence are based on the Current State. In other words, the CX Tool provides clear schema for decision trees. The full Tier 1 CX Tool, is depicted in Appendix.

One example of how a CX Tool decision tree could be used would be to discover that some system elements are not, in fact, present. For instance, on the Performance Management side in terms of vertical congruence, a system may have clearly delineated one element, the Essential Deliverable(s), such as, "Providing uninterrupted generation of nuclear power for 500,000 residents." However, there may not be Essential Standard(s) in place, such as "100 % quality assured compliance with all safety/risk metrics for the generation of nuclear power." Or there may not be an Essential Actions element in place, such as, "Funds will be allocated to ensure safety standards are met and employees are trained for safe nuclear power generation." These missing elements will cause system dysfunction.

Another example of how decisions are informed is through improvement of the metrics of the system. For instance, if there is a state of horizontal congruence between the Organizational Intelligence element, Essential Assessment(s), Audit (s) and the Essential Deliverable(s), but the metrics of congruence fall short, the system will experience dysfunction. If the benchmark metrics for a nuclear power plant are efficiency related, then an Essential Assessment/Audit element that requires a year-long testing cycle and the Essential Deliverable that there must be an annual safety/quality update, then these congruent elements are not efficient enough. These elements need to be improved so that shorter time spans are used and more efficiency can be introduced into the system. These missing metrics will also cause system dysfunction.

4 Case Studies of the CX Tool

The Tier 1 version of the CX Tool has been applied in different contexts through student projects at Midwestern and northern pacific universities. Presentations on the findings from these studies have been shared at 2014 ISSS [24, 25], 2014 ASME IMECE [53], and during workshops at the 2014 and 2015 AIAA CASE Academic Roundtables.

4.1 Midwestern University Tier 1 CX Tool Case Studies

Empirical testing of the Tier 1 CX Tool has taken place where practitioners have been asked to apply the CX Tool to a variety of systems in order to analyze the current state and to make suggestions for the future state. The tool has received positive comments from this testing, related to how informative it was to the current state, how helpful it was to collective improvement work, and how it provided a clear picture of the system. In these empirical tests, the suggested decision tree for using the CX Tool was to examine the systems' current state using this protocol as:

1. To check for the presence of all system elements as being "Essential." The decision tree poses this inquiry: Are the elements widely shared, succinctly understood, easy to articulate, and aligned with mission, vision, strategic plans, and goals?
2. To identify the same metrics of interest to use on all Essential Elements. The decision tree poses this inquiry: Are the system metrics widely shared, succinctly understood, easy to articulate, and aligned with mission, vision, strategic plans, and goals?
3. To qualitatively and quantitatively measure the state of congruence throughout the system. The decision tree poses this inquiry: Is there a state of vertical, horizontal, diagonal congruence between and among the CX Elements with acceptable metrics that are widely shared, succinctly understood, easy to articulate, and aligned with mission, vision, strategic plans, and goals?

For example, the Tier 1 CX Tool was used in educational leadership and management. One case study was conducted for a school district using the tool for a data management solution. It was determined that several system elements could be improved, such as two Essential Deliverables to, "*Highly effective data backup and retrieval*" *and* "*Cross—trained staff.*" Another case study was conducted at a school district using the tool for an instructional problem. In spite of the purchase of new, highly-vetted curriculum, reading scores were not improving. After a series of meeting, the CX Tool was used to view the problem from a systems perspective. The executive team was able to quickly visually ascertain that this curricular system was missing the Essential Assessment/Audit Element. Another series of case studies was conducted by students in a graduate leadership program as an observational and analytical exercise in systems thinking. Students in the program were tasked with observing any educational system, such as an instructional program system, a professional development system, a curricular system, an instructional technology system, or a collaboration system in the current state and conducting a future state analysis on it. The goal of this activity was to enable these current and practicing educational leaders to grasp the complexity and complicatedness of a selected educational system in order to generate well-founded improvement decisions. The experiences using the tool were very positive. One student practitioner reported having an enhanced leadership perspective on system complexity, such as realizing that visualizing system elements based on congruence measures, "*High degrees of congruence can be achieved when the structure of development choice is driven by ideas, actions, standards, processes, assessments, and deliverables that relate to the fluid nature of educational needs.*" Another student leader shared the results of the using the tool as, "*For 2 ½ years I have been trying to mandate Guided Reading and purchase a Guided Reading library. I was so excited to see a light bulb click with my boss as he recognized how our expectation of differentiation was not congruent to the whole group teaching of reading. When he looked up and said, 'That makes sense,' my heart skipped a beat! The tool not only helped me plan for Guided Reading, but it was used to help me demonstrate the lack of congruence in our current situation and show the congruence in Guided Reading!*"

4.2 Northwest Pacific University Tier 2 CX Tool Case Studies

The Tier 2 CX Tool was also used by students in a graduate engineering program as a workplace application of system-based problem solving. The goal for using the CX Tool was to allow a hematology and oncology care provider clinic to more effectively serve patients over the phone, resulting in an increase in patient satisfaction. The clinic has five different locations, seven physicians, and team of registered nurses, and many patients. It strives to provide patient-centered care at every step of treatment and patient ratings are then used to determine the reimbursement that the provider will receive from insurance companies [54]. The clinic ascertained the telephone system, the most common type of non-face-to-face interaction was inefficient with long response times, being consistently unable to respond to patients within a single business day. The volume of calls, complexity of the system, and lack of standardization caused long delays between initial call and final response. When patients are concerned about their health and need answers quickly, significant stress results if expectations are not met. The problem had been intensified by the recent adoption of a new electronic medical records system. Response times to patients have been delayed as care providers learn how to navigate the new system. By conducting a current state analysis of various stakeholders: doctors and nurses, patients, and management, an enriched view of system complexity was obtained under three weltanschauungen (world views). The stakeholders were identified in alignment with the CATWOE approach as defined by Checkland [55]. This way, a more holistic analysis was completed.

In CATWOE, each letter stands for:

- **C** lients—customers or the people who use the process to benefit from it
- **A** ctors—the people who make the process work
- **T** ransformation—process that transforms the inputs into outputs
- **W** eltanschauung (World view)—impact of solution or process
- **O** wner—who owns the process
- **E** nvironmental constraints—broader constraints to solution

For example, from the patient's weltanschauung, there were two main process flows. The first is shown in Fig. 3.

Figure 3 shows the process map for the phone system process from the point of view of the patient. The patient starts by calling the department. The decision that they have to make is what number to press. The process ends with the patient waiting and finally receiving the final response from the nurse. This process flow is the patient's view of how the phone system works. The first major breakdown that the patients encounter is the phone tree. If they do not press the correct number initially then they can be transferred from person to person before reaching the appropriate nurse. The next problem that the patients see is the act of waiting for a response. They do not know how long it takes to receive a response. The second process map is shown in Fig. 4.

Figure 4 shows the current phone tree that is presented when a patient places a call. In order to connect with the nurses, the patient needs to press three, four, five, or six. Any of the other options can result in multiple transfers. Pressing zero connects the patient with the answering service, who decides where to transfer the patient. The answering service is not always accurate with their choice; thus, the patient is transferred multiple times. The same analysis for doctors/nurses and management was conducted. And in then these results are combined based on the CATWOE elements and examined.

In summary, the Tier 2 CX Tool is capable of handling a higher level of complicatedness in a system because congruency scores between and within perspectives are analysed, allowing for a series of stakeholders' Tier 1 templates to be input into the analysis and capturing system complicatedness. Table 1 depicts this visually.

Notice how, in the Tier 2 template, the Organizational Intelligence and Performance Management have been re-arranged vertically, while the left column now only contains current state elements and the right column contains future state elements. The complexity of the Tier 2 templates makes it difficult to use them as visual management tools. So, Figure 5 presents a matrix structure used as a visual management for a CX Tool application. This example contains several cells blacked out and cells not filled out at all. The blacked out cells indicate interactions not considered in the analysis, while the blank cells indicate the analysis was only focusing on one-way interactions, as complexity of the system remains relatively low (Fig. 5).

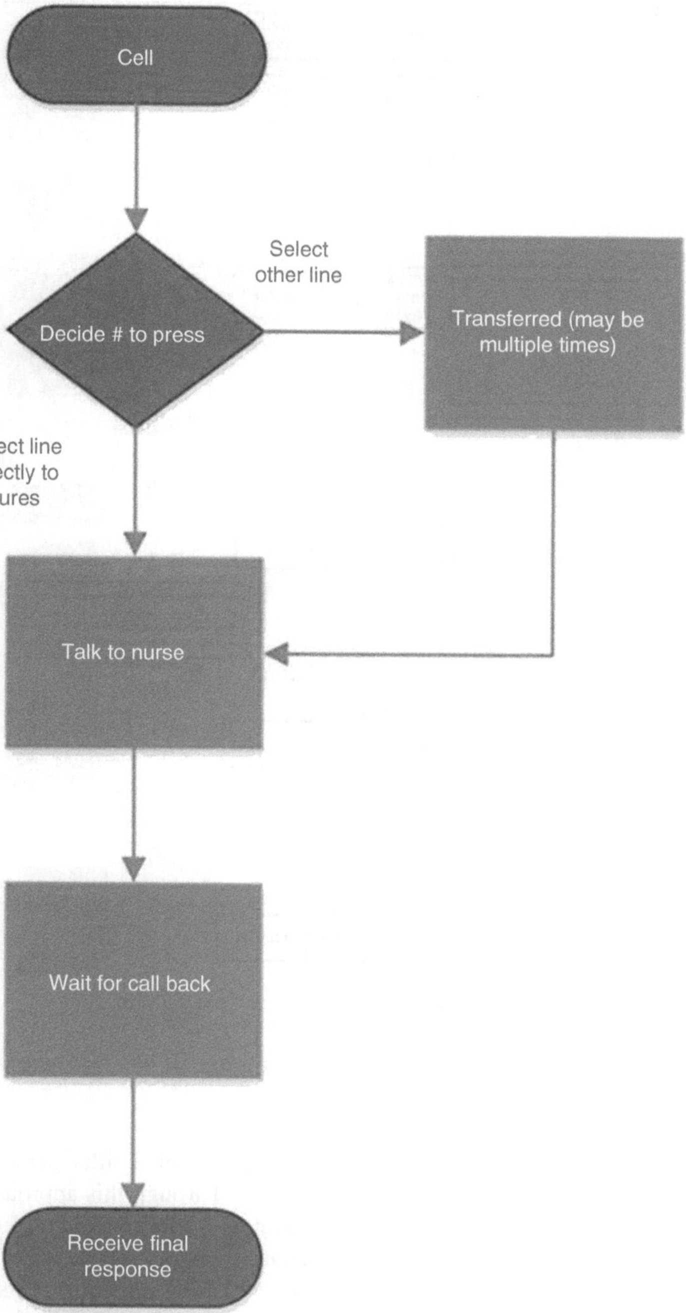

Fig. 3 Patient weltanschauung process flow diagram

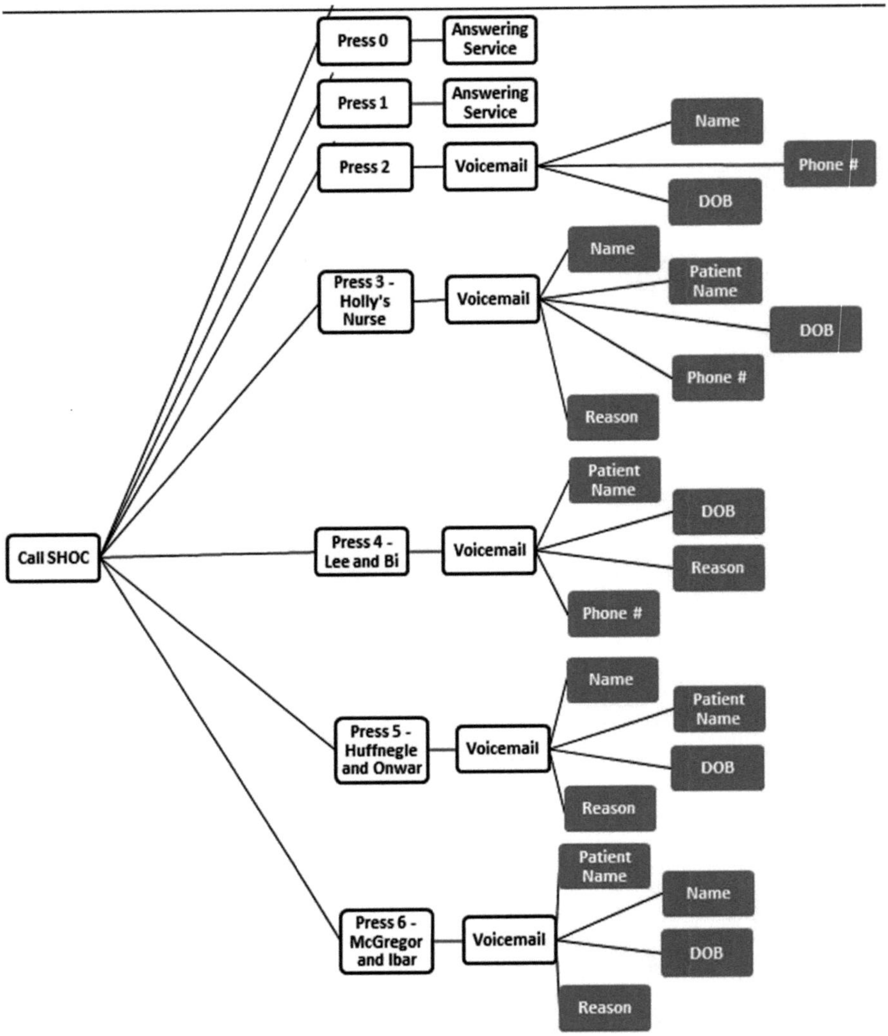

Fig. 4 Current phone tree process

The Tier 2 approach provided insight into three stakeholder perspectives and provided valuable findings for better congruency. Through this approach, the processes under various perspectives are correlated with selected and different system metrics, such as patient satisfaction, efficiency, patient cooperation or nurses' ability to adjust questions to each situation, providing an informed solution for better system management. In the end a series of solutions, all interrelated and accounting for system complexity were deployed, providing a clear outline of distinctions between the current and future state of this medical system.

Table 1 CX Tool Tier 2 template—congruency between stakeholders' perspectives

CX Tool Tier 2 Version							
Current state			**Congruency scores**		**Future state**		
Organizational intelligence					**Organizational intelligence**		
Essential Ideas					Essential Ideas		
Stakeholder 1	Stakeholder 2	Stakeholder 3	**C**	**F**	Stakeholder 1	Stakeholder 2	Stakeholder 3
Essential Processes, Protocols, Structures					Essential Processes, Protocols, Structures		
Stakeholder 1	Stakeholder 2	Stakeholder 3	**C**	**F**	Stakeholder 1	Stakeholder 2	Stakeholder 3
Essential Assessments					Essential Assessments		
Stakeholder 1	Stakeholder 2	Stakeholder 3	**C**	**F**	Stakeholder 1	Stakeholder 2	Stakeholder 3
Performance management			**Congruency scores**		**Performance management**		
Essential Actions					Essential Actions		
Stakeholder 1	Stakeholder 2	Stakeholder 3	**C**	**F**	Stakeholder 1	Stakeholder 2	Stakeholder 3
Essential Standards					Essential Standards		
Stakeholder 1	Stakeholder 2	Stakeholder 3	**C**	**F**	Stakeholder 1	Stakeholder 2	Stakeholder 3
Essential Deliverables					Essential Deliverables		
Stakeholder 1	Stakeholder 2	Stakeholder 3	**C**	**F**	Stakeholder 1	Stakeholder 2	Stakeholder 3

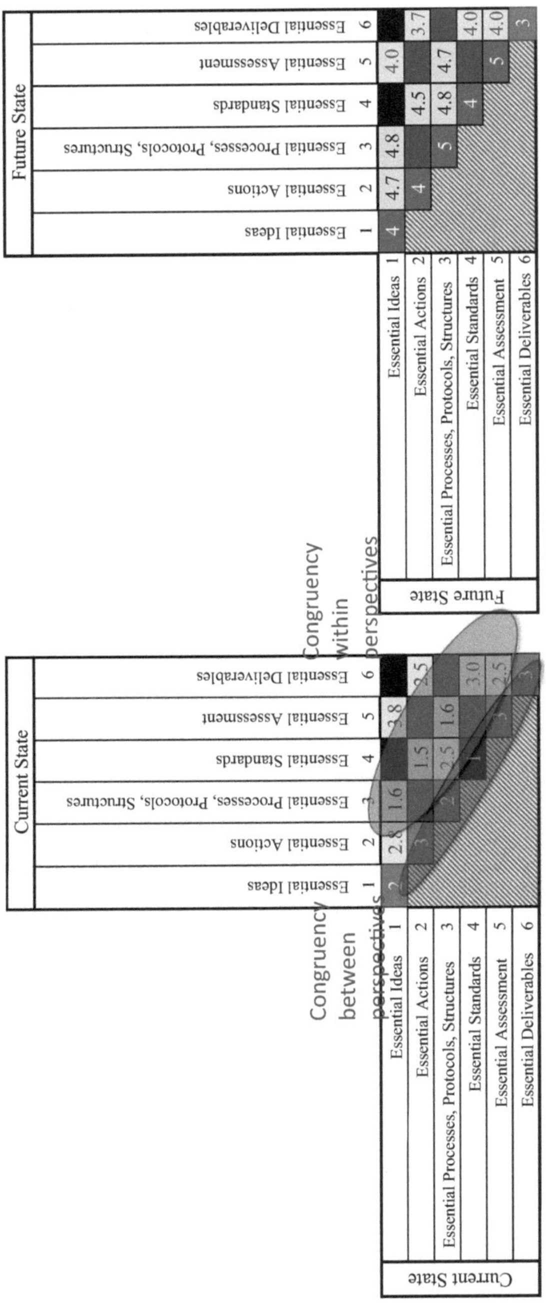

Fig. 5 CX Tool Tier 2 visual management summary (adapted from Calvo-Amodio and Flumerfelt [56])

5 Conclusion

The management of any system may seem befuddling due to the oblique attributes of systems themselves. While the CX Tool is still under testing, this tool allows for the identification of systems-based problems in organizations. Because there is a tendency for what people think and what people do in organizations to get disconnected, system functionality and resiliency suffers. Therefore, the state of equality or congruence between Organizational Intelligence and Performance Management must be tended to. The CX Tool offers three strategies to improve systems by: (1) ensuring that system elements representing thinking and doing are present, (2) identifying and testing valued system metrics on the system elements, and (3) testing for congruence in thinking and doing individually and collectively. In turn, without proactive management of any system's congruence condition without the CX Tool, three dissonance problems will potentially exist: (1) a lack of essential system elements' presence, (2) poor or missing system metrics, and (3) organizational dissonance.

As seen in the Tier 1 and Tier 2 case studies, the CX Tool provided roadmaps to identify systems' dissonance in their current states. In addition, one case study demonstrated the use of different system stakeholders' weltanschauungen to illustrate how the CX Tool can generate a holistic solution, based on some basic KPIs. The holistic analysis from the CX Tool facilitates the ability to "see" the system changes as the system incorporates these changes. Developing multi-faceted and coordinated potential solutions without harming the system has been demonstrated.

In summary, a theoretical and empirical basis for the CX Tool as a visual management method or holistic method that allows people from different functional departments to inform an interdisciplinary view of any system in the organization has been described. By working from a holistic understanding of the current state of the system, brainstorming and adjusting system impacts are possible to map out before deployment. The proposed future state of the system can be set out and decisions made as to whether it should be emerge as the new current state of the system. The iterative learning loop of moving the current state into improvement mode and then designing the future state ensures that organizational learning is ongoing and that the system is always getting better. The use of congruence to guide and measure systems performance and to leverage complexity is a tenet of systems management presented here. It is intended that the theory and practice of the CX Tool will, in fact, enable better organizational sense making by understanding more precisely where complex system failure is inevitable or occurring. The costs and consequences of system failure are too great. And, finally, it is hoped that the CX Tool will provide critical information for systems-based decision making.

Appendix: The CX Tool

ORGANIZATIONAL INTELLIGENCE				PERFORMANCE MANAGEMENT		
	Essential Ideas			Essential Actions		
	Future State	Current State		Current State	Future State	
12345 12345			1 2 3 4 5 1 2 3 4 5			12345
	Essential Processes, Protocols, Structures			Essential Standards		
	Future State	Current State		Current State	Future State	
12345 12345			1 2 3 4 5 1 2 3 4 5			12345 12345
	Essential Assessments			Essential Deliverables		
	Future State	Current State		Current State	Future State	
12345 12345			1 2 3 4 5 1 2 3 4 5			12345 12345

Description of the CX Tool

The CX Tool is a mental model "test kit" that can be used for analysis of any current or new system, process or project. "C" in the name stands for "congruence" or "equal state" and "X" stands for the myriad of ways that congruence can be developed or improved in a system, resulting in up to 21 points of congruence. This tool fully employs the Plan-DoCheck-Adjust organizational learning loop, a foundational concept of Lean.

The CX Tool identifies two Areas, Organizational Intelligence and Performance Management, with three Elements each. The three Elements for Organizational Intelligence are *Essential Ideas, Essential Processes/Protocols/Structures/Taxonomies and Essential Assessments/Audits*. The three Elements for Performance Management are *Essential Actions, Essential Standards, and Essential Deliverables*. The interrelationships between the six Elements represent points of Congruence.

When using this CX Lean Tool, a visual depiction of the current state of a system or process is created for an easier examination and prioritization of points of improvement needed in a system. By improving a system to include all six Elements, to place all Six Elements in the appropriate Area and to create Congruence between Elements, immediate benefits can be realized. The CX Tool helps managers test for three states in a system. First, the model tests for the presence of six essential Elements. Second, the model tests for the placement of six Elements into Organizational Intelligence activities and Performance Management activities. Third, the model tests for the Congruence of these Elements within and between using three Metrics, Efficiency, Effectiveness, and Relevance.

The CX Tool has been used in organizations to improve systems and processes that include strategic planning, core service programming, leadership development, technology systems, budgeting processes, support services, employee evaluation systems, and professional development/training. Directions for using the CX Tool and definitions of the Areas, Elements and Metrics follow.

Directions for the CX Tool

1. Consider any process, system or initiative in your organization and read the attached definitions of the two Areas, six Elements and three Metrics to understand the scope of the CX Lean Tool.
2. Identify a specific system, process or initiative, whether strategic or tactical, in your organization for analysis in its Current State.
3. Beginning with any of the six system Elements, identify one that is present in your system and enter a short description of it in the boxed space provided.
4. Then consider which, if any, of the other five system Elements are present in your current system and write short descriptions of those in the boxed spaces provided. You may decide there are some missing system Elements, so leave those spaces blank. The absence of any of the six Elements is a point of improvement.
5. Last, consider the three Metrics of Congruence both vertically and horizontally between present Elements. Circle those ratings numbers for Efficiency, Effectiveness, Relevance and then Congruence. Given the presence of all six Elements, it is possible to identify the state of Congruence between three horizontal pairs (*Essential Ideas to Essential Actions, Essential PPSTs to Essential Standards and Essential Assessments/Audits to Essential Deliverables*) and among

two vertical quadruplets (*Essential Ideas to Essential PPSTs to Essential Assessments/Audits to Essential Ideas and Essential Actions to Essential Standards to Essential Deliverables to Essential Actions*). In the end, there may be Congruence between all Elements, a possible 21 states of Congruence. For beginners, it is most helpful to examine the horizontal pairs and vertical quadruplets first.
6. Based on the absence of an Element or based on the worst Congruence rating, identify areas for improvement. Develop a description of the Future State and reassess that state after initiating improvements using this CX Lean Tool.

Definitions for the CX Tool

Two System Areas

Organizational Intelligence is the "shared smarts" or "IQ" of a corporate group based on its capacity to create and manage knowledge well.

Performance Management is the "shared work" of a corporate group based on its capacity to create and manage activity well.

Six System Elements

Organizational Intelligence

Essential Ideas are key concepts, major theories, big ideas or "meaning" that exists collectively. Examples are shared vision or benchmarks of practice, like the need to diversify, the benefit of new market development, the impact of going "green," or the value of collaborating to co-innovate. *Essential Process-Protocol-Structural* understandings are identified steps, forces and systems, both tacit and explicit, that are commonly understood. Examples are governance structures, business architecture, policies and practices, like budgeting processes, HR protocols, IT systems, Customer Records Management systems, training and development, manufacturing processes, grievance processes, social networks or strategic planning.

Essential Assessments-Audits are corporate data collection and analysis activities using quantitative and/or qualitative data and formative or summative assessments. Examples are analytics of operations or business intelligence, such as resource allocation effectiveness, performance activity results, ROI, scalability data, contextualized results, test or performance scores or achievement data.

Performance Management

Essential Actions are core values, initiatives, corporate ethos, and traditions that demonstrate what is highly valued in the culture by what is done. Examples are customer service, guarantees of conduct or quality, implementation of new initiatives, use of e-learning for training, product recalls, new marketing initiatives based on globalization, resource allocation to support an initiative, or capital expenditures.

Essential Standards in action are compliance-based or professionally-based codes designed to produce a result valued in the culture. They can be explicitly defined or tacitly understood behaviors. Examples are risk management compliance, key performance indicators or operational standards, like SOX compliance, globalization standards, expectations for creativity, common syntax, professional codes of conduct, organizational citizenship behavior, commonly shared practice or behavior shaped by culture. *Essential Deliverables* in action are the valued tangible and intangibles results of corporate effort aligned with mission. Examples are realized transactional or transformational value, thought leadership, societal improvement, quality products, outstanding service delivery, customer satisfaction, research and development breakthroughs or community service.

Three Congruence Metrics

Congruence is an equal state between Elements. Congruence, as opposed to equivalence or approximation, is a relation which implies a parallel existence. Congruence is not the same as alignment, which is more linear in nature. Congruence includes the concept of alignment, but is more expansive by including intensity, pace and interdependency of the interrelationship between Elements as being equal.

The metrics of Congruence in the CX Lean Tool are Efficiency, Effectiveness, and Relevance. *Efficiency* is a metric of Congruence that demonstrates the ability to achieve a desired result without wasted resources, energy or effort such as minimizing wasted time, motion or funds.

Effectiveness is a metric of Congruence causing an intended result that is favorable to organizational mission and performance such as facilitating sales objectives, employee productivity or financial stability.

Relevance is a metric of Congruence that is the sensible or logical connection that an activity has to a bigger scope such as department to organizational objectives, corporate to real-world issues, research and development to the marketplace, and executive goals to stakeholder value.

References

1. Skinner, B. F. (1965). *Science and human behavior*. New York: Free Press.
2. Glasser, W. (1994). *The control theory manager*. New York: HarperCollins.
3. Flood, R. L., & Jackson, M. C. (1991). *Creative problem solving*. Chichester: Wiley.
4. Felder, W. N., & Collopy, P. (2012). The elephant in the mist: What we don't know about the design, development, test and management of complex systems. *Journal of Aerospace Operations, 1*(4), 317–327.

5. Miller, J. H., & Page, S. E. (2007). *Complex adaptive systems: An introduction to computational models of social life*. Princeton, NJ: Princeton University Press. Princeton Studies in Complexity.
6. Jones, K. H., Parker, P. A., Detweiler, K. A., McGowan, A. R, Dress, D. A., & Kimmel, W. M. (2013). *Analysis and perspectives from the Complex Aerospace Systems Exchange (CASE)* (pp. 1–12). Hampton, VA: NASA Langley Research Center, American Institute of Aeronautics and Astronautics.
7. Senge, P. (1990). *The fifth discipline: The art and practice of the learning organization*. New York: Doubleday/Currency.
8. Drucker, P., Collins, J., Kotler, P., Kouzes, J., Rodin, J., Kasturi-Rangan, V., et al. (2008). *The five most important questions you will ever ask about your organization. Leader to Leader Institute*. San Francisco: Jossey-Bass.
9. Wheatley, M. (2002). We are all innovators. In F. Hesselbein, M. Goldsmith, & I. Somerville (Eds.), *Leading for innovation and organizing for results* (pp. 11–22). San Francisco: Jossey-Bass, The Drucker Foundation.
10. Burke, J. (2002). When 1 + 1 = 3. In F. Hesselbein, M. Goldsmith, & I. Somerville (Eds.), *Leading for innovation and organizing for results* (pp. 185–196). San Francisco: Jossey Bass, The Drucker Foundation.
11. Bánáthy, B. H. (2000). *Guided evolution of society: A systems view*. New York: Springer.
12. Madni, A. M. (2012). Elegant systems design: The fusion of simplicity and power. *Systems Engineering, 15*(3), 347–354.
13. Felder, W. N. (n.d.). *Interactions among components in complex systems* (pp. 1–5). Hoboken, NJ: Stevens Institute of Technology, American Institute of Aeronautics and Astronautics.
14. US News & World Report Health. (2014). *Mayo clinic* [online report]. Retrieved from http://health.usnews.com/best-hospitals/area/mn/mayo-clinic-661MAYO
15. Mannion, J. (2005). *Essential philosophy: Everything you need to understand the world's greatest thinkers*. Avon, MA: Adams Media Corporation.
16. Templeton, G. F., & Dowdy, J. F. (2011). CASE-mediated organizational and deuteron learning at NASA. *Information Systems Frontiers, 2012*(14), 741–764.
17. Axelrod, R., & Cohen, M. D. (1999). *Harnessing complexity: Organizational implications of a scientific frontier [ibook]*. New York: The Free Press.
18. Sterman, J. D. (2001). System dynamics modeling: Tools for learning in a complex world. *California Management Review, 43*(4), 8–25. Retrieved April 11, 2011, from ABI/INFORM Global (Document ID: 82668655): http://proquest.umi.com/pqdlink?vinst=PROD&fmt=6&startpage=1&ver=1&vname=PQD&RQT=309&did=82668655&exp=04-09-2016&scaling=FULL&vtype=PQD&rqt=309&TS=1302576861&clientId=43939
19. Checkland, P. (2000). Soft systems methodology: A thirty year retrospective. *Systems Research and Behavioral Science*, 17: S11–S58. doi: 10.1002/1099-1743(200011). http://citeseerx.ist.psu.edu/viewdoc/download?doi=10.1.1.133.7381&rep=rep1&type=pdf
20. Ackoff, R. (2004, May). *Transforming the systems movement*. Opening Speech at 3rd International Conference on Systems Thinking in Management, Philadelphia, PA. Retrieved from http://www.acasa.upenn.edu/RLAConfPaper.pdf
21. Amen, H., Flumerfelt, S., Halada, G., & Kahlen, F. J. (2011, December). Complexity and consequence: [Crowdsourcing Project]. *Mechanical Engineering*.
22. Amen, H., Flumerfelt, S., Halada, G., & Kahlen, F. J. (2012, March). Complexity by design: [Crowdsourcing Project]. *Mechanical Engineering*.
23. Felder, R. M., Woods, D. R., Stice, J. E., & Rugarcia, A. (2000). The future of engineering education: Teaching methods that work. *Chemical Engineering Education,* 34, 26–39. Retrieved from http://citeseerx.ist.psu.edu/viewdoc/down-load?doi=10.1.1.34.1082&rep=rep1&type=pdf
24. Calvo-Amodio, J., Patterson, P. E., Smith, M. L., & Burns, J. R. (2014). A generalized system dynamics model for managing transition-phases in healthcare environments. *Journal of Industrial Engineering and Management Innovation, 1*(1), 13–30. Retrieved from http://www.atlantis-press.com/php/download_paper.php?id=14481

25. Calvo-Amodio, J., Flumerfelt, S., & Hoyle, C. (2014). *A complementarist approach to lean systems management*. 2014 International Society for the Systems Sciences Conference, Washington, DC.
26. Calvo-Amodio, J., Patterson, P. E., Smith, M. L., & Burns, J. R. (2015). Application of transition-phase management model for an electronic health record system implementation: A case study. *Engineering Management Journal, 27*(3), 131–140.
27. Calvo-Amodio, J. (2013). Transition-phase management model as an action research tool for healthcare managers. In *Proceedings of the 2013 Industrial and Systems Engineering Research Conference, San Juan, Puerto Rico*.
28. Vitiello, P. F., & Kalawksy, R. S. (2012). *Visual analytics: A sense making framework for systems thinking in systems engineering*. IEEE International Systems Conference SysCon 2012, pp. 1–6.
29. Shewhart, W. (1939). *Statistical method from the viewpoint of quality control*. Washington, DC: Department of Agriculture.
30. Deming, W. E. (2000). *Out of the crisis*. Boston: MIT Press.
31. Bahill, A. T., & Gissing, B. (1998). Re-evaluating systems engineering concepts using systems thinking. *IEEE Transactions on Systems, Man and Cybernetics—Part C, 28*(4), 516–527.
32. Snowden, D. J. (2000, July). *Cynefin, a sense of time and place: An ecological approach to sense making and learning in formal and informal communities*. Conference proceedings of KMAC at the University of Aston. CiteSeerX: 10.1.1.196.3058.
33. Bolman, L. G., & Deal, T. E. (2008). *Reframing organizations: Artistry, choice and leadership* (4th ed.). San Francisco: Jossey-Bass.
34. Drucker, P. (1998). *The coming of the new organization*. Boston: Harvard University Press.
35. Jensen, E. (1996). *Brain-based learning*. Del Mar, CA: Turning Point Publishing.
36. Gardner, H. (2000). *Intelligence reframed: Multiple intelligences for the 21st century*. New York: Basic Books.
37. Arygis, C., & Schön, D. A. (1974). *Theory to practice: Increasing professional effectiveness*. San Francisco: Jossey Bass.
38. Moss-Kanter, E. (1989). *When giants learn to dance* (p. 21). New York: Simon & Schuster.
39. Frank, M. (2000). Thinking and systems thinking. *Systems Engineering, 3*(3), 163–168.
40. Pfeffer, J. (1998). *The human equation: Building profits by putting people first*. Boston: Harvard Business School Press.
41. Argyis, C. (1982). *Reasoning, learning and action: Individual and organizational XXX*. San Francisco: Jossey-Bass.
42. Argyis, C., & Schon, C. (1978). *Organizational learning: A theory of action perspective*. Reading, MA: Addison-Wesley.
43. Ward, D. (2010, March–April). *Faster, cheaper, better revisited: Program management lessons from NASA*. Defense AT&L.
44. Bennis, W. (1989). *On becoming a leader*. New York: Addison-Wesley.
45. Bennis, W. (1989). *Why leaders can't lead: The unconscious conspiracy continues*. San Francisco: Jossey-Bass Publishers.
46. Pfeffer, J., & Sutton, R. (2000). *The knowing-doing gap: How smart companies turn knowledge into action*. Boston: Harvard Business School Press.
47. Hurwitz, J. (2012, November 20). Re: Hewlett-Packard: It only gets worse. *Business week, The management blog*. Retrieved from http://www.businessweek.com/articles/2012-11-20/whyyou-still-can-t-bet-on-a-hewlett-packard-comeback
48. Taylor, W. C. (2011, September 11). How Hewlett Packard lost the HP Way. *Business week: The management blog*. Retrieved from http://www.businessweek.com/management/how-hewlettpackard-lost-the-hp-way-09232011.html
49. Associated Press. (2005, May 31). Enron's auditors edict reversed: Supreme Court rules June '02 obstruction decision flawed. *Corporate Scandals on NBC News.com*. Retrieved from http://www.nbcnews.com/id/8046535/ns/business-corporate_scandals/t/enron-auditors-verdict-reversed/#.VJjMBJ0FwA

50. Rapoport, M. (2014, September 2). Tax firm to revive Arthur Andersen name: Enron scandal brought down firm; consultant is buying rights to use name. *The Wall Street Journal*. Retrieved from http://www.wsj.com/articles/tax-firm-to-revive-arthur-andersen-name1409626508
51. Markman, A. (2012). *Smart thinking: Three essential keys to solve problems, innovate, and get things done*. Toronto: Penguin Group.
52. Womack, J. P., Jones, D. T., & Roos, D. (1990). *The machine that changed the world*. New York: Rawson Associates.
53. Flumerfelt, S., Kahlen, F.-J., Alves, A. C., Calvo-Amodio, J., & Hoyle, C. (2014, November 14–20). Systems competency for engineering practice. In *Proceedings of the ASME 2014 International Mechanical Engineering Congress & Exposition (IMECE2014), Montreal, PQ*.
54. Geiger, N. (2012, July). On tying Medicare reimbursement to patient satisfaction surveys [online]. *American Journal of Nursing, 112*(7), 11. Retrieved from LWW Journals: http://journals.lww.com/ajnonline
55. Checkland, P. (1981). *Systems thinking, systems practice*. Chichester: Wiley.
56. Calvo-Amodio, J., & Flumerfelt, S. (2015). *Sense-making between and across stakeholder perspectives*. 2015 International Society for the Systems Sciences Conference, Berlin, Germany.

Additive Manufacturing: A Trans-disciplinary Experience

Paul Witherell, Yan Lu, and Al Jones

1 General Principles of Additive Manufacturing

1.1 Overview of Additive Manufacturing

Additive manufacturing (AM) is a layer-by-layer fabrication process that is initiated from a 3D digital geometric model. AM has demonstrated considerable potential for fabricating geometrically complex, value-added, and customer-oriented products [1]. In addition, when compared with more traditional manufacturing processes, AM has several advantages, which are described below.

Since its inception in 1982 (as Stereolithography [2]), AM been known has been known by several "trendy" names including freeform fabrication, rapid prototyping, and 3D printing technology. It has also become an economically attractive manufacturing method. As of this writing, AM is generally recognized to consist of seven types of layering processes [3]. Each AM process, in essence, provides a way of fabricating customized parts from polymers, metals, and/or ceramics, without any of the geometric constraints of traditional manufacturing processes.

The recurring emergence of AM can be attributed to three major technology advances associated with layer-by-layer processes. Early excitement started with the first 3D printer in 1984 [2], was revived with the introduction of self-replicating machines in 1995 [4, 5], and then again with the RepRap Project in 2005 [6]. Projects such as the RepRap began to open the doors for personal desktop printers, a main enabler of the so-called maker movement [7], and a primary driver of AM's most recent reemergence. Another important driver in the latest wave of excitement is the fact that key patents have expired or are expiring [8]. This makes it more cost

P. Witherell (✉) • Y. Lu • A. Jones
National Institute of Standards and Technology, Gaithersburg, MD 20899, USA
e-mail: pwithere@nist.gov

F.-J. Kahlen et al. (eds.), *Transdisciplinary Perspectives on Complex Systems*,
DOI 10.1007/978-3-319-38756-7_6

effective for vendors to introduce new, technologically and economically attractive, AM technologies to users and industry alike.

Most recently, government, academia, and industry have recognized that AM is one of the key enablers for a U.S. economic renaissance in the near future [9, 10]. When discussing production in AM, it has been said that "complexity is free," meaning that the AM process does not care how complex the geometry is [11]. However, with that freedom come new technological constraints associated with material properties and the layering processes used to create that geometry. Many of those constraints negatively affect planning, testing, certification, qualification, validation, and verification. These constraints are a result of the inherent uncertainty, due to process, material and geometry variabilities, of AM; and, consequently, satisfying these constraints can drive production costs extremely high. Yet, even so, at least one manufacturer, GE, has estimated that it will be profitable to manufacture over 85,000 nozzles for a new line of jet engines with additive manufacturing.[1]

There are two reasons for this estimation. First, there is the ever-increasing understanding of the material-process-structure relationships in AM. That understanding is predicated on the continued use of methodologies and tools most often associated with systems engineering. Systems principles are being adopted to support design, process planning, fabrication, measurement, control, and qualification in AM. This chapter will review the role of systems engineering in advancing the understanding of fundamental AM principles.

Second, there is the recent progress of AM technologies at both the research and development stages. Research achievements have been made in key fundamental issues in AM. Those achievements have led to developments in in-situ monitoring/controlling systems, new qualification and certification methods, innovative designs, availability of more powder material and so on. Using these developments, vendors are finding ways to realize the potential for practical utilization of AM technologies in various sectors including manufacturing, biomedical, energy, and sustainability [12]. This chapter will explore the implications of expanding manufacturing technologies across traditional boundaries and will review the challenges presented when systematically managing and integrating information across multiple AM domains and beyond.

1.2 Systems in AM

In 2013, NIST published the report "Measurement Science Roadmap for Metal-Based Additive Manufacturing," [13] the result of an AM-focused workshop held at NIST and attended by various leaders in the field. The technology development

[1]http://www.bloomberg.com/news/articles/2013-11-12/ge-printing-engine-fuel-nozzles-propels-6-billion-market

challenges and needs identified include (1) robust and in-situ process-monitoring techniques including sensors for monitoring and measuring AM processes and products, (2) a database for storing material and process information, (3) expert systems for AM design, and (4) validated physics-based models.

Previously, in 2010, Frazier et al. [14] arrived at similar findings, including the need to (1) understand and reduce machine-to-machine variability through integrated, in-process monitoring and control systems, (2) develop a new paradigm of qualification methods to reduce costs of qualification, (3) develop new, innovative, structural design paradigms to save energy and weight, and (4) develop physics-based models to relate microstructure, material properties, and performance for process optimization.

Though worded differently, these two reports describe the same challenges. Moreover, both reports conclude that underlying all these challenges is a significant need for integrated information systems across all phases of the AM-product realization process. Addressing this need requires three things. First, the information flow between the various phases of that process must be defined. Second, the related technologies that implement these phases must interoperable and interchangeable. Third, the feedback and feedforward information needed to predict and ensure the desired quality of the final AM products must be identified, collected, and available to be used by those technologies.

1.3 The Bigger Picture

Addressing the aforementioned challenges has some urgency. In their consumer-based forms, AM technologies have empowered the general population with the ability to create simple, on-demand products. As a result, AM technologies have played a large role in the current maker movement [7]. In their more advanced forms, AM technologies are opening possibilities for those with specialized needs, such as doctors or the Warfighter, to make significantly more complex, on-demand products. It is clear, however, from the preceding discussions, that realizing those possibilities will require new system approaches that integrate all phases of the AM-product realization process. In this chapter, we will discuss what we mean by these systems approaches and implementation challenges in AM.

In this chapter, we will refer to activities in this systems approach as intra- inter- and trans-disciplinary with each signifying a progression on the previous. Intradisciplinary activities can be performed in isolation. Interdisciplinary activities involve multiple interactions between several disciplines. Transdisciplinary activities transcend traditional systems principles, extending beyond disciplinary perspectives [15]. In the next section, we discuss some of these disciplines. We will use multidisciplinary to signify that the discussion is irrespective of intra, inter or transdisciplinary implications.

2 Defining the Disciplines: Design, Materials, Processes, Qualification

According to both the NSF and NIST roadmaps [1, 13], AM research can be separated into four major categories: AM design, AM materials, AM processes and equipment, and AM qualification and certification. This section will discuss the fundamentals of each research discipline and introduce the interactions that lead to multidisciplinary systems engineering opportunities.

2.1 AM Design

Producing AM parts introduces new design opportunities and issues over and above those encountered with traditional manufacturing and basic CAD modeling. Such issues are often related to topological optimization and the designs of internal lattice structures, support structures, internal cooling channels, heat sinks, and part thickness. Sometimes these issues are resolved in the original, process-independent, CAD design; other times, they are resolved in AM—specific software applications. For example, internal cooling channels can be designed in CAD software; but minimum feature sizes, which may depend on the 3D printing process, cannot.

Earlier we mentioned the phrase "complexity is free" in AM. This means that it does not "cost" anymore to create a complex geometry with AM than it would to create a simpler geometry that may be typical of subtractive, milling processes. The phrase is associated with the free-form design opportunities offered by the process, such as the topological optimization of AM products. Topological optimization is achieved by (1) designing geometry, (2) minimizing unneeded geometry, and (3) maximizing the load carrying capacity of the part. In other words, add material only where it is needed for a part to perform its function. Topological optimization has been around for a long time, but the nature of subtractive manufacturing constrains the possible final geometries. AM removes those constraints and can realize complex geometries to fulfill the load carrying requirements of a part.

Figure 1a shows different designs from the GE engine bracket challenge in 2013.[2] The challenge was to optimize the bracket design for additive manufacturing while still maintaining its desired functionality and load-carrying characteristics. In this GE challenge, students with little or no experience in AM were able to successfully solve a practical problem using advanced optimization methods and a basic understanding of additive manufacturing. Figure 1b shows an extreme example of topological optimization. The main features of the part are the base and top plates, the two ducts on the top, and one duct on the side. The two ducts on

[2] http://grabcad.com/challenges/ge-jet-engine-bracket-challenge

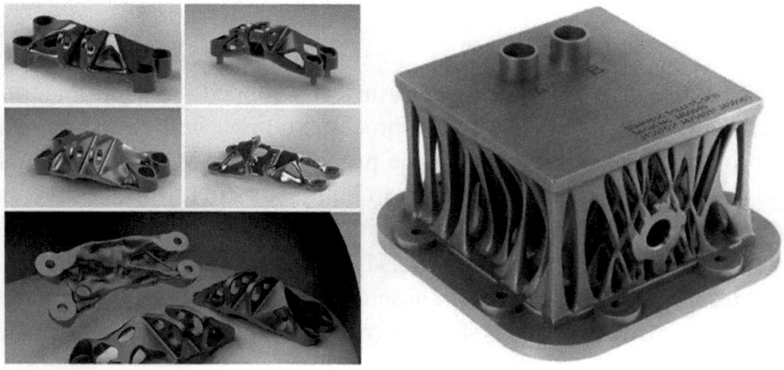

Fig. 1 (**a**) and (**b**) Examples of topological optimization

the top merge into the one duct. With conventional subtractive manufacturing processes, this might be an assembly of several pieces. With additive manufacturing, it is one piece with a very organic structure carrying the load between the top and bottom plates and supporting the internal duct.

Outside of free-form activities, internal lattice structures are a commonly used means for incorporating topological optimization principles into an AM build. Such structures can reduce significantly both the weight of a part and the amount of materials required while maintaining structural strength. The mechanical and thermal impact from lattice structures depends on the lattice configuration, porosity, build orientation, and constituent materials. Oftentimes lattice structures can be designed into a part automatically. For instance, some vendors support the generation of lattice structures as a honeycomb structure to reduce material consumption—rather than completely fill the interior of a structure. The user can control the density of the honeycomb, which is the percentage of the interior to be filled. This is a feature unique to AM processes.

Similar to the design of lattice structures, the incorporation of complex design requirements into the process itself provides the ability to create digitally enhanced materials. Graded materials allow for the ratio of multiple (usually two) printing materials to vary along the part. Materials with different mechanical properties can be combined to generate new properties for the part as a whole.

Using some AM processes can introduce new types of specific design requirements. For instance, support structures are needed in many AM processes to facilitate the build process. Supports are highly dependent on the build orientation. Optimizing the build orientation can reduce the amount of support material that needs to be removed. Support structures are mechanically removed after the build process. The build orientation can also be optimized to minimize the contact between the part and support structures in order to reduce the amount of post-building finishing. As an example of AM's multidisciplinary nature, another purpose for a support structure is to act as a heat sink. It can control the thermal characteristics of regions of the part during the build process.

2.2 Materials

AM materials and AM processes are intrinsically related—certain materials may only work with certain processes and process settings will determine the final material properties. Materials that can be processed by AM include metals, polymers, and ceramics. Feedstock may come in liquid or solid form. The diversity seen in AM materials is a reflection of the diversity seen in AM processes. This diversity creates challenges when "designing for AM," since the material selection is highly dependent on the process selection. These challenges highlight the importance of understanding, from a systems perspective, design-material-process interdependencies.

In coming to this understanding, it is important to realize that these AM interdependencies are significantly different from those that exist in traditional manufacturing. There are two important reasons. The first is the variance in processes and raw materials. The second is that the final composition of those materials and the final geometry are being created at the same time. To explain, in subtractive manufacturing, we know the final constituent material properties of a metal billet before milling takes place. In AM, however, the resultant material properties of sintered AM materials, fused plastic filaments, or light-cured polymers—are not completely known until after the geometry of the part has been formed.

As an example, we look at metals, specifically metal powders. Metal powders can be AM-processed using a binder, a laser, or an electron beam, to name a few technologies. The characteristics of metal powders respond differently to each type of processing. To identify the optimal type of metal powder for a given process, we must know (1) the pertinent material characteristics, (2) the pertinent process characteristics, and (3) the interactions between the process and the materials. Currently, none of this is known with certainty; and this creates an interesting systems optimization problem.

Characterizing the properties of raw materials in AM requires a significant amount of details. Different methods are needed to obtain those details. For instance, consider the morphology of powder particles. Characterizing the morphology requires two kinds of details and two kinds of methods: qualitative and quantitative. The qualitative method characterizes the shapes of the powder particles by assigning adjectives to them such as acicular, flake, granular, irregular, needles, nodular, platelet, plates, and spherical. Quantitative methods may include single-number classifications, in which a shape is defined by using only one number associated with a feature of a particle. Some standards efforts address sampling techniques and methods that can be used in determining size, flow, and density characteristic.[3] Other standards efforts are addressing how to characterize and to measure the morphology, the chemical composition, and the thermal properties of metal powders.

[3]http://www.astm.org/COMMIT/SUBCOMMIT/F4205.htm

As noted, the meticulous characterization (i.e. morphology, thermal, density) is necessary to understand completely how AM processes alter material characteristics. Currently, a complete understanding does not exist. To arrive at such an understanding, a thorough characterization of the prominent AM processes is also needed.

2.3 Process

As noted earlier, AM processes can be classified into seven types: Material Jetting, Binder Jetting, Powder Bed Fusion, Directed Energy Deposition, Vat Photopolymerization, Sheet Lamination, and material extrusion [3]. Each AM process is unique and requires a vastly different characterization. Most processes can handle more than one type of material, but the definition of "type" may vary. Despite process differences, the main principles of each process are based on the characteristics of a layer-by-layer process. Different processes create these layers differently—such as by sintering, by curing, by lamination, or by melting. The way the layers are joined depends on both the feedstock material and the process. The way the layers are formed will also play a vital role in determining the characteristics of the final part.

Continuing with our metal powder example, we now discuss some of the parameters that factor into the manufacture of a metal part. When categorized by layer type, metal AM systems can be divided into three broad categories [16]: powder bed systems, powder feed systems, and wire feed systems (Fig. 2). When we include the different energy sources (electron beam, laser, arc, etc.), we get additional characterization. When categorized by layer type and energy source, metal powder can be processed using several different AM processes [16]: EBM (Electron Beam Melting), SLM (Selective Laser Melting), DDM (Direct Digital Manufacturing), DMLS (Direct Metal Laser Sintering), SLS (Selective Laser Sintering), LENS (Laser Engineered Net Shaping), and so on.

Clearly, metal processing systems get more complex with each additional relationship. The physical phenomena that occur during the processing of metal powder become very complex sub-systems. High-powered beams will heat the powder to a temperature above its melting point to fuse it together. This is done at enormously fast speeds, and in very focused laser spots. Let us look a bit deeper into one such process: powder bed fusion (PBF).

Typical PBF processes for metal powder processing include four major sub-processes: heat radiation, heat absorption by powders, melt pool formation, and solidification. Heat radiation is a process where a beam (laser or electron) emits a predetermined amount of energy, at a predetermined traveling speed, and, with a fixed spot size. Heat absorption is a sub-process where powders absorb beam energy while scattering and diffusing energy at the same time. After powders are heated to a molten state, a melt pool is formed. As the beam moves away, the melt pool temperature decreases and molten liquid solidifies, causing solidification.

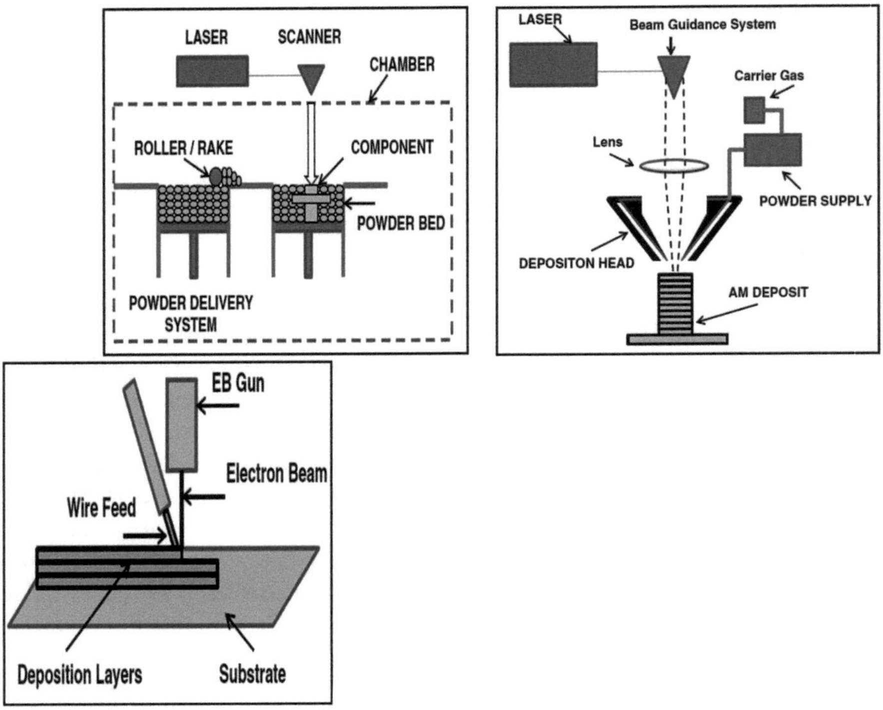

Fig. 2 Generic illustration of metal AM machine: powder bed system (*left*), powder feed system (*middle*), and wire feed system (*right*) [16]

During solidification, defects can occur and residual stresses can develop, creating imperfections in manufactured parts.

The effort involved in measuring these sub-processes is substantial and the data generated is significant. This makes AM processes very difficult to control and to repeat, and it creates complex systems problems. To understand the process-material interactions, research has been conducted on many of the processes by monitoring the environment and developing empirical models. For instance, many models have been developed to analyze and simulate melt pools [17]. These empirical models, however, do not easily translate. Because of the variability in, and the uncertainty of, AM processes, developing testing methods and tools to qualify the final part is an important part of AM production.

2.4 *Qualification*

Repeatability is a challenge during the production of AM part, from material preparation to part build. For this reason, qualification procedures are often put in place, especially for industry-manufactured components. These procedures can be applied to the feedstock, the process, and the part itself. Qualification procedures allow users to create a level of confidence in the materials and the processes they are using, and, the parts they are creating. Qualification of a part requires successful interactions among design, feedstock and process, embodying the notion of multi-disciplinary systems.

Without standardized methods, current empirical methods for qualification are stand-alone, which means customized methods are utilized. Given the complexity of the measurements, qualification can be cumbersome and impractical. Complete qualification of new materials and manufacturing processes can involve thousands of individual tests, cost millions of dollars, and take 5–15 years to complete [18]. One of main reasons is that the data necessary for material and process qualification are expensive to collect and difficult to analyze. Here we discuss the qualification of a material, process, and part in the context of metal powders and power-based fusion (PBF).

Given the importance of qualification, its dependence on repeatability, and the PBF process parameters, various research efforts for characterizing metal powders are underway [19–21]. Qualification of powders requires identifying important powder characteristics and identifying their desired qualities. For instance, powder particle size is one of the most important properties, since it determines the minimum layer thickness and the minimum geometric feature sizes of a part. There are several methods for size determination; they include sieving, gravitational sedimentation, and microscopy-based techniques.

Unlike traditional subtractive manufacturing, qualification methods for AM processes are not well developed, and are therefore far from standardized. Challenges in developing such methods come from the complexity of the process and the resulting variations. The qualification of the process often comes down to the repeatability of developing a part. Common means for machine/process qualification are performing mechanical testing (e.g., tensile strength), thermal testing (e.g., heat deflection), electrical testing (e.g., volume resistivity), chemical testing (e.g., corrosion test), microstructure analysis (e.g., SEM), and other testing (e.g., cleanliness).

Part dimensional accuracy provides an indication of how closely the physical dimensions of an AM part meet the design specifications. The mean of the dimensional errors is often used as a measure of accuracy. The standard deviation of the dimensional errors is a measure of the precision for a process. If both the accuracy and precision were high, this would indicate a certain level of repeatability in the AM process. Dimensional error is often dependent on the build orientation and part size. The accuracy, precision, variability, and surface finish of AM metal parts are generally poorer than their counterparts in traditional machining processes are. When the same design is built on consecutive days using the same machine, same powder, and same process parameters, the resulting parts have typical dimensional differences larger than the dimensional errors found in a computer numerically controlled (CNC) removal processes [22].

Another reason for the difference in quality control requirements between traditional and AM processes is the complexity of AM parts. Traditional, destructive test methods require the building of coupons. Each build requires multiple coupons in various orientations, which can quickly become expensive. Non-destructive tests can provide an alternative means for qualifying a part [23]. NDE methods can be selected based on the governing properties for a particular qualification. Regardless of the testing technique used, qualification in general creates large amounts of data that must be managed. Given the complex interactions, collecting and storing the data for later reuse requires methodical organization and deliberate systems approaches.

3 Exploring the Interactions: Systems Principles in AM

Various roadmaps and reviews [24] have noted that AM data and information—key elements of the so called AM digital spectrum [25]—can be used to improve AM processes. This section investigates the roles information infrastructure can play across the value chain of AM design-to-product transformation. We highlight interdisciplinary systems challenges and opportunities that will later be related to advancing AM in transdisciplinary settings.

3.1 The Digital Spectrum

The AM design-to-product transformation process involves various data-driven activities. Activities include design, tessellation and/or data generation, model fixing and cleaning, deciding the best support structures and orientation, generation of machine code, AM machining using processes and materials, post processing, and finally testing for qualification. It is important to recognize these as data-driven activities, since they can be leveraged to drive systems opportunities. To understand the feedforward and feedback opportunities, we must understand the flow of information and systems interactions during the design to product transformation.

The design-to-product transformation process can be separated using six different constraint-inducing "controllers," each controller creating the need for a transition in how AM information is represented and managed. These transitions can be enabled by a supporting infrastructure that (1) involves a number of AM-related standards, methods, techniques, hardware, and software and (2) facilitates the exchange of the feedforward and feedback information needed to change from a set of individual transitions into a system of transitions. To effect such a change and to understand its potential benefits, we must have a clear understanding of each transition, its boundary, and its interactions with other transitions. The six transitions are design for AM, machine independent process planning, machine dependent process planning, build, post process, and qualification [25]. We define these further below.

Design for AM—creates a file that specifies the design geometry of the part to be manufactured by AM, independent of the process that will be used.

Machine Independent Process Planning—incorporates process dependencies into the design. Once a process has been identified, factors such as optimal orientation and support structure requirements can be added.

Machine Dependent Process Planning—addresses the generation of machine-specific information. This is needed because there are multiple machine manufactures for a single process. Furthermore, each such manufacturer uses its own "recipe" for creating a part.

Build—uses the specified parameters from process planning to control the layering of the material during runtime.

Post Process—involves other, more traditional manufacturing processes to complete the fabrication of the part. Examples include heat treatment to improve stress relief and polishing to improve surface finish.

Qualification—uses a variety of testing and inspection procedures to qualify the part. This may include destructive testing of coupons or non-destructive evaluation of the part.

Figure 3 depicts an activity model of these transitions using IDEF0.[4] The boundaries of these transitions are dictated by their control inputs (feeding in from top) and the equipment they rely on (feeding in from the bottom). These six transitions result in the creation of six different phases of AM information. These six phases of information are represented as inputs and outputs in Fig. 3. Individually, information interfaces within these phases provide the foundations for functional feedforward and feedback interactions.[5] Collectively, these phases constitute information from the "end-to-end digital spectrum" that can be associated with a single build—sometimes referred to as the "digital thread."

The concept of a 'digital thread' is not new. The term 'thread' has been used when speaking about improving interoperability across the supply chain where all partners are now expected to create feedforward digital information when designing, manufacturing, and testing products. The same term has been used when speaking about improving the quality of manufacturing processes in multiple industries. In each case, the digital thread ties together a number of activities.

We use the same 'thread' concept to tie together the activities within and across the six phases defined in AM. By 'tie together', we mean that the thread provides both the path that information will use to be exchanged and the information that comprises that exchange. The information captured and maintained in this digital thread facilitates rapid design-to-product transformations by establishing real-time provenance during production. This provenance allows us to manage and trace "what, how, where, who, when, and why to get" during the whole AM process.

---

[4]http://www.idef.com/idef0.htm

[5]It is important to realize that the exact nature of these interactions is not specified in the model.

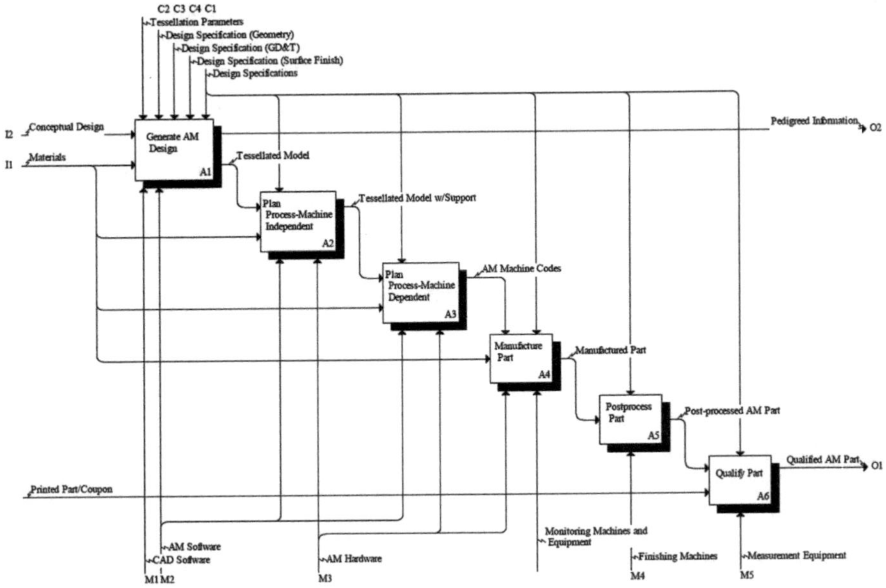

Fig. 3 IDEF0 Representation of AM design to product transformation

The information that makes up the digital thread is created, sorted, exchanged and used dynamically by activities throughout the whole transformation processes. That information is frequently transmitted back and forth among those activities. More importantly, it is one of the keys to the successful execution of those activities. For example, feedforward information can be used to build a simulation to assess and optimize process parameters, before the actual build process begins. During build, in-situ measurements can then provide the feedback information needed to control the process and ensure a quality part. Understanding and taking advantage of the feedforward and feedback opportunities is critical to enabling not only interdisciplinary, but also transdisciplinary activities.

Figure 4 details information that may be associated with each of the six phases defined from Fig. 3. With a clear understanding of these phases, and their relationships to the transitions, we are now to view the entire design-to-product transformation process as a system. It is through this view that we can identify opportunities to replace physical implementations of activities with highly accurate digital replicates. These replicates, together with the digital thread, enable us to develop a virtual design-to-product transformation process.

This virtual design-to-product transformation process has several potential benefits in terms of both utility and quality. In terms of quality, the inter-disciplinary replicates—usually simulations or models—allow us to test and optimize all aspects of the build and post process domains before anything happens in the real world. This should improve the quality of the part significantly. This, in turn, will increase utility by reducing production costs, including those associated with

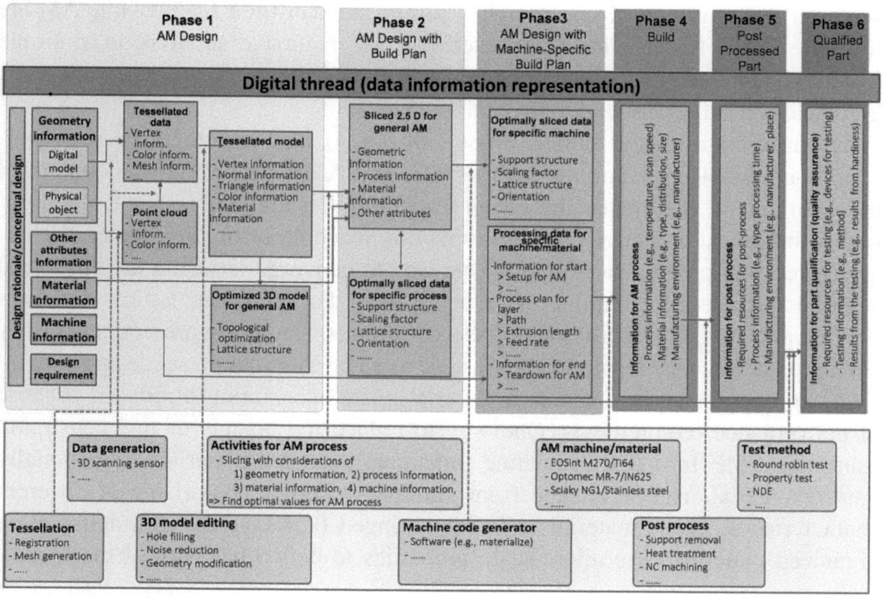

Fig. 4 NIST end-to-end digital spectrum in AM based on [22]

finishing and failed builds. In the following sections, we discuss this new virtual design-to-product transformation process in more detail, from the perspective of the six phases of AM.

3.2 *The Design Phase*

3.2.1 Designing Simple Parts

The wide, almost limitless, range of geometries that can be built using AM processes is one of AM's most appealing attributes. AM processes are often used to manufacture parts that could be easily built with more traditional manufacturing process, such as machining or injection modeling. This is often the case when prototyping or building "one of" parts. However, using AM to build parts that could be easily built using traditional manufacturing processes does not take full advantage of what AM processes offer. This section discusses what it means to "Design for AM," and how inputs from multiple phases should be considered.

While almost any geometry can be created using AM processes, the usable geometry representations are limited. Most AM systems use the stereolithography (.stl) from CAD solid model [26], though Additive Manufacturing File Format (AMF) has become a more recent alternative. Both formats approximate geometry using tessellations, which provide the basis for generating slices and assigning process parameters for a particular AM machine. While the STL format primarily supports geometry, the AMF format has introduced many new features into AM file formats. Those features include curved patches, recursive subdivision, multiple materials, graded materials, internal structures, material properties, colors, graphics, constellations, and metadata.

Downstream design activities (Phase 1 of Fig. 4), including topological optimization, are enabled by formats such as STL and AMF. The richer the content of these formats is, the greater the systems opportunities are. For instance, the topological optimization activity is constrained by machine resolution, material characteristics, and process capabilities. One way to converge faster on the optimized solution is to reduce/simplify those constraints. For example, lattice structures provide a means for simplifying process constraints, since they are not necessarily a function of machine resolution but instead a can be a used as preset design pattern.

As another example, consider a material-related constraint introduced because of performance requirements. One way to reduce the bounds of this constraint, using the AMF format, is to define functionally graded materials. Functionally graded materials can be created from multiple materials comprising of different characteristics. These materials are then arranged (and layered) in a deliberately organized pattern to use micro-scale properties to define behavior at the macro-scale.

3.2.2 Designing Complex Parts

In this subsection, we discuss the design of parts that are unique to AM and inherently face greater system-imposed constraints. For instance, in addition to fabricating single AM parts, additive manufacturing can also create an assembled part in one build. This creates unique opportunities, where benefits are still being explored, and unique systems challenges. For instance, NASA has designed and fabricated an AM wrench.[6] Fabricating such assemblies can be a challenging endeavor, since the assembled pieces (1) will determine the required orientation of the part and (2) allow for moving parts to be created within solid structures (Fig. 5).

Another example of a complex AM part is one that incorporates embedded conductors. As the layering process proceeds, it is possible to embed fully enclosed and fully functional 3D electronics. This capability suggests the possibility of using AM processes to manufacture printed circuit boards (PCBs). The technology of designing and manufacturing such parts is still limited by; but with advances being made in the printing of conductive materials [27] this technology is quickly emerging [28, 29]. Such a capability embodies the systems challenges that arise in AM processing, as new dimensions can be added to the already-complicated designs of PCBs.

[6]http://www.nasa.gov/mission_pages/station/research/news/3Dratchet_wrench/

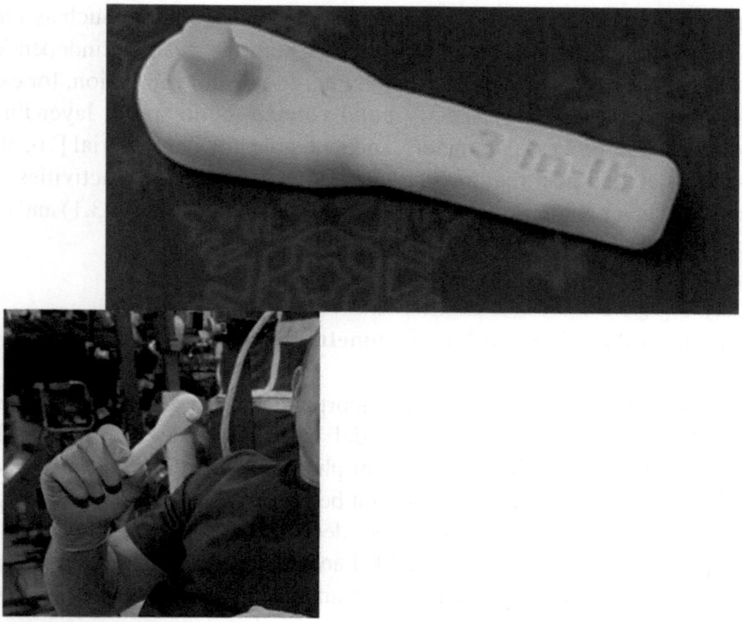

Fig. 5 NASA wrench assembly built with AM

This section highlighted some of the unique design-related opportunities available in AM. To take advantage of these opportunities, progress must be made elsewhere in the digital spectrum as well.

3.3 Process Planning Phases

Fundamentally, designing the geometry of AM parts is no different from designing the geometry of molded or machined parts. Realizing the geometry as a part, however, often requires planning for a specific AM process and AM machine. It is well known that different AM processes have different strengths and weaknesses [12]. The process planning phases, both machine independent and machine dependent, are where such process considerations are taken into account. For example, considerations such as orientation and support structures can be made once a process has been identified. On the other hand, machine-specific process parameters, such as those for slicing and scanning, are not decided until the machine has been identified.

Process planning requires three major informational inputs: the design, the process, and the materials. These inputs must be synthesized in order to find the process plan that optimizes the likelihood that the finished AM part matches the AM design. To identify an optimal plan, a great deal of insight into how materials, processes, and machines interact is necessary, because the outputs of that plan depend highly on those interactions. For example, considerations such as shrinkage and residual stresses should be incorporated at both the machine independent and machine dependent phases of process planning. In powder bed fusion, for example, an optimal process plan may specify build volume, build speed, layer thickness, surface roughness, process parameters, and available powder material [16, 30]. Creating such a detailed plan requires the integration of several sub-activities, such as part setup and layer configuration, based on part geometry (Sect. 3.3.1) and material properties (Sect. 3.3.2).

3.3.1 Process Planning and Part Geometry Models

Process planning begins with, and must incorporate, a part geometry represented as a 3D CAD model. That 3D CAD model can affect several process-planning decisions during the machine-independent phase. One of those decisions is "build orientation." Build orientation is important because it affects part tensile properties and surface roughness. Another of those decisions is "Nesting." Nesting allows multiple parts to be built within the build chamber, a means for reducing production times and creating assemblies in a single build. A third decision that can be made using the 3D models involves the design of support structures.

Machine-dependent process planning results in a 2D slicing model, a path plan, and all the necessary process parameters. All of these decisions require an understanding of the capabilities and the limitations of the AM machine and the 3D geometry model. For instance, in order to minimize both stair stepping and build time, a 2D adaptive slicing model must be developed based on the geometry curvature at nearly horizontal surfaces. In choosing the path plan (represented in G-code, STEP-NC [31], or something similar) and the process parameters, a thorough understanding of the material properties is necessary.

In both machine dependent and independent process planning, it is clear that downstream production implications can influence upstream decisions. These interactions become even more intricate with complex parts. Interdisciplinary systems approaches are taken by designers to account for process-related nuances, however the methods used are often built into the software and require design expertise. Transdisciplinary endeavors must incorporate similar approaches, creating challenges without the proper domain expertise.

3.3.2 Process Parameters and Material Properties

Various research efforts have focused on establishing relationships between processes and materials. Characterizing these relationships is key to determining process and material combinations that will produce a quality AM part. By relating process parameters with material characteristics, mappings can be established to support process selection, material selection, and process planning in the early stages of the design-to-product transformation.

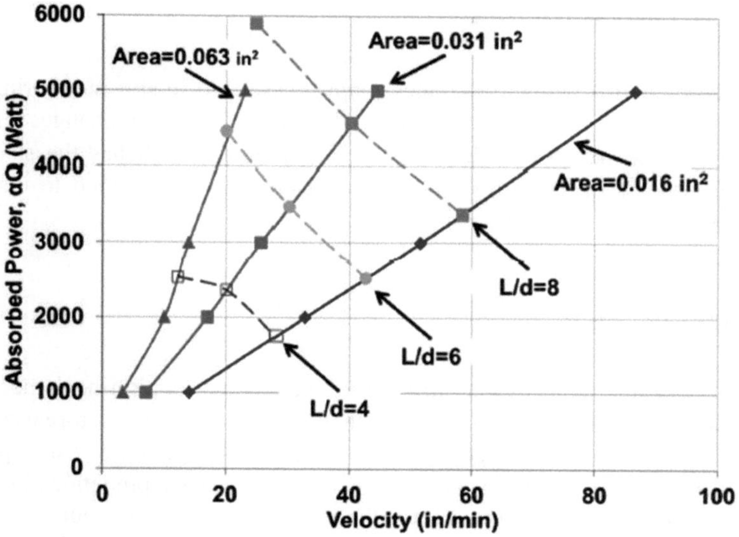

Fig. 6 Process map for controlling melt pool dimensions for single bead Ti64 deposition in wire feed e-beam AM (Sciaky) [33]

Beuth et al. [32–35] have presented such a process mapping approach and identify five primary process variables of importance in any heat-based AM process. They are heat source power (P), heat source travel speed (velocity, V), material feed rate (MFR or other variable defining MFR), temperature profile of the part (T0), and feature geometry (the local geometry of the part). While recognizing that other process variables (such as average powder particle size or beam spot size) may also affect process outcomes, they emphasize five primary process variables first. Figure 6 is an example process map for controlling melt-pool dimensions for single-bead Ti64 deposition in wire feed e-beam AM (Sciaky), where d is the maximum effective melt pool depth [defined as $d = \sqrt{(2A/\pi)}$] and L is the melt pool length from the point of maximum depth to the melt pool trailing edge.

The complexities of the mappings discussed above highlight the systems-related challenges faced by AM users. While efforts such as the mapping in Figure 6 provide insight into how to account for variances in parameters, it remains a challenge to account for all possible variances. As the interactions between disciplines become more detailed, the systems requirements (and applications) become more intricate. Though this intricacy is paramount to some AM systems opportunities, such process control applications, it again creates significant challenges when supporting trans-disciplinary efforts.

3.4 AM Process Control

AM process control is about (1) anticipating, based on in-situ monitoring, how process parameters will influence the build and (2) making changes in those parameters when necessary. Unlike process planning, which has the same first objective, process control uses real time measurements to respond to unwanted changes in the build of a part.

3.4.1 How Does It Work

To make real-time adjustments during a build, process monitoring is necessary. Sensor data is collected and processed in real time to determine the current state of a process. The state can include geometric accuracy and temperature change profiles. When differences between actual and predicted states reach a predefined threshold, corrective actions can be taken to force the required changes to occur.

When available, most AM machines use what is called open-loop control, meaning that feedback is received but not processed in real-time; hence, there is no automated response. Commonly found issues with open loop control include (1) defects cannot be timely detected, (2) part variability is high, and (3) good part quality is difficult to obtain. Open loop control monitors and collects data that can be acted on only later for quality traceability. Alternatively, close-loop control relies on sensors and feedback loops to adjust operating parameters to attempt to compensate for perceived errors during the build of a part.

3.4.2 Process Control and Modeling

Let us continue our discussion with our previous example: a powder bed fusion (PBF) process. PBF processes present significant process control challenges principally because of the problems in (1) accurately predicting the variability of the powders, even from the same manufacturer and (2) accurately tracking a number of dynamically changing variables including the melt pool, the heat radiation, the heat absorption, the scattering, and the diffusion. Powder variability comes from the uncertainties in several material properties including particle size distribution, powder density, and other powder coefficients. These all impact the melt pool formation, which in turn influences the grain size, grain morphology, residual stress distribution, and porosity of the final part [33, 36–38].

To account for the intricate dynamics associated with AM, AM process controls rely heavily on high-fidelity models. Predictive models can be derived from many sources [39] and are used to anticipate errors and initiate feedback. These models can be derived based on physics or using empirical data and statistical derivations; however, they come with significant developmental challenges. As noted on many occasions in this chapter, the intricate relationships inherent to AM builds create many measurement, representation, and interoperability challenges. The more complex the physical phenomena, the greater the modeling challenge. For these reasons, closed-loop process control has been a difficult to attain goal.

Researchers have had some recent successes in closed-loop control of DMD (Direct Metal Deposition) processes [40–42] where it is possible to monitor and control geometry, temperature, composition and microstructure with the collected light data at the laser material interaction zone. Models of these systems, however, do not have to account for heat dispersion throughout a powder bed, therefore simplifying the models.

The management of key parameters, and related process-structure-material relationships, is the primary driver of process control. The difficulties in transitioning from open-loop to closed-loop controls embody the systems challenges in AM. While monitoring processes have been implemented, and feedback is available, the systems characterizations and interactions are not yet complete enough to automate the process. Interdisciplinary advancements in the areas of modeling and process control are essential to enabling trans-disciplinary opportunities.

3.5 AM Qualification

The immediate goal of qualification is to certify parts built using an AM technology against design specifications. The longer-term goal is to improve repeatability, reliability, and reproducibility in material, machine/process, and part development. Regardless of short or long-term goals, the concept of qualification builds on validating, to some extent, the fundamental relationships and characterizations discussed in the previous sections. Material qualification addresses the characterization of raw materials (e.g., powder shape, size, distribution, etc.). Material/process qualification establishes relationships between the material, process parameters, and their corresponding built part properties.

The qualification of materials and processes are helpful in establishing repeatability in production; but, ultimately, it is the part that must be qualified. Part qualification occurs in two steps (1) In-situ process qualification through monitoring and control and (2) post process part inspection. In-situ process qualification looks to monitor a build and validate collected data against expected results, similar to process control. However, this approach is limited by the available measurement techniques. For off-line part inspection, there can be many inspection tests on characteristics such as geometric tolerances, surface roughness, porosity, crack, residual stress analysis, and microstructure analysis. To gain insight into all of these, measurement methods such as mechanical testing, round-robin testing [43], and non-destructive evaluation (NDE) techniques are used.

A new direction of high significance in AM qualification is the shift from physical and NDE testing to alternative qualification methods such as modeling and simulation. As with process control, modeling for qualification is inherently an interdisciplinary problem, incorporating aspects from material, process, and geometry, and often requiring correlations with statistical or empirical data. However, while modeling for control requires high-fidelity models to compensate for build discrepancies, the same is not necessarily true for models used in qualification. Modeling for process control was subjected to design, material, and process constraints. While modeling for qualification must still incorporate constraints from each of these areas, the focus is directed at the functional requirements established during design. Therefore, the effectiveness of models for qualification can be directly correlated with the uncertainty of the models and the room for error established by initial design requirements. This observation can create unique systems opportunities for part qualification, including the development of design rules, a key enabler of any transdisciplinary opportunities.

Section 3 reviewed the AM design-to-product transformation process and highlighted numerous systems issues that arise in the execution of that process. It also related AM process characteristics through systems engineering principles on multiple occasions. It is through these principles that AM advancements can continue to be made, especially from a transdisciplinary viewpoint.

4 Beyond the Plant Floor: The Tran-disciplinary Nature of AM

Trans-disciplinary systems cross traditional disciplinary boundaries, extending into the many different aspects of human life. The continued emergence of AM applications has embodied the notion of trans-disciplinary systems. Scientists, doctors, designers, artists, teachers, and entrepreneurs are all embracing the new opportunities AM technology has presented. Additive Manufacturing has enabled the "thinker" to become the "doer" when it comes to making an idea into a reality.

At its core, AM is a manufacturing process that can be represented as the interdisciplinary system described in Sect. 3 (Fig. 7). It allows a design idea to be transformed into a physical entity through controlled layering processes. What makes AM different from other manufacturing processes is not only its ability to create customized, one of a kind parts, but also its ability to be contained within a single machine, making the technology easily accessible. The range of capabilities offered by AM makes the technology suited for many different needs, from industrial applications, to medical applications, to tools for the soldiers on the front line, to the hobbyist that needs to make a repair. This is the trans-disciplinary essence of AM. This section will discuss how various communities have been successful in harnessing the already complex, multidisciplinary nature of AM and the challenges that remain for full integration.

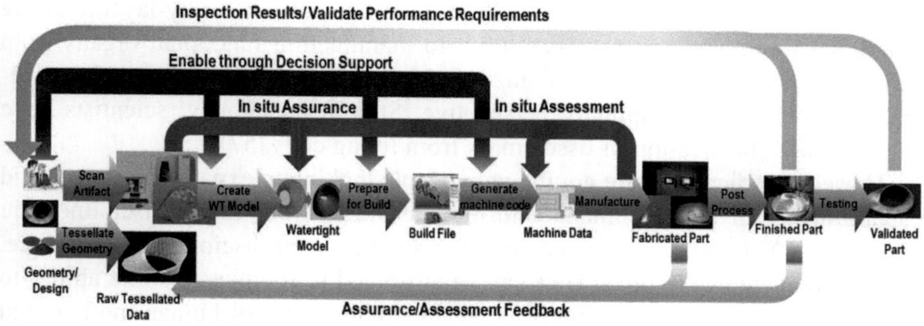

Fig. 7 The AM Process as an interdisciplinary system. Derived from [25]

4.1 In the Lab

One of the early adopters of AM technology has been the medical community. The customization capabilities of AM are well suited for individualized needs. AM has allowed doctors to create customized bone structures to replace damaged or deteriorating bones [44]. While this can be achieved to some extent with more traditional processes such as machining, AM allows for the creation of truly individualized parts. AM also provides a means to more closely simulate the porosity of the bone structure [45], and promote more consistent tissue and cellular ingrowth within the body [46]. While doctors are not yet able to create Adamantium skeletal structures (Sorry X-Men fans [47]), AM has enabled creativity in skeletal system repair more than ever before. AM has given the doctors the ability to use data collected from MRIs, and CT scans [48, 49] to build replacement parts for the human body, maybe someday even improving upon what biology is able to do.

Beyond skeletal repairs, doctors have begun to look at AM as a means for replicating soft-tissue body parts. As prosthetics require precise balancing to simulate the mass distribution of body parts, the desired traits of a prosthetic are weighted far more on functionality than aesthetics. Current prosthetic designs that are manufactured using composites or machining are unable to visually simulate the body parts they are meant to replace. AM allows doctors to not only come closer to simulating the physical appearance of the body part, but also customize to the shape and build of the person it was meant for. As prosthetics become more advanced and dynamic in nature, AM will play a key role in allowing doctors to mimic the biological parts in look, function, and feel as close as possible [50, 51].

Dentists and oral surgeons have also been able to exploit AM capabilities. Tooth replacement has long been an area where customization has been applied, where dentists have made molds [52] to reconstruct and repair jawlines. AM has provided a means to streamline and simplify dental processes, significantly cutting turn-around times [53]. Given the materials available to AM, AM has provided dentists with in-house capabilities for practicing their profession [54].

Though not yet realized in a practical manner, the AM processing of biomaterials [55, 56] is pushing the bounds of what we understand as manufacturing.

Scientists are investigating ways to build cellular structures layer-by-layer to create functional tissue and organs. The ability to manufacture functional organs from cells would be the pinnacle breakthrough for medical applications in AM, especially from a trans-disciplinary perspective. Still in its infancy, scientists have demonstrated the creation of tissue mass from living cells [57].

Those in medical labs are not the only people looking to expand their reach and capabilities with additive manufacturing. Scientists from many disciplines are exploiting AM to progress the state-of-the-art within their discipline. For instance, many advancements in prosthetics can be attributed to robotics, and the ability to replicate motion. Feedback control is an important aspect of human motion, and AM has been used to support the sensory aspects of the control loop [58]. Soft exoskeletons have been developed using AM, bringing new perspectives on how machines can advance human capabilities [59]. AM has been used to replicate the biological features of animals used by forensics experts, such as canines, to simulate their evolved sensing capabilities [60].

Though many scientists and researchers outside of manufacturing communities have embraced AM, their capabilities are still limited by their understanding of the technology and tools they are provided. The expertise of doctors is needed to understand how to duplicate optimal bone porosity when repairing skeletal structures. Biologists must provide insight when attempting to duplicate the functionality of an animal organ, such as a snout. Input from kinesiologists and physical therapists must be solicited when creating exoskeletons for mobility support. New approaches are needed to better integrate the manufacturing capabilities of AM with the unique requirements of scientific communities. While scientists have indeed begun to take advantage of AM capabilities, many others have had similar success outside the lab.

4.2 In the Field

Many other professions have also embraced AM layering concepts as a means for expanding their horizons. Fashion designers have found AM as a way to create never-before seen clothing items, perhaps as a taste of what the future may hold [61, 62]. While some have used AM as a means to create truly unique designs, others have seen AM as a way to incorporate functionality into what we wear [63]. Jewelers are able to customize jewelry like never before, in terms of materials, size, and design [64]. One of the more creative applications of AM has been in the food industry [65]. Bakers have applied AM concepts to create customized pastries [66]. Though not yet at "star trek" levels [67], others in the food industry have also experimented with the "printing" of food, including pizza, burgers, and chocolate [68].

Perhaps one of the most important drivers of AM technology is improving the capabilities of the Warfighter. Militaries are exploring ways of deploying AM to the front lines to support the Warfighter, bringing millions of dollars of research into

AM as means to support defense efforts [69]. When in the field, it is often difficult, if not impossible, to replace parts that have failed on equipment. Unless the part is readily available, part failure can lead to equipment inoperability, and wasted assets [70, 71]. For specialists with unique missions, AM offers the ability to create unique tools to accomplish these missions. With the ability to customize, and manufacture on demand, AM has the potential to enable the Warfighter to new levels [72].

As similarly stated in Sect. 4.1, further integration is needed between the technology-provider (AM) and the technology-user (professionals) to fully realize AM's potential. As a manufacturing alternative, AM technologies are more accessible than any other production options. AM technologies can be made available in the office and in the field, providing on-demand manufacturing capabilities. These unique traits offered by AM make the technology extremely attractive to those that need practical solutions without the having the luxury of a "go-to" manufacturer.

When discussing trans-disciplinary challenges with the scientists, the focus was how to incorporate disciplinary knowledge into the design. For the general user, a significant the trans-disciplinary challenge is how to communicate manufacturing limitations when producing a functional part. With users expecting to build parts that meet specified functional requirements, scientific-type constraints must be communicated to those with non-scientific backgrounds.

4.3 *The Maker Movement*

One of the most unique aspects of AM is its appeal to all, independent of professional applications. Hobbyists, for one, have seen AM as a way to bring new life to their hobbies, using AM to do things such as repair their RC cars [73], build pieces for their replica railroad towns [74], or bring their photography to new heights [75]. People are using AM as a means to further their passions in ways that have not been possible before.

The Maker Movement [76] has stirred the interests of groups around the globe. AM machines have been placed in schools around the world, from elementary schools to universities. The White House has moved to take advantage of the momentum and promote the interests of the next generation of builders and inventors, hosting maker faires[7] to show their support for the maker movement. Across disciplines, professions, and generations, AM has reignited the imaginations of many, providing new outlets for "making things." As people gain experience with AM, and capabilities are added to the processes and the machines, AM has the potential to become a part of everyday lives. With it, many skills and tools often associated with manufacturers and engineers will also become a part of everyday lives. To realize this potential, various interdisciplinary and trans-disciplinary challenges must be overcome.

---

[7] http://www.whitehouse.gov/nation-of-makers

4.4 At the Core

True to the trans-disciplinary principles, this chapter presents the use of AM outside of the traditional engineering boundaries observed by AM practices. Doctors, scientists, teachers and hobbyists were mentioned as beneficiaries of AM technology, where manufacturing has been made available at levels previously unrealized. Though each of the different disciplines has applied AM in different ways, there are similarities that are shared by all. At the core, these applications all shared a similar characteristic, their use as a problem-solving tool.

As a developer of critical thinking skills and a problem solving tool, AM has vast potential. Problem solving skills can be "brought to life" through AM technologies, where the defining characteristic is the ability to transform a mass into infinite design possibilities. With that understanding of AM, it is easy to see how AM can translate so seamlessly to different disciplines. In practice, however, the "critical thinking skills" of AM requires some familiarity to be put to use. While the concept is generally well understood, transforming the shape of raw material, the implementation is dependent on equipment capabilities and user know-how.

Though the data and information pertaining to AM may be standardized with neutral formats and data transformations, cross-discipline perspectives (Fig. 8) create their own systems challenges in how knowledge is captured and communicated. The definition of a "digital spectrum," as defined in Sect. 3, is no longer adequate. Geometry-process-material relationships must be expanded to geometry-process-material-function relationships. In a trans-disciplinary setting, design rules are no longer a process planning technique but a guideline for problem-solving in a

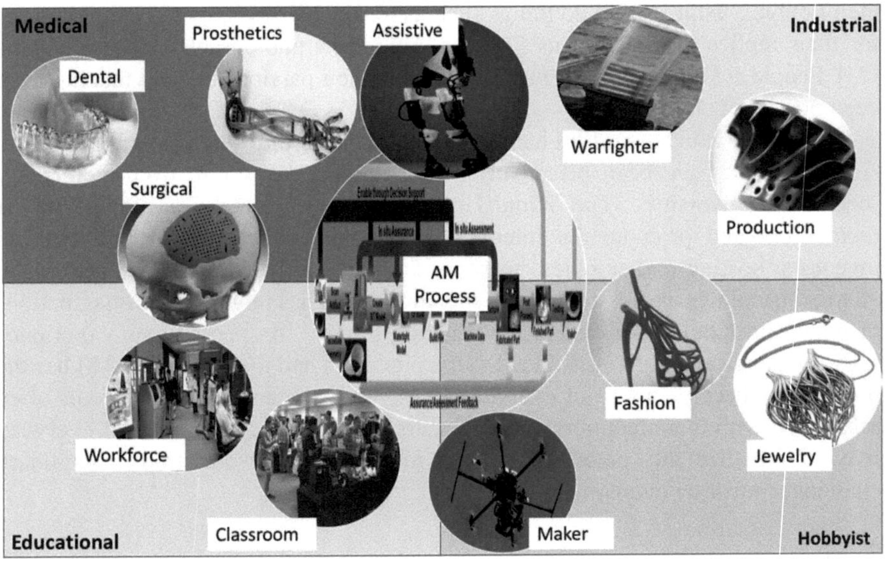

Fig. 8 Trans-disciplinary perspectives on AM processes

particular field of application. Support for these needs have not been adequately addressed with current approaches.

5 Systems for Trans-disciplinary Perspectives in AM: Moving Forward

For the average "maker," one of the main attractions of AM as a creative pathway and a problem-solving tool is the simplicity of its applications. Software such as Minecraft and Mineways [77] are available to create simple geometries. Various repositories are available for collecting and sharing existing designs. The low-cost 3D printer selection is continuously expanding, and most work on very similar technologies. For a maker, these traits are each appealing in their own right. However, as these users learn to push boundaries beyond toys and "trinkets," towards more functional pieces, available tools are no longer sufficient. As such, the same qualities that make AM so intuitive and simplistic to many users also begin to inhibit its applications.

The complexities and controls discussed in the previous sections are left out of the sightlines of most adopters, who have been able to get by with a more basic understanding of the process. Without understanding the many of intricacies of AM processes, user capabilities become limited, as do user group applications. Users are very much subject to constraints imposed by AM design tools and equipment capabilities. There are limited design tools available to non-expert users of AM, and the best alternatives can be difficult to learn as they are directed towards skilled engineers. While the impact may be lessened for those who are familiar with design tools, (jeweler, fashion designer), it limits the possibilities, and therefore the creativity, of those with lesser design skills. Systems principles provide a foundation for synthesizing design challenges with process capabilities, allowing users to more precisely understand and take advantage of AM capabilities.

In Sect. 3, Fig. 4 provided key data-elements of the multidisciplinary systems problem presented by AM. Recognizing each of the phases as independent, we must create and leverage systems interactions between them to open doors for feedforward and feedback control, as shown in Fig. 6. By addressing interdisciplinary challenges, we can facilitate the development of rules and models to improve decision-making support in each of the domains described in Sect. 3. Resolving interdisciplinary challenges, however, does not directly translate to meeting the needs of the specialized and general user bases. To further integrate AM capabilities with the general user bases, multi- and trans-systems principles must be adopted.

In AM, rules and models have been used to support the simplification of systems interactions. These rules and models have demonstrated success as a means for crossing disciplines in interdisciplinary settings. We propose that a similar approach can be used to bridge disciplines in trans-disciplinary settings, enhancing

AM applications and making the technology more attractive and accessible to general user bases. We believe such an approach can be specialized to satisfy specific trans-disciplinary requirements as well.

Section 4 discussed challenges faced at various degrees of trans-disciplinary integration, from the scientist to the hobbyist to the student. Though the applications differed, the challenges faced by the users, and need for further integration, were consistent. As we discuss trans-disciplinary principles in the context of design intent and design constraints, here we associate improved integration with better communication across disciplines. Communication can happen at different levels of detail, and enhanced communication is an essential component in addressing any trans-disciplinary challenge. The communication of data and information may be sufficient for interdisciplinary systems challenges, but additional perspective is needed to address trans-disciplinary challenges. This perspective can be achieved through the communication of knowledge [78–80].

Knowledge management (KM) is an area of research that has been addressed with varying degrees of intensity over recent years. Knowledge can be captured, stored and retrieved through various methods, including through annotations or formal relationships. Formal languages can provide the relationships necessary to exchange knowledge, and have been explored as a means for systems development [81–83]. These languages may include ontologies [84–86] and other logics [87]. Formal mathematical structures, such as category theory [83], are emerging as alternatives as well [88, 89]. Regardless of the technique used, we propose that further integration of AM systems must be based on more expressive formalisms. To advance state-of-the-art, trans-disciplinary approaches must communicate knowledge to enable the AM user base.

6 Conclusion

While the technology has been around since the early 1980s, AM's most recent emergence has captivated the masses at a previously unmatched level. The publicity has been widespread, and the placement of 3D printers in the workplace, in schools, and in homes has driven interest in the technology. However, despite AM's current success, AM is still seen as a novelty by many [90–92]. To sustain interest, AM, and especially the casual 3D printer, must prove to be more than a novelty.

The multidisciplinary nature of AM requires greater systems emphasis for sustained growth. There are many systems opportunities at the interdisciplinary level, including:

1. innovative design and processing planning paradigms to exploit the advantages offered by AM
2. developing relationships between microstructure, processing, and properties in support of the development of an AM material-process database,

3. developing models to support closed-loop feedback control and process monitoring and the sensing systems that supply the data to build those models,
4. developing models to support qualification of parts with higher confidence and reduced sensitivity

These interdisciplinary challenges are a first step in addressing the greater challenge of satisfying trans-disciplinary perspectives in AM.

To take advantage of transdisciplinary opportunities in AM, we must expand our definition of the AM digital spectrum. While interdisciplinary systems are able to capture essence of AM, transdisciplinary approaches are necessary to fully realize the potential of AM processes. Knowledge of the process and from the specialized user must be integrated into our definition of the AM digital spectrum. Expanding our use of design rules could lead to new sets of rules that would incorporate basic user needs with geometry design, material selection, and process parameters.[8] Though various technological and standards challenges must be overcome before this and other similar opportunities can be realized, it is a position that needs to be furthered.

This chapter reviewed the knowledge required from different disciplines to successfully manufacture a product using AM technologies. We discussed emerging opportunities, from the manufacture of assemblies to the printing of electronics. We explored the trans-disciplinary nature of additive manufacturing and discussed how additive technologies have transcended the reach of traditional manufacturing and brought design and manufacture directly to the consumer. Finally, we explored information barriers in additive manufacturing, and discussed how systems applications can help open new doors.

References

1. Bourell, D. L., Leu, M. C., & Rosen, D. W. (2009). *Roadmap for additive manufacturing: Identifying the future of freeform processing*. Austin: The University of Texas.
2. Lipson, H., & Kurman, M. (2013). *Fabricated: The new world of 3D printing*. Indianapolis, IN: Wiley.
3. International Organization for Standardization, ASTM International. (2015). ISO/ASTM 52900:2015 Additive manufacturing—General principles—Terminology.
4. Lackner, K. S., & Wendt, C. H. (1995). Exponential growth of large self-reproducing machine systems. *Mathematical and Computer Modelling, 21*(10), 55–81.
5. Bass, T. (1995). Robot, build thyself. *Discover*.
6. Sells, E., Smith, Z., Bailard, S., et al. (2009). RepRap: The replicating rapid prototyper: Maximizing customizability by breeding the means of production. In F. T. Piller & M. M. Tseng (Eds.), *Handbook of research in mass customization and personalization*. New Jersey: World Scientific.
7. Economist. (2012). The new maker rules. *Economist Newspaper*.

[8] Similar to Design for Manufacturing (DFM), Design for Additive Manufacturing (DFAM) can provide a more complete design solution.

8. Horn, T. J., & Harrysson, O. L. (2012). Overview of current additive manufacturing technologies and selected applications. *Science Progress, 95*(3), 255–282.
9. Ford, S. L. (2014). Additive manufacturing technology: Potential implications for US manufacturing competitiveness. *Journal of International Commerce and Economics*.
10. Markillie, P. (2012). A third industrial revolution: Special report manufacturing and innovation. *Economist Newspaper*.
11. Friedman, T. L. (2013). When complexity is free. *The New York Times*.
12. Gibson, I., Rosen, D. W., & Stucker, B. (2010). *Additive manufacturing technologies*. New York: Springer.
13. NIST. (2013). *Measurement science roadmap for metal-based additive manufacturing*. National Institute of Standards and Technology. http://www.nist.gov/el/isd/upload/NISTAdd_Mfg_Report_FINAL-2.pdf
14. Frazier, W. (2010, August). Digital manufacturing of metallic components: Vision and roadmap. *Solid Free Form Fabrication Proceedings, Austin, TX, pp. 9–11*.
15. Hadorn, G. H., Hoffmann-Riem, H., Biber-Klemm, S., Grossenbacher-Mansuy, W., et al. (2008). *Handbook of transdisciplinary research*. New York: Springer.
16. Frazier, W. (2014). Metal additive manufacturing: A review. *Journal of Materials Engineering and Performance, 23*(6), 1917–1928.
17. Raghavan, A., et al. (2013). Heat transfer and fluid flow in additive manufacturing. *Journal of Laser Applications, 25*(5), 052006.
18. Brice, C. A. (2011). Unintended consequences: How qualification constrains innovation. In *Proceedings of the 1st World Congress on Integrated Computational Materials Engineering (ICME)*. Hoboken, NJ: Wiley
19. Cooke, A., & Slotwinski, J. (2012). *Properties of metal powders for additive manufacturing: A review of the state of the art of metal powder property testing*. US Department of Commerce, National Institute of Standards and Technology.
20. Slotwinski, J., & Moylan, S. (2014). *Applicability of existing materials testing standards for additive manufacturing materials NISTIR 8005*. Gaithersburg, MD: NIST Technical Publications.
21. Slotwinski, J. A., et al. (2014). Application of physical and chemical characterization techniques to metallic powders. *AIP Conference Proceedings, 1581*(1), 1184–1190.
22. Slotwinski, J., & Moylan, S. (2014). Metals-based additive manufacturing: Metrology needs and standardization efforts. In *Proceedings of the 2014 ASPE Spring Topical Meeting—Dimensional Accuracy and Surface Finish in Additive Manufacturing, Berkeley, CA, pp. 11–12*.
23. Slotwinski, J. (2014). *Additive manufacturing: Overview and NDE challenges*. In 40th Annual Review of Progress in Quantitative Nondestructive Evaluation: Incorporating the 10th International Conference on Barkhausen Noise and Micromagnetic Testing, AIP Publishing.
24. Herderick, E. (2011). Additive manufacturing of metals: A review. *Materials Science and Technology*, 1413–1425.
25. Kim, D. B., Witherell, P., Lipman, R., et al. (2015). Streamlining the additive manufacturing digital spectrum: A systems approach. *Additive Manufacturing, 5*, 20–30.
26. Moylan, S., & Slotwinski, J. (2014). Assessment of guidelines for conducting round robin studies in additive manufacturing. In *Proceedings of the 2014 ASPE Spring Topical Meeting—Dimensional Accuracy and Surface Finish in Additive Manufacturing, Berkeley, CA, pp. 82–85*.
27. Calvert, P. (2001). Inkjet printing for materials and devices. *Chemistry of Materials, 13*(10), 3299–3305.
28. Sun, K., et al. (2013). 3D printing of interdigitated Li-Ion microbattery architectures. *Advanced Materials, 25*(33), 4539–4543.
29. Gross, B. C., et al. (2014). Evaluation of 3D printing and its potential impact on biotechnology and the chemical sciences. *Analytical Chemistry, 86*(7), 3240–3253.

30. Vayre, B., Vignat, F., & Villeneuve, F. (2012). Metallic additive manufacturing: State-of-the-art review and prospects. *Mechanics and Industry, 13*(2), 89–96.
31. ISO 10303-238. (2007). *Industrial automation systems and integration—Product data representation and exchange—Part 238: Application protocol: Application interpreted model for computerized numerical controllers*. Geneva: International Organization for Standardization (ISO).
32. Beuth, J., et al. (2013). *Process mapping for qualification across multiple direct metal additive manufacturing processes*. In SFF Symposium.
33. Gockel, J., & Beuth, J. (2013). Understanding Ti-6Al-4V microstructure control in additive manufacturing via process maps. In *Solid Freeform Fabrication Proceedings*.
34. Soylemez, E., Beuth, J. L., & Taminger, K. (2013). Controlling melt pool dimensions over a wide range of material deposition rates in electron beam additive manufacturing. In *Solid Freeform Fabrication Symposium. Solid Freeform Fabrication Proceedings, Proc., Austin, TX, pp. 571–581*.
35. Fox, J., & Beuth, J. (2013). *Process mapping of transient melt pool response in wire feed e-beam additive manufacturing of Ti-6Al-4V*. Solid Freeform Fabrication Symposium, Austin, TX.
36. Yadoitsev, I. (2009). *Selective laser melting—Direct manufacturing of 3D-objects by selective laser melting of metal powders*. Germany: Lambert Academic Publishing.
37. Strohm, P., et al. (2011). 2.5-controlling laser material processing with real-time algorithms on cellular neural networks. *Proceedings OPTO, 2011*, 60–65.
38. Craeghs, T., Bechmann, F., Berumen, S., et al. (2010). Feedback control of layerwise laser melting using optical sensors. *Laser Assisted Net Shape Engineering 6, Proceedings of the Lane 2010, Part 2, 5*, 505–514.
39. Witherell, P., et al. (2014). Toward metamodels for composable and reusable additive manufacturing process models. *Journal of Manufacturing Science and Engineering, 136*(6), 061025.
40. Mazumder, J., & Song, L. (2010). Advances in direct metal deposition. In S. Hinduja & L. Li (Eds.), *Proceedings of the 36th international MATADOR conference* (pp. 447–450). London: Springer.
41. Song, L., & Mazumder, J. (2011). Feedback control of melt pool temperature during laser cladding process. *Control Systems Technology, IEEE Transactions on, 19*(6), 1349–1356.
42. Song, L., et al. (2012). Control of melt pool temperature and deposition height during direct metal deposition process. *The International Journal of Advanced Manufacturing Technology, 58*(1–4), 247–256.
43. Moylan, S., Slotwinski, J., Cooke, A., et al. (2012). Proposal for a standardized test artifact for additive manufacturing machines and processes. In *Proceedings of the 2012 Annual International Solid Freeform Fabrication Symposium*.
44. Murr, L. E., et al. (2012). Next generation orthopaedic implants by additive manufacturing using electron beam melting. *International Journal of Biomaterials, 2012*, 14.
45. Heinl, P., et al. (2008). Cellular Ti–6Al–4V structures with interconnected macro porosity for bone implants fabricated by selective electron beam melting. *Acta Biomaterialia, 4*(5), 1536–1544.
46. Ryan, G., Pandit, A., & Apatsidis, D. P. (2006). Fabrication methods of porous metals for use in orthopaedic applications. *Biomaterials, 27*(13), 2651–2670.
47. Universe, M. (2015). *Adamantium*. Retrieved from http://marvel.com/universe/Adamantium
48. Knight, M. (2014). *3-D printing is revolutionizing surgery*. Chicago: Crain's Chicago Business.
49. Rengier, F., et al. (2010). 3D printing based on imaging data: Review of medical applications. *International Journal of Computer Assisted Radiology and Surgery, 5*(4), 335–341.
50. Bibb, R., Eggbeer, D., & Evans, P. (2010). Rapid prototyping technologies in soft tissue facial prosthetics: Current state of the art. *Rapid Prototyping Journal, 16*(2), 130–137.

51. Terdiman, D. (2010, March 5). *3D printing changing prosthetics forever*. CNET. Retrieved from http://www.cnet.com/news/3d-printing-changing-prosthetics-forever/
52. Strub, J. R., Rekow, E. D., & Witkowski, S. (2006). Computer-aided design and fabrication of dental restorations: Current systems and future possibilities. *The Journal of the American Dental Association, 137*(9), 1289–1296.
53. van Noort, R. (2012). The future of dental devices is digital. *Dental Materials, 28*(1), 3–12.
54. King, L. (2013). *3-D printing—The death of the dental lab?* Retrieved from http://www.dentistryiq.com/articles/2013/08/3-d-printing-the-death-of-the-dental-lab.html
55. Melchels, F. P., et al. (2012). Additive manufacturing of tissues and organs. *Progress in Polymer Science, 37*(8), 1079–1104.
56. Mironov, V., et al. (2003). Organ printing: Computer-aided jet-based 3D tissue engineering. *TRENDS in Biotechnology, 21*(4), 157–161.
57. Murphy, S. V., & Atala, A. (2014). 3D bioprinting of tissues and organs. *Nature Biotechnology, 32*(8), 773–785.
58. Cranny, A., et al. (2005). Thick-film force and slip sensors for a prosthetic hand. *Sensors and Actuators A: Physical, 123*, 162–171.
59. Majidi, C. (2014). Soft robotics: A perspective—Current trends and prospects for the future. *Soft Robotics, 1*(1), 5–11.
60. NIST. (2015). *NIST additive manufacturing*. Retrieved from http://www.nist.gov/public_affairs/factsheet/additive-manufacturing.cfm
61. Taylor, A., & Unver, E. (2014). *3D printing-media hype or manufacturing reality: Textiles surface fashion product architecture*. Textiles Society Lecture, 17th February 2014. Huddersfield UK: Textile Centre of Excellence.
62. Flaherty, J. (2014). *This dress is made from 3-D printed plastic, but flows like fabric*. Retrieved from http://www.wired.com/2014/12/dress-made-3-d-printed-plastic-flows-like-fabric/
63. McCann, J., & Bryson, D. (2009). *Smart clothes and wearable technology*. Burlington: Elsevier.
64. Wannarumon, S., & Bohez, E. L. (2004). Rapid prototyping and tooling technology in jewelry CAD. *Computer-Aided Design and Applications, 1*(1–4), 569–575.
65. Periard, D., Schaal, N., Schaal, M., et al. (2007). Printing food. In *Proceedings of the 18th Solid Freeform Fabrication Symposium, Austin, TX*.
66. Willett, M. (2014). This is the best 3D food printer we've seen yet—And it makes stunning desserts. *Business Insider*.
67. Birtchnell, T., & Urry, J. (2013). 3D, SF and the future. *Futures, 50*, 25–34.
68. Prisco, J. (2014). *'Foodini' machine lets you print edible burgers, pizza, chocolate*. Retrieved from http://www.cnn.com/2014/11/06/tech/innovation/foodini-machine-print-food/
69. Wolhers, T. (2013). *Wohlers report 2013*.
70. Gaereminck, R. (2011). *Strategic alliances: Making a difference one warfighter at a time*. DTIC Document.
71. Zimmerman, B. A., & Allen, E. E., III. (2013). *Analysis of the potential impact of additive manufacturing on Army logistics*. Monterey, CA: Naval Postgraduate School.
72. Pettus, E. L. (2013). *Building a competitive edge with additive manufacturing* (Doctoral dissertation). Air War College, Air University.
73. (2014). *This 3D printable R/C Car Truggy is testing the limits of 3D printing*. Retrieved from http://www.3ders.org/articles/20140910-truggy-car-a-3d-printable-rc-car-that-is-testing-the-limits-of-3d-printing.html
74. Krassenstein, E. (2014). *The openrailway project launches: Bringing 3D printing & railroad modeling together for hobbyists*. Retrieved from http://3dprint.com/24721/open-railway-project-3d/
75. Golson, J. A. (2014). *Military-grade drone that can be printed anywhere*. Retrieved from http://www.wired.com/2014/09/military-grade-drone-can-printed-anywhere/
76. Anderson, C. (2010). The new industrial revolution. *Wired Magazine* 18, 2.

77. Humphries, M. (2012). *Mineways lets you export Minecraft models for 3D printing*. Retrieved from http://www.geek.com/games/mineways-lets-you-export-minecraft-models-for-3d-printing-1454333/
78. Mylopoulos, J., et al. (1990). Telos: Representing knowledge about information systems. *ACM Transactions on Information Systems (TOIS), 8*(4), 325–362.
79. Piateski, G., & Frawley, W. (1991). *Knowledge discovery in databases*. Menlo Park, CA: MIT Press.
80. Liu, A. (2004). *The laws of cool: Knowledge work and the culture of information* (552 pages). University of Chicago Press. ISBN-10: 0226486990 and ISBN-13: 978-0226486994.
81. *OMG Systems Modeling Language (OMG SysML™) Version 1.4* (OMG Document Number: formal/2015-06-03). http://www.omg.org/spec/SysML/1.4/
82. Johnson, T., et al. (2012). Integrating models and simulations of continuous dynamics into SysML. *Journal of Computing and Information Science in Engineering, 12*(1), 011002.
83. Spivak, D. I. (2014). *Category theory for the sciences*. Cambridge, MA: MIT Press.
84. Witherell, P., Krishnamurty, S., Grosse, I.R., et al. (2008). *FIDOE: A framework for intelligent distributed ontologies in Engineering*. In 2008 ASME IDETC/CIE Conferences, New York.
85. Witherell, P. (2009). *Semantic methods for intelligent distributed design environments*. Amherst, MA: Department of Mechanical and Industrial Engineering, University of Massachusetts Amherst.
86. D'Alessio, A., Witherell, P., & Rachuri, S. (2012). Modeling gaps and overlaps of sustainability standards. In D. A. Dornfeld & B. S. Linke (Eds.), *Leveraging technology for a sustainable world* (pp. 443–448). Berlin: Springer.
87. Witherell, P., et al. (2010). Improved knowledge management through first-order logic in engineering design ontologies. *Artificial Intelligence for Engineering Design, Analysis and Manufacturing, 24*(2), 245–257.
88. Williamson, K., Healy, M., & Barker, R. (2001). Industrial applications of software synthesis via category theory—Case studies using Specware. *Automated Software Engineering, 8*(1), 7–30.
89. Tse, T. H. (2009). *A unifying framework for structured analysis and design models: An approach using initial algebra semantics and category theory* (Vol. 11). Cambridge: Cambridge University Press.
90. Osborne, C. (2014). *3D printing: Don't believe all the hype*. Retrieved from http://www.zdnet.com/article/3d-printing-dont-believe-all-the-hype/
91. Mims, C. (2012). *Why 3-D printing will go the way of virtual reality*. Retrieved from http://www.technologyreview.com/view/426702/why-3-d-printing-will-go-the-way-of-virtual-reality/
92. Banker, S. (2014). *3D printing's ability to transform supply chains is years away*. Retrieved from http://www.forbes.com/sites/stevebanker/2014/02/26/3d-printings-ability-to-transform-supply-chains-is-years-away/

Expanding Sociotechnical Systems Theory Through the Trans-disciplinary Lens of Complexity Theory

Lisa Troyer

1 Introduction

The aim of this chapter is to trace the history of sociotechnical systems theory and the relevance of this research in contemporary organizations. I propose that the key to the ongoing utility of this influential theory lies in recognizing the complexity of both social and technical systems, individually and jointly. As I will describe, the term "sociotechnical systems" refers to the interface between social systems (including interpersonal relations, as well as how groups and group members solve problems, design work, allocate tasks, analyze and present information) and technical systems (i.e., the complex interdependencies among innovation, design, engineering, and the manufacturing and marketing of materials and processes). Sociotechnical systems theory was primarily developed by researchers at the Tavistock Institute beginning in the 1940s, integrating insights from a wide range of disciplines, including psychology, group dynamics, management science, sociology, and engineering. It began as an attempt to create more humane work environments and has evolved since its beginnings in the human relations movement to contribute to complexity theory through its recognition that human and social systems interact in unique ways with complex technical systems. The result is that, when combined, seemingly linear relationships between discrete elements of a system (e.g., linear relationships between parts, linear relationships between people and between processes) can engender non-linear complex dynamics. These dynamics, in turn, create complex interdependencies, which present unexpected challenges and potentials for catastrophic failures that are difficult to predict. To complicate matters, neither social systems nor technical systems are consistently

L. Troyer (✉)
Army Research Office/Army Research Laboratory, Research Triangle Park, NC, USA

Duke University, Durham, NC, USA
e-mail: lisa.l.troyer.civ@mail.mil

F.-J. Kahlen et al. (eds.), *Transdisciplinary Perspectives on Complex Systems*,
DOI 10.1007/978-3-319-38756-7_7

organized in linear relationships. These empirical realities make it difficult to design risk-free systems. Yet, all is not lost. Emerging advances in agent-based modeling, epidemiological methods of tracing complex causal interdependencies, and advances in complex systems analysis offer promise to identify risky points of interdependence within and across systems. Importantly, the trend is toward trans-disciplinary orientations to enable a holistic approach to understanding systems (For additional discussion of the importance of holistic approaches to systems analysis, see Hester et al. [1]). These advances may enable managers to identify potential points of failure and intervene to preclude such failures.

2 Background: Early Research from the Sociotechnical Systems Perspective

In the 1950s the engineering, physical, natural, and social sciences were transformed by the insights of Ludwig von Bertalanffy [2], who introduced the transcendent perspective of general systems theory. The Bertalanffy thesis is that foundational elements of the physical, natural, and social world could be conceptualized as interrelated systems. Key to this is the notion of “interdependence”—that these systems are not discrete and closed from one another, but rather that what occurs in one system affects others. Initially, general systems theorists proposed that technical systems were “tightly coupled,” meaning that relations between material parts are predictable, constrained, and well-understood (a premise, however, which is no longer considered valid as I will take up later). In contrast, early general systems theorists viewed social systems as more complex, variable, and less predictable. That is to say, how people behave and interact with one another and with materials was generally viewed as less constrained even when formal rules and codified processes exist. As recognition of the interdependence between social and technical systems and their unique features began to take hold among organizational scientists, it became apparent that (1) organizations are not necessarily rational systems; and perhaps more importantly, (2) efforts to impose rationality could undermine efficiencies and engender greater risks of loss of control (e.g., [3]). It was out of this intellectual tradition that sociotechnical systems theory emerged.

Shortly after World War II ended, researchers in the United States and England began focusing increasing attention on human relations research, which had begun to build a compelling body of research on how human and social processes affect and are affected by work environments. Much of the work grew out of a growing appreciation for the importance of the group as a fundamental unit of analysis (e.g., [4]). This research aimed at uncovering and addressing problematic aspects of the work environment that undermined employee efficiency and adversely affected morale. Perhaps most representative of this early human relations work was research by Elton Mayo and his colleagues at the Western Electric Company’s Hawthorne Plant outside of Chicago that began in the late 1920s. Mayo was a student of scientific management, which aimed to imbue rationality into work design by breaking down work into its smallest tasks and creating the most efficient protocols to complete the tasks (e.g., [5]). In a field application of scientific management, however, Mayo and his colleagues (e.g., [6, 7]) found that the changes they introduced created unexpected effects. Perhaps most famously, in the so-called "lighting experiment," they increased lighting in one room for workers, while holding it at a lower constant level in the other room. As expected, production increased in the well-lit room; but unexpectedly, production also increased in the less well-lit room. More surprisingly, when lighting was decreased in the well-lit room, production again increased. Through discussions with workers, the researchers found that their attention to the workers led the workers to be more conscientious about their efforts, regardless of which lighting condition they experienced. Their conscientiousness enhanced productivity. This has become famously known as the "Hawthorne Effect"—i.e., when people are observed, their work tends to improve. From this, the human relations school was born. In a string of research emanating from this foundation, researchers found, for instance, that individual worker productivity wasn't so much motivated by material rewards. In fact, some workers would slow their production to match their peers' expectations. This occurred even if it meant forgoing bonuses so as not to disrupt the average productivity of the work group and risk social ostracism from co-workers if they out-paced them (a phenomenon known as "rate-busting"). It became clear that the design of work required a shift away from thinking of workers as solely and rationally motivated by external material gains and instead considering them as members of social groups whose work depended on the responses of not only superiors, but also (and perhaps more so) of their colleagues.

This shift in thinking took hold at a psychotherapy clinic in England, the Tavistock Clinic, and led the Clinic to re-organize around human relations research and rename itself the Tavistock Institute for Human Relations. As an action-oriented clinical institute, the Tavistock Institute generated a number of influential studies demonstrating the importance of employee discretion in the execution of work and the superior efficiencies that could emerge by designing technical systems in which work groups were cultivated and encouraged to act autonomously. Importantly, Tavistock researchers shifted focus in operations research from a sole emphasis on designing superior technical systems to jointly optimizing both social and technical systems, treating them as a single interdependent *sociotechnical* system.

As the research progressed, however, scientists found that attempts to allow greater discretion at the work-group level were not successful unless changes were made with respect to the structures in which those groups were embedded, including the organizational structure and the broader culture in which the organization is situated. In other words, the organization is embedded in a political, economic, and social environment itself, introducing additional complexities in building intersystemic compatibilities (for a history of the Tavistock Institute's role in the development of sociotechnical systems theory, see Trist [8]). Consequently, researchers increasingly turned to a view of organizations as complex systems comprising sociotechnical systems and embedded in a web of environmental sociotechnical interdependencies. This open systems view of organizations [9] sets the stage for recognizing the true complexity of organizational processes. As the open systems perspective has taken hold, researchers have increasingly adopted methods and ideas from complexity theory to attempt to understand and model sociotechnical systems.

Early studies emerging from the Tavistock Institute examined sectors like mining and found that workers were resisting attempts to create mechanistic, repeated roles for miners based on a division of labor, which was thought to create superior efficiencies (i.e., having the same person do the same task repeatedly). Instead, the workers created autonomous groups that rotated tasks and shifts. The Tavistock researchers found that the autonomous work group design increased productivity *and* worker satisfaction (e.g., [10, 11]). Subsequently, their insights on participatory work re-design were adopted by other organizations in different industrial sectors (mostly in manufacturing) and rapidly began to influence management science. One key observation that emerged from sociotechnical systems theory is that human groups are the most adaptive and agile unit of the organization.

The early work in sociotechnical systems theory emphasized the looser coupling of social systems vis-à-vis a tighter coupling of technical systems in which the technical requirements of production and manufacturing could be represented in more linear terms. Today, however, it's widely recognized that technical systems are often not so tightly coupled. For example, with the advent of nanoscience (and the exponential rate of discovery in the engineering and physical sciences, in general), new materials and processes are being developed and introduced into manufacturing, which sometimes exhibit unexpected behaviors and interactions with other materials. Moreover, the effects of these interactions may be initially subtle, but those subtleties can become magnified across the parts in unexpected ways, leading to unanticipated behaviors of the overall technical system.

Additionally, contemporary organizations are not always independent. They often rely on a dynamic global network of suppliers and customers that contribute to the development and deployment of the products they generate. The challenges from an engineering perspective are further exacerbated because throughout the life cycle of the development and deployment of complex technologies, different subsystems may be constantly adjusted to improve their effectiveness, draw on new innovations, or conform to new (and sometimes competing or contradictory) customer requirements and supplier capabilities. These subsystems, are of course, interrelated (often in recursive and non-linear ways) creating an ongoing spiral of adjustment and changes in the ways the technical subsystems co-operate.

Consequently, the challenges of conceptualizing technical systems alone are enormous. Add to this, the difficulty of analyzing and understanding social systems (the focus of this chapter), and our view of the terrain of contemporary engineering becomes highly complex and unpredictable. Complexity theory emphasizes how subtle (or not-so-subtle) effects in one component of an interrelated system can flow across other components to generate unpredictable and unexpected effects in other parts of the system. This is captured by the oft-cited example of how the flutter of a butterfly's wings in one region of the world could theoretically change global meteorological patterns in ways that may create conditions for a hurricane in another region. While critics contest this kind of dramatic effect, arguing that there is an important distinction between random influences and "predictable chaos" (e.g., [12]), complexity theory sensitizes us to the unexpected ways that social and technical systems may interact, as well as the challenges to both in predicting and managing these complex interactions. In the next section, we will consider some of the human and social processes that undermine the effectiveness of individuals, groups, and organizations and the difficulty of intervening in those processes to create more efficient social systems.

3 Human and Social Behavioral Challenges to Organizational Design and Performance

As we've seen, sociotechnical systems theory places particular emphasis on the importance of work groups and on enabling their ability to adjust to changes in the work environment, particularly changing demands. Yet it's also important to note the biases that may adversely affect the performance of groups and especially their adaptive capacity, as well as the prospects for intervening in these processes.[1] As already noted, humans are subject to an array of social pressures, including peer pressure to conform to the norms and expectations of group members. This creates substantial non-linearity and unpredictability in social interaction. In the context of work, such pressures can arise from either members of one's work group or the organization itself. The most influential pressures, however, are those that arise locally from the individuals with whom one shares a common identity, which emerges from shared values, goals, and interdependent roles among group members (e.g., [14]). Indeed, as discussed above and exemplified in the Hawthorne Studies, subordinate workers tend to forego the incentives (and punishments) imposed by superiors, if conforming to the directives of superiors requires action that will deviate from local work group expectations. This is because (1) the most immediate source of reward/punishment arises from one's in-group members; and (2) in most cases, social acceptance is a far more important and enduring motivator of human behavior than material rewards/punishments (e.g., [15]). Thus, groups have a profound effect on individuals and on the outcomes of group work. The pressures to achieve a common identity and conform to group norms will lead individual members to suppress their own opinions, evaluations, and even skills to conform to the social expectations of the group. In a striking study of this phenomenon known

[1] In a provocative alternative to the framework on human and social biases set forth in this section, Manuse and Sniezek [13] suggest that the very notion of "complexity" itself reflects a deep-rooted assumption, or mental construction, regarding work and engineering that generate biases in how system design is approached.

as "groupthink," Irving Janis [16] examined historical instances of poor decision-making, most notably the executive decision-making surrounding the failed Bay of Pigs Invasion in 1961. As is well-documented, the U.S.-backed invasion, aimed at overthrowing the Castro regime was defeated in 3 days. Janis undertook a comprehensive analysis of memos, reports, and insider journal entries documenting the days leading up to and following the decision to support a rebel invasion of Cuba. Much to his surprise and that of others, his analysis showed that several risks and shortcomings related to the planned invasion were widely recognized by the President's advisors. Yet, these concerns were either dismissed as inconsequential or never raised. Interviews with and journals of advisors revealed that members of the advisory group felt strong pressures to go along with the President's espoused opinion that the invasion should proceed. In addition, the advisory team was basking in the glow of an improbably successful presidential victory. This engendered a sense of invulnerability among the team—i.e., the sense that their actions could withstand even the most formidable of oppositions. Consequently, advisors downplayed or withheld pertinent information from the group, including doubts regarding the level of rebel support in Cuba, awareness that the Castro regime had been tipped-off to the invasion plans, and concerns that the effectiveness of Cuban military forces were being underestimated. The perception was that voicing such doubts and concerns would undermine the planning, adversely affect the team's cohesiveness, and be dismissed by the President and his other advisors. Instead, the documentation of the discussions suggests that the advisory group spent much of its time reinforcing the view that US aims and resources were superior to those of Castro (both operationally and morally) and quelling any emerging contradictory opinions.

Subsequent studies have been conducted by Janis and an array of other researchers on both high-level political decision-making and corporate declines. Among the cases exhibiting symptoms of groupthink are the case of the collapse of Swissair [17] and the near-demise of British Airways in the 1990s [18]. With respect to the former, Hermann and Rammal documented how the Swissair board downsized in the face of board member criticisms and questioning of some members about operations and financial health. Rather than confront the concerns of these board members, executives jettisoned them from the group by changing the board membership. In hindsight, this demonstrated the lack of tolerance for dissent—a clear symptom of groupthink. The result, however, left the board without much-needed expertise. In the case of British Airways, Eaton analyzed press releases to document the manifestation of groupthink. His research revealed that the organization repeatedly turned a blind-eye to serious indications of a financial decline and engaged in repeated attempts to demonize competitors, which ultimately led to severe legal and financial sanctions that nearly destroyed the company.

One important process identified by groupthink researchers is the role of social influence on group dynamics. On the one hand, we expect experts, superiors, and group members with specific skills relevant to a group's work to be more influential—that is, to exert a greater ability to guide the group's decisions and work strategies. And, in fact, groups are often designed to ensure that the appropriate range of knowledge, skills, and leadership abilities are embodied in the group. Yet, on the other hand as social psychologists have repeatedly demonstrated, influence also arises from members who possess traits that are valued in general ways throughout an organization and a society, even when those traits are not germane to the task at hand (e.g., [19]). That is, general characteristics, such as sex, race, and even physical attractiveness affect a person's ability to influence others, even when the persons possessing higher status states of such characteristics lack task-relevant knowledge and skills.

For instance, male members of groups tend to speak more, be listened to more, and gain more support for their positions compared to female group members, even if female group members possess superior knowledge and skill that is relevant to the group's tasks (e.g., [20]). This phenomenon isn't limited to gender, however. In a striking demonstration of the powerfulness of such status generalization processes, Cohen and Roper [21] showed that even when it is publicly known that racial minorities in a group possess superior task knowledge and expertise, they, at best, are as equally influential as racial majorities who are known to *lack* the same level of task knowledge and expertise. And Webster and Driskell [22] demonstrated this same outcome for individuals who are perceived as more physically attractive. In fact, even when groups have been sensitized to this phenomenon, the effects remain pervasive. Research by Silver et al. [23] examined decision-making in a heterogeneous organizational team that had undergone extensive team-building training in an effort to eliminate its susceptibility to processes like groupthink and influence that isn't grounded in member knowledge or expertise. Within a few meetings, however, symptoms of groupthink and influence arising from status characteristics irrelevant to the group task (such as sex, age, race) were clearly evident and grew in the strength of their effects over the course of the team's work across several months. This research demonstrated that it is not enough to understand these status generalization dynamics. They are deeply embedded in the repertoires of individual action, interact in complex ways with task demands, and are difficult to both detect and hinder.

Exacerbating these kinds of group dynamics are the cognitive limitations and biases that characterize individual decision-making and problem-solving. As Nobel Prize winner, Herbert Simon famously documented, individuals may strive to be rational in their approach to problem-solving, but fall woefully short and are at best "boundedly rational" [24, 25]. That is, individuals lack certain cognitive capabilities, including perception, memory, and the ability to rapidly process large amounts of information, necessary to make perfectly rational decisions. In addition, they are swayed in seemingly illogical ways by emotion and irrelevant personal preferences in making decisions. Consequently, rather than generating optimal solutions to problems, individuals tend to converge on satisfactory solutions that minimally meet requirements.

The realization that individuals are cognitively limited has led psychologists to try to uncover patterns in the biases. One particularly relevant one when it comes to group decision-making (which is also likely to contribute to groupthink) is the phenomenon of loss aversion [26]. In a series of provocative experiments, Tversky and Kahneman showed that individuals are highly resistant to giving up material goods already in their possession, even if by giving up those goods they would fall into possession of more valuable goods (as in the case of a beneficial trade). In fact, these researchers showed that psychological effects of anticipated losses are about twice as powerful as gains. Extending this work, researchers have found that it applies to attitudes and opinions, not just material goods (e.g., [27]). Individuals work much harder to justify current points of view than they do at trying to understand and accept alternatives. The result is that groups tend to converge upon and hold fast to a "status quo bias" [28]. Thus, the biases set into motion via group processes like groupthink and status generalization tend to become rapidly and deeply entrenched in both individual and group psyches.

The combined effects of biases arising through group processes and biases reflecting individual cognitive limitations suggest that the social dimensions of sociotechnical systems are themselves very complex, even at the small group level. Compounding this is the fact that organizations commonly comprise multiple groups, often with cross-cutting memberships. For example, an employee may simultaneously serve on multiple project teams and even answer to different supervisors, as is the case in matrix organizations. Matrix structures have many theoretical advantages, including the potential capacity to facilitate communication and coordination across different components of an organization or a particular set of interrelated projects. Yet, if we conceptualize the web of individuals' work relations with one another within and across work groups, it becomes clear that as these social networks also become channels through which the kinds of biases described above can flow. That is, the more interconnected an organization's members are, the greater the potential is that a bias at one point (e.g., within one group) may "infect" other individuals, groups, and thus, organizational outcomes. Moreover, a small (perhaps even innocuous) bias in one part of an organizational process can become magnified as it flows across the organizational network, leading to disastrous outcomes elsewhere, just as theorized by complexity theory's "butterfly effect." By way of example, a decision to continue to use particular materials for one component of a complex technology such as a car, airplane, ship (even when potentially superior new materials may be available), necessarily impacts the performance of materials in other components, particularly when the technical system is tightly coupled. When one part of technology directly depends on the performance of another technology, then the technical system is said to be "tightly coupled" (e.g., [29]). In such instances, the flow of effects across the system may be more easily detected. But as we will take up later, in many instances, complex technologies may be comprised of more loosely coupled technical components (e.g., [30]). The point is that human and social dynamics occurring within one area of a social system have effects on other areas, including other social areas and other technical areas. Again, it's important to recognize the profound effects that biases like loss aversion can play—it is much harder to convince ourselves to switch to new processes, materials, or technologies even in the face of rational information, because we work much harder cognitively to avoid losing that which we know and we're generally unaware of the operation of these kinds of biases.

4 The Organization as an Open System: Setting the Stage for Normal Accidents

As already noted, the general systems perspective that emerged in the 1950s with Bertalanffy's work represented an intellectual revolution of sorts, affecting biological, physical, and social sciences, including management and organizational sciences. Contemporary organizational scientists have long abandoned the conceptualization of organizations as closed systems, dependent only on the workers they hire or even customers they serve. Rather, organizations today are increasingly embedded in networks of relations with other organizations, including governmental organizations, citizen/community organizations, professional organizations (including unions, professional associations), not to mention the dynamic value chains that comprise suppliers and customers. Moreover, increasingly organizations are embedded in a global network of individual and corporate actors. Recognition of this complex social environment reflects the "open systems perspective" on organizations [9].

The open systems perspective treats the boundaries of an organization as fluid and changing according to demands placed on the organization by the environment, which of course, includes other organizations. The extent to which organizations are tightly coupled with one another can change over time and space. For instance, at one point in time an organization may be deeply engaged with governmental and citizen organizations as it confronts legislation that may impede or improve operations, and legislation may be undergoing potential change at local, national, or global levels. Once determinations are made regarding the legislation, the organization may become less engaged with those groups and re-focus its efforts and resources (including human capital) elsewhere.

An important insight that the open systems perspective reveals is that organizations are constantly adjusting to the demands of their environments and thus, their interrelations with other organizations are constantly changing. For instance, as product lines expand or change, new suppliers may enter the organization's network. In some cases, organizations may even become integrated with one another or subsume others (e.g., through mergers and acquisitions, respectively). In so doing, their networks are likely to further shift. Moreover, as Buckley [30] argued the more interrelations there are (i.e., the larger the organization's network), the more organizations come to depend on the transmission of information (not simply material resources) to successfully adapt to and attempt to coordinate new demands. And, as we've seen, the formulation, adoption, deployment, and transmission of information occurs as the result of human and group dynamics in organizations, which are subject to both loose coupling and profound biases.

Consequently, the open systems perspective on organizations sensitizes us to the fact that organizations are at once highly adaptive, but also and in part as a result of that adaptive capacity, highly fragile. That is, as networks within and between organizations change and grow increasingly complex, managing and coordinating flows of resources—particularly information—across the network becomes increasingly challenging. Furthermore, each node in a network through which resources travel corresponds to a point within the system where a failure may occur. Sometimes, those failures can cascade over the system (often undetected for a long period of time) until they result in a catastrophic failure. This is what Perrow referred to as a "Normal Accident" [31].

Adopting an open systems view of organizations, Charles Perrow launched a perspective that revealed the complexity of sociotechnical systems. Over several years, he conducted careful case studies of major catastrophes and also served as a consultant on investigative teams examining the causes and consequences of such major disasters as the Three Mile Island nuclear power plant failure, the space shuttle Challenger tragedy, the Exxon Valdez catastrophe, and disasters that later followed, including Chernobyl and the 2008 financial crisis [32]. Perrow's incisive analyses of these and other cases resulted in the conclusion that accidents (or more descriptively, catastrophic failures) are "normal" (i.e., to be expected) because of the intense complexity of both social systems (which embody human limitations) and technical systems (involving non-linear chains of spatial-temporal interdependencies between materials and process). The genius of Perrow's research is that it recognizes that *both* social systems and technical systems can be highly complex and interdependent. Furthermore, Perrow notes that the more complex that systems are, the greater the tendency to try to impose control systems. At the same time, however, formalization can set the stage for a "normal accident" (For a more thorough discussions of Perrow's framework, see Grieves and Vickers [33]).

In 2007, researchers at the Human Factors Defence Technology Centre examined how sociotechnical systems theory might inform and be informed by growing state and organizational capacity for network enabled capabilities (NECs) [34]. NEC involves attempts to better recognize, analyze, and deploy information systems to generate more flexible and responsive state and military systems. As these researchers recognized, simply deploying new information technologies within a hierarchical control system, without recognizing human and group advantages and limitations could be highly ineffective. Existing strategies attempt to use a more linear framework to represent sociotechnical systems (e.g., X $\rightarrow$ Y $\rightarrow$ Z) and manage them within a hierarchical model of command and control. In reality as this chapter has noted, the relationships between parts, processes, and people is more often non-linear, recursive, and difficult to predict. Small shifts in decisions regarding the use of materials, design of the process, timing of production, which often arise from non-rational individual and group dynamics, can engender major speed bumps and even failures in production and manufacturing well into the innovation and production life cycle. Walker et al. [34] advocate a return to sociotechnical systems theory, which would recognize the importance as well as limitations of group dynamics and the need to detect the complex social and technical interdependencies, which have been discussed in this chapter.

Challenging sociotechnical systems theory to incorporate a conceptualization of organizations as open systems, populated by individuals, groups, and rapidly changing technologies necessarily poses challenges to managerial and engineering science. Perhaps the biggest challenge is how to conceptualize, analyze, and model sociotechnical systems. Fortunately, recent advances in computational social science, network science, and physics provide a template for helping us move forward with new approaches and tools for the analysis of complex sociotechnical systems. In the next section, we will briefly outline some promising new strategies.

5 New Methodological Approaches to Address Sociotechnical Complexity

Substantial advances are being made in the computational social sciences to enable improved ways to represent complex sociotechnical systems. In particular, advances in network science and network analytic methods coupled with increasingly more sophisticated agent-based models can help engineers, organizational analysts, and social scientists better understand the complex interrelationships between people, groups, parts, processes, and outcomes. Additionally, emerging automated strategies for mining data from the communications of key organizational actors allow analysts and researchers to rapidly generate models that are amenable to statistical prediction and forecasting. Together, these advances offer promising ways to wrestle with the complexity of sociotechnical systems.

One of the biggest challenges for computational scientists who seek to represent sociotechnical complexity is the fact that the actors, materials, processes, and interdependencies operate on different spatial-temporal scales. This makes it difficult to connect and model how shifts in one network (say, a network of individual team members and their communications surrounding a project) might interconnect with another network (say, a network of different materials needed for the production of a new product). For instance, status structures and influence may crystallize quite rapidly in groups determining decision objectives for a production process. Suppliers, relying on one another for the materials to create a component, however, may operate at a much slower pace (e.g., slowed down by administrative requirements, contractual demands across their organizations, and externally imposed regulatory requirements). Furthermore, networks are dynamic. A snapshot of the network at one time may be very different than a snapshot at a different time as individuals, groups, and organizations come-and-go, relationships evolve, alongside changes in materials and requirements. Moreover, all may operate in different geographical spaces, complicated by transit issues (e.g., transportation requirements, including import/export relationships across nations and partners). And as Baldwin and Felder [35] demonstrate complex systems are perhaps better captured and modeled as "systems of systems," with each system in the set of systems characterized by a degree of autonomy, each with distinct functions, each with different degrees of connectivity. Developing network models that can accurately represent these complexities and their interdependencies has been computationally challenging, although Baldwin and Felder offer an elegant strategy. Additionally, recent advances drawing on models from mathematics and physics that accommodate different spatial-temporal behaviors of interrelated systems, and particularly non-linear dynamics within and between these systems, are proving promising for adaption to engineering/production analyses (e.g., [36, 37]).

In a different approach, West and colleagues [38] use mathematical models that transcend network content to demonstrate how inverse power laws can be used to represent a variety of non-intuitive, yet commonly observed social phenomena. These approaches are amenable to modeling the emergence of consensus on minority opinions and the failure of individuals to abandon self-interests and conform to behavior that advances a group outcome, even at great personal risk.

Another area gaining steam that can help in the development of models to detect complexities in sociotechnical systems arises from agent-based modeling to represent and analyze organizations confronting rapidly changing high-stress environments. Agent-based models (ABMs) integrate network analyses and conceptualizations of individuals, groups, and organizations within interdependent networks as decision-making actors who are subject to constraints, including limits on rationality. Examples include work by Helbing and Carley and colleagues (e.g., [39–41]). An important benefit of this line of research is its ability to use scenario-driven models to depict shifts in relations between individuals, groups, and organizations, driven from content-based data. ABM research suggests that "command and control systems" (a common approach for managing technical systems) need to account for the unique dynamics of social systems, which includes explicitly accounting for the kinds of individual and group biases discussed in this chapter.

The "Digital Twin" strategy that Grieves and Vickers [33] describe is suggestive of a variant of agent-based modeling to capture complexities and interdependencies within and across systems over a product lifecycle. This modeling approach portrays systems as interrelated information components and the model itself evolves as new pieces of information come to light. In this regard the Digital Twin approach enables an evolutionary strategy for capturing complexity and simulating various outcomes far beyond the capacity of human cognition and enabling the kind of structured approach to experimentation to cope with complexity that is suggested by Holt et al. [42]. Moreover, the Digital Twin is agnostic regarding the kinds of information that can be incorporated and so is amenable to incorporating complexity as it arises from both social and technical sources.

Complementing these advances are emerging methods of analyzing the transmission of information, including how meaning flows and morphs over social networks. For instance drawing on epidemiological models of disease, researchers have empirically demonstrated that whether and how the content that individuals contribute to a group survive to be "heard" and acted upon in a complex sociotechnical system depends on the group ecology (e.g., [43]). In this approach, researchers use event history models, drawn from epidemiological methods to study disease dynamics. The application of these models to social dynamics demonstrates that status processes and group identity dynamics determine whether an idea or the critique of an idea survives and becomes the basis for decisions, independent of the empirical evidence supporting or refuting the idea. This approach documents how models adapted from the life sciences can inform understandings of group dynamics within complex sociotechnical systems. Indeed, as demonstrated by Berglund [44], survival analyses promises an effective strategy for analyzing information transmission over multiple complex heterogeneous networks that correspond to contemporary organizations and the innovation and manufacturing life cycles that play out within and across organizations. This work provides provocative insights on how information "survives" in complex social networks. It reveals both empirical confirmation of insights discussed in this chapter regarding the suppression of key insights from individuals as a result of group dynamics (like groupthink, status processes, and loss aversion), as well as strategies for detecting the likelihood that a group is vulnerable to such risks.

Finally, an interesting recent development in the study of sociotechnical systems is the emergence of the field of social physics (e.g., [45]). The aim of social physics is to formally model the relationships between the flow of information (esp. ideas) and social behavior. In doing this, the social physics framework attempts to capture the complexity of social learning, productivity, and the evolution of social groups and organizations. Importantly, drawing on theory and models from physics, social physics recognizes that the relationships between people, information, and groups is often non-linear and recursive. One of the most interesting insights emerging from this nascent field is that optimizing innovation and minimizing risk is best accomplished not by changing material technologies, but rather by changing social networks—that is, re-designing the group—an inference that is consistent with the earliest tenets of sociotechnical systems theory. At the same time, however, the family of theory and methods in this domain is deeply sensitive to the limitations of group re-design. For instance, in explicating the approach of sociophysics, Galam [46] suggests that the seemingly disordered, individualized nature of human behavior can be modeled using techniques from the physics of disordered matter. Yet, operationalizing these models in terms that afford opportunities for intervention and group re-design may be challenging.

With these new advances from social network analysis, agent-based modeling, adaptation of epidemiological models, and the emergence of social physics, organizational analysts are much better equipped to leverage insights from complexity theory and capture the complexity of sociotechnical systems to advance sociotechnical systems theory. To be sure, there are still many computational, analytical, and operational challenges. Also, validating and refining models remains a challenge, although the increasing availability of data archives (including communication, production, financial, and marketing records) are providing important test beds to help analysts validate and refine different theoretical representations of complex sociotechnical systems.

6 Conclusion

This chapter has focused on the history and ongoing utility of sociotechnical systems theory. Initiated over 60 years ago, this perspective has had a profound impact on our understanding of how individuals, groups, and organizations interact around technical opportunities and demands to innovate and perfect new products and systems. As well, it is one trans-disciplinary perspective among other theories and methods that reveals the fragility of systems and has sensitized organizational analysts to the ease with which a small flaw in materials, failure in processes, or human error can (1) occur; and (2) cascade through a system leading to catastrophic disaster. In this regard, it may offer a sensitizing framework to aid in the development of risk assessment models that might prevent or mitigate adverse effects of disasters. As discussed in this chapter, sociotechnical systems theory is ripe for integration with emerging insights and methods from complexity theory. When coupled with complexity theory, sociotechnical systems theory can further leverage psychology, sociology, and engineering, as well as mathematics and physics to advance the development of a trans-disciplinary lens for understanding and analysis of the complex systems in which innovation, work, and organizational performance occur. Adding insights from complexity theory to better understand sociotechnical systems and their impact on individual and organizational outcomes is enlightening. Yet what is perhaps most remarkable about sociotechnical systems theory is how robust it is as a framework for analyses, regardless of the complexity of the social or technical system, the point in time in which organizational systems are being analyzed, or the extensiveness of the global environment in which organizations are embedded.

Acknowledgements I am grateful for the thoughtful and constructive comments of the editors and anonymous reviewers of this chapter, as well as from the visionary research of other authors contributing to this volume who gave me new insights on complex systems. Any opinions expressed in the paper or errors should be attributed to me and not the Army Research Office/Army Research Laboratory, or Duke University.

References

1. Hester, P. T., Akpinar-Elci, M., Shaeffer, J. M., & Shaeffer, M. B. (2016). On complementarity and the need for a transdisciplinary approach in addressing emerging global health issues. In F.-J. Kahlen, S. Flumerfelt, & A. Alves (Eds.), *Interdisciplinary perspectives on complex systems*. Heidelberg: Springer.
2. von Bertalanffy, L. (1956). General systems theory. In L. von Beralanffy & A. Rapoport (Eds.), *General systems: Yearbook of the society for the advancement of general systems theory* (Vol. 1, pp. 1–10). Ann Arbor, MI: The Society.
3. Emery, F. E., & Trist, E. L. (1965). The causal texture of organizational environments. *Human Relations, 18*, 21–32.
4. Lewin, Kurt. (1947). Frontiers in group dynamics. *Human Relations, 1*, 5–41; *2*, 143–153.
5. Taylor, F. W. (1911). *The principles of scientific management*. New York: Harper.

6. Mayo, E. (1945). *The social problems of an industrial civilization*. Boston: Graduate School of Business Administration, Harvard University.
7. Roethlisberger, F. J., & Dickson, W. J. (1939). *Management and the worker*. Cambridge, MA: Harvard University Press.
8. Trist, E. L. (1981). The evolution of sociotechnical systems as a conceptual framework & as an action research program. In A. H. Van de Ven & W. F. Joyce (Eds.), *Perspectives on organization design & behavior* (pp. 19–75). New York: Wiley Interscience.
9. Scott, W. R. (1998). *Organizations: Rational, natural, & open systems* (4th ed.). Upper Saddle River, NJ: Prentice-Hall.
10. Bion, W. R. (1961). *Experiences in groups*. London: Tavistock Publications.
11. Trist, E. L., & Bamforth, K. (1951). Some social & psychological consequences of the longwall method of coal getting. *Human Relations, 4*, 3–38.
12. Levy, D. L. (2000). Applications & limitations of complexity theory in organization theory & strategy. In J. Rabin, G. J. Miller, & W. Bartley Hildreth (Eds.), *Handbook of strategic management* (2nd ed., pp. 67–87). New York: Marcel Dekker.
13. Manuse, J. E., & Sniezek, B. (2016). On the perception of complexity and its implications. In F.-J. Kahlen, S. Flumerfelt, & A. Alves (Eds.), *Interdisciplinary perspectives on complex systems*. Heidelberg: Springer.
14. Kramer, R. M., & Brewer, M. B. (1986). Social group identity and the emergence of cooperation in resource conservation dilemmas. In H. Wilke, C. Rutte, & D. M. Messick (Eds.), *Experimental studies of social dilemmas* (pp. 205–234). Frankfurt: Peter Lang Publishing Company.
15. Turner, J. C., & Oakes, P. (1986). The significance of the social identity concept for social psychology with reference to individualism. *Interactionism and Social Influence, British Journal of Social Psychology, 25*, 237–252.
16. Janis, I. L. (1972). *Victims of groupthink*. Boston: Houghton-Mifflin.
17. Hermann, A., & Rammall, H. G. (2010). The grounding of the 'flying bank'. *Management Decision, 48*, 1048–1062.
18. Eaton, J. (2001). Management communication: The threat of groupthink. *Corporate Communication, 6*, 183–192.
19. Berger, J., Cohen, B. P., & Jr, M. Z. (1972). Status characteristics & social interaction. *American Sociological Review, 37*, 241–255.
20. Ridgeway, C. (1988). Gender differences in task groups: A status & legitimacy account. In M. Webster & M. Foschi (Eds.), *Status generalization: New theory & research* (pp. 188–206). Stanford, CA: Stanford University Press.
21. Cohen, E. G., & Roper, S. (1972). Modifications of interracial interaction disability: An application of status characteristics theory. *American Sociological Review, 37*, 643–665.
22. Jr Webster, M., & Driskell, J. E. (1983). Physical attractiveness as status. *American Journal of Sociology, 43*, 220–236.
23. Silver, S. D., Cohen, B. P., & Troyer, L. (2000). Effects of status on the exchange of information in team decision-making: When teambuilding isn't enough. In M. M. Beyerlin, D. A. Johnson, & S. T. Beyerlin (Eds.), *Advances in interdisciplinary studies of work teams: Team development* (Vol. 7, pp. 21–51). New York: Elsevier.
24. Simon, H. A. (1976). *Administrative behavior: A study of decision-making processes in administrative organization*. New York: Free Press (orig. published 1947).
25. Simon, H. A. (1991). Bounded rationality & organizational learning. *Organization Science, 2*, 125–134.
26. Tversky, A., & Kahneman, D. (1981). The framing of decisions & the psychology of choice. *Science, 211*, 453–458.
27. Kahneman, D. (2002). *Maps of bounded rationality: A perspective on intuitive judgment & choice*. Nobel Prize Lecture, December 8, 2002, Aula Magna, Stockholm University, Stockholm, Sweden.

28. Kahneman, D., Knetsch, J. L., & Thaler, R. H. (1991). The endowment effect, loss aversion, & status quo bias. *Journal of Economic Perspectives, 5*, 193–206.
29. Weick, K. (1979). *The social psychology of organizing* (2nd ed.). Reading, MA: Addison-Wesley.
30. Buckley, W. (1967). *Sociology & modern systems theory*. Englewood Cliffs, NJ: Prentice-Hall.
31. Perrow, C. (1999). *Normal accidents: Living with high-risk technologies* (Rev. Orig. 1984). Princeton, NJ: Princeton University Press.
32. Perrow, C. (2010). The meltdown was not an accident. In M. Lounsbury & P. M. Hirsch (Eds.), *Markets on trial: The economic sociology of the U.S. financial crisis: Part A* (Research in the sociology of organizations, Vol. 30, pp. 309–330). Bingley: Emerald Group Publishing.
33. Grieves, M., & Vickers, J. (2016). Digital twin: Mitigating unpredictable, undesirable emergent behavior in complex systems. In F.-J. Kahlen, S. Flumerfelt, & A. Alves (Eds.), *Interdisciplinary perspectives on complex systems*. Heidelberg: Springer.
34. Walker, G., Stanton, N., Salmon, P., & Jenkins, D. (2007). *A review of sociotechnical systems theory: A classic concept for new command & control paradigms*. Somerset: Human Factors Integration Defence Technology Center.
35. Baldwin, W. C., & Felder, W. N. (2016). Mathematical characterization of system-of-systems attributes. In F.-J. Kahlen, S. Flumerfelt, & A. Alves (Eds.), *Interdisciplinary perspectives on complex systems*. Heidelberg: Springer.
36. Brummitt, C. D., D'Souza, R. M., & Leicht, E. A. (2011). Suppressing cascades of load in interdependent networks. *Proceedings of the National Academy of Sciences of the United States of America, 109*. doi: 10.1073/pnas.1110586109.
37. D'Souza, R. M. (2013). Complex networks: A winning strategy for interdependence. *Nature Physics, 9*, 212–213.
38. West, B. J., Turalska, M., & Grigolini, P. (2014). *Network of echoes: Imitation, innovation, & invisible leaders*. Switzerland: Springer International Publishing.
39. Helbing, D. (Ed.). (2012). *Social self-organization: Agent-based simulations & experiments to study emergent social behavior*. Berlin: Springer. doi:10.1007/978-3-642-24004-1.
40. Lanham, M. J., Morgan, G. P., & Carley, K. M. (2014). Social network modeling & agent-based simulation in support of crisis de-escalation. *Systems, Man and Cybernetics: Systems, IEEE Transactions, 44*, 103–110. doi:10.1109/TSMCC.2012.2230255.
41. Moon, I.-C., Carley, K. M., & Kim, T. G. (2013). *Modeling and simulating command and control for organizations under extreme situations*. London: Springer.
42. Holt, S., Collopy, P., & Deturris, D. (2016). So it's complex, why do I care? In F.-J. Kahlen, S. Flumerfelt, & A. Alves (Eds.), *Interdisciplinary perspectives on complex systems*. Heidelberg: Springer.
43. Troyer, L., Watkins, G., & Silver, S. D. (2007). Time dependence in micro social interaction: An elaboration of information exchange theory & initial empirical test. *Sociological Focus, 40*, 161–181.
44. Berglund, P. A. (2011). *An overview of survival analysis using complex sample data*. Paper #338-2011, SAS Global Forum, Institute for Social Research, University of Michigan, Ann Arbor, MI.
45. Pentland, A. (2014). *Social physics: How good ideas spread—The lessons from a new science*. New York: Penguin Press.
46. Galam, S. (2012). *Sociophysics: A physicist's modeling of psycho-political phenomena (understanding complex systems)*. New York: Springer.

On Complementarity and the Need for a Transdisciplinary Approach in Addressing Emerging Global Health Issues

Patrick T. Hester, Muge Akpinar-Elci, James M. Shaeffer Sr., and Margaret B. Shaeffer

1 Introduction

Complexity is a term used all too-often in literature and everyday conversation with little specificity. How might we define complexity? What makes a situation complex? While researchers such as noted psychologist Joachim Funke provide a list of characteristics of complex problems [1], it may be best to think of complexity as existing along a continuum and not as a binary characteristic of a problem. One such continuum is presented by Jackson [2], who summarizes the work of Boulding [3] in creating a nine-level hierarchy for real-world complexity, as shown in Table 1.

Each of these levels is of increasing complexity and each contains emergent properties not found in the levels below. Thus, in seeking to understand a given level, we must also understand those levels beneath it. The key for us, as problem solvers, is to understand that increasingly complex problems necessitate increasingly complex methods to address them.

Significant literature has been devoted to matching problem type (e.g., simple vs. complex) to appropriate methods for addressing them. Jackson [4] portrays systems methods using a typology. Jackson and Keys [5] and Jackson [6] proposed a relationship between a problem's inherent complexity and its participants (i.e., stakeholders) and the type of methodology that could be used to address the problem. Many of these taxonomies exist within a larger meta-methodology, such as Flood and Jackson's [7] Total Systems Intervention (TSI) or Hester and Adams' [8] systemic thinking. Hester and Adams [8] speak of the need for a "discipline-agnostic" (p. ix) approach to solving complex problems. Such an approach is

P.T. Hester, Ph.D. (✉) • M. Akpinar-Elci, M.D., M.P.H. • J.M. Shaeffer Sr., Ph.D. • M.B. Shaeffer, Ph.D.
Old Dominion University, Norfolk, VA, USA
e-mail: pthester@odu.edu

F.-J. Kahlen et al. (eds.), *Transdisciplinary Perspectives on Complex Systems*,
DOI 10.1007/978-3-319-38756-7_8

Table 1 A summary of Boulding's [3] hierarchy of complexity ([2], p. S25)

Level	Description	Example
1	Structures and frameworks which exhibit static behavior and are studied by verbal or pictorial description in any discipline	Crystal structures
2	Clockworks which exhibit predetermined motion and are studied by classical natural science	The solar system
3	Control mechanisms which exhibit closed-loop control and are studied by cybernetics	A thermostat
4	Open systems which exhibit structural self-maintenance and are studied by theories of metabolism	A biological cell
5	Lower organisms which have functional parts, exhibit blue-printed growth and reproduction, and are studied by botany	A plant
6	Animals which have a brain to guide behavior, are capable of learning, and are studied by zoology	An elephant
7	People who possess self-consciousness, know that they know, employ symbolic language, and are studied by biology and psychology	Any human being
8	Socio-cultural systems which are typified by the existence of roles, communications and the transmission of values, and are studied by history, sociology, anthropology and behavioral science	A nation
9	Transcendental systems, the home of 'inescapable unknowables', and which no scientific discipline can capture	God

predicated on the need to go beyond traditional discipline boundaries to solve problems; such problems require a transdisciplinary approach.

In order to discuss the value of a transdisciplinary perspective, we invoke the concept of complementarity. The main goal of this chapter is to demonstrate the value of a transdisciplinary approach in complex problems and to understand how a transdisciplinary approach works beyond traditional discipline boundaries. Following a discussion of complementarity, we discuss emerging health issues such as infectious diseases as a case study for the use of a transdisciplinary perspective. This problem is addressed first from an engineering perspective, then from a global health perspective, and finally from an education perspective, each taking a unitary perspective in reasoning about this complex problem. These unique perspectives are driven by the authors' varied educational and professional backgrounds, with each author hailing from one of these unique disciplines. An approach for integrating these perspectives using cognitive mapping is then presented, and this approach is demonstrated on an example problem. Finally, some conclusions are drawn regarding the utility of adopting a transdisciplinary approach to solving complex problems.

2 On Perspectives in Complex Problems

Examining the many perspectives on complexity, one may conclude that it is difficult to succinctly define complexity in a unified, authoritative manner. That is to say, there are numerous ways we can look at a problem and no one perspective is truly complete. This notion is captured by the systems principle known as complementarity [9]. Complementarity addresses the aspect that no single perspective or view of a system can provide complete knowledge of the system. Niels Bohr coined this term during his experiments on particle physics. Bohr stated that if two concepts are complementary, an experiment that clearly illustrates one concept will obscure the other complementary one. For example, an experiment that illustrates the particle properties of light will not show any of the wave properties of light ([10], p. 128). The complementarity principle states:

> Two different perspectives or models about a system will reveal truths regarding the system that are neither entirely independent nor entirely compatible. ([8], p. 26)

Further complicating this matter is the fact that individuals often have unique worldviews which drive their analytic techniques for addressing a problem. Aerts et al. [11] define world view as "....a system of co-ordinates or a frame of reference in which everything presented to us by our diverse experiences can be placed" (p. 9). Consistency in world view is what Checkland [12] refers to as *weltanschauung*, the image or model of the world that provides meaning. Each of these definitions hints at the idea of a world view as a shared perspective or frame of reference for understanding the world. As problems become more complex, it is necessary, for example, to adopt the complementary advantages afforded by using both a constructivist (assembling systems into larger and larger constructs to infer meaning) and reductionist (reducing the world to its barest elements to provide meaning) worldview to decompose and understand problems.

Each discipline, then, provides a unique perspective regarding a problem, likely informed by its own unique worldview. Adams and Meyers [13] offer a mathematical representation of contextual understanding for a limited number of perspectives (n):

$$Contextual\ Understanding = \sum_{i=1}^{n} p_i \qquad (1)$$

where i is the number of perspectives (P_i) utilized to represent a problem under study.

The more perspectives we incorporate during the assessment of a problem, the greater our understanding is. An infinite number of perspectives provide a theoretical upper limit on understanding known as *perfect understanding*. This notion is depicted in Eq. (2) below [13]:

$$Perfect\ Understanding = \sum_{i=1}^{\infty} p_i \qquad (2)$$

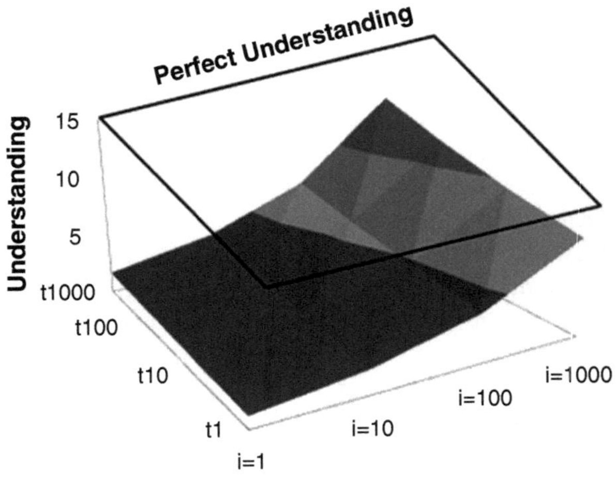

Fig. 1 Depiction of increased understanding as a function of time (t) and perspectives (i) (adapted from Hester and Adams [8], p. 27)

These concepts are shown graphically in Fig. 1. This figure shows that as both time (*t*) and the number of perspectives increases, our understanding increases dramatically. Perfect understanding is depicted as an asymptotic plane that we attempt to attain but cannot reach no matter how much time passes or how many perspectives we consider.

By definition, we cannot consider an infinite number of perspectives; we must strive to consider as many as is practical in the understanding of our problems. This necessitates that we adopt a transdisciplinary perspective in reasoning about complex problems and thus, reason across disciplines. To illustrate this point, we will discuss a problem relating to emerging diseases from three unique perspectives (engineering, global health, and education) and then present an approach for integrating these perspectives in a transdisciplinary fashion.

3 Emerging Diseases

In order to understand the efficacy of a transdisciplinary perspective in addressing emerging diseases, a brief introduction to fundamental concepts of disease is necessary. Crowley [14] defines disease as "any disturbance of structure or function of the body" (p. 1). Crowley makes an important point that health and disease should be considered as two ends of the same continuum. Health can be defined using the World Health Organization's seminal definition as "a state of complete physical, mental and social well-being and not merely the absence of disease or infirmity" ([15], p. 100). The specific focus of this chapter's case study is on communicable diseases, described by Crowley [14] as "an infectious disease that is readily transmitted from person to person" (p. 149). There are both *endemic* diseases that persist in small, but continually present, factions of the population, while *epidemics* represent diseases that affect a large proportion of the population. One of the goals of proper disease control is to prevent the outbreak of a epidemic from what was previously an endemic.

Communicable diseases can be transmitted "from person to person by either direct or indirect methods. Direct transmission is either by direct physical contact or by means of droplet spread, such as by coughing or sneezing. Indirect transmission of an infectious agent is accomplished by some intermediary mechanism, such as transmission in contaminated water or by means of insects. A few communicable diseases are primarily diseases of animals and are transmitted to humans only incidentally" ([14], p. 149). Crowley [14] points out that effective control requires knowledge of both the cause of the disease and the method of the disease. In the absence of this information, infection control will be problematic. Further, education of the general public regarding what is fact and what is hearsay becomes an issue. Each of engineering, global health, and education have a unique perspective to bring to examining the problem of emerging diseases. In the following sections, discussion is provided on addressing an emerging global health crisis through the lens of each of these disciplines. The specific disease is not important; it could be conceptualized using a recent example such as the Ebola virus disease, H1N1, Avian flu, or using a more historic example such as the bubonic plague.

4 Viewing the Outbreak of Emerging Diseases Through an Engineering Lens

Engineering is defined as "the art or science of making practical application of the knowledge of pure sciences, as physics or chemistry, as in the construction of engines, bridges, buildings, mines, ships, and chemical plants" [16]. Thus, engineering is not a pure science such as chemistry or biology but rather an applied science. The branch of engineering concerned with the design and execution of systems is systems engineering. The IEEE standard for systems and software vocabulary defines systems engineering as the "interdisciplinary approach governing the total technical and managerial effort required to transform a set of customer needs, expectations, and constraints into a solution and to support that solution throughout its life" ([17], p. 361). Thus, although it is an engineering sub-discipline, per se, nothing about systems engineering is engineering-centric. It is all about creating a solution to a problem.

The two earliest books on engineering for systems were written by Harry H. Goode and Robert E. Machol [18] and Arthur D. Hall [19]. Both texts were focused on general systems and paid significant attention to the basics of understanding systems. Hall also introduced a formal methodology for the analysis and synthesis of large scale systems. Hall's methodology continues to serve as a basis for many systems engineering models currently in use. Hall [19] spoke of general systems principles as:

> It happens that certain properties apply to systems in general, irrespective of the nature of the systems or of the fields in which they are normally studied. While it is true that not all of the most general properties are useful in an operational sense for applied work, they have considerable conceptual value in more deeply understanding creative or developmental processes. This fact is the real justification for including them in this chapter. (p. 59)

Hall [19] acknowledges the notion of a general systems theory and states that "…sometimes very difficult systems problems are greatly illuminated by looking at them in the light of the appropriate generalized property" (p. 65).

Both texts were very focused on the aforementioned definition of engineering as applied problem solving. As such, and with their focus on systems, they were very much advocating a transdisciplinary perspective, albeit with a different terminology. Hester and Adams [8] elaborate:

> It is worth noting that both Hall's and Goode and Machol's texts on the subject of systems engineering are substantially more general in nature than their successors and do not predicate their discussion on engineering disciplines. To them, engineering is problem solving; there is no distinction. To think about engineering systems was to think about how systems interacted with one another, how they functioned, how they could be understood, designed, and improved. (pp. 36–37)

Systems engineering is deeply rooted in the concept of holism, the understanding that a system must be viewed as whole rather than in terms of its constituent parts. In simpler terms, the whole is greater than the sum of the parts.

"The *principle of holism* [20] requires that multiple perspectives be included as part of the development of problem context" ([8], p. 157). Holism [20] is the philosophical position which holds that understanding a system is based not solely in terms of the functions of the component parts, but by viewing the system as a whole. Indeed, the incorporation of multiple perspectives is a fundamental tenet of traditional systems engineering.

Understanding the interactions between systems and their components and one another requires us to think in a transdisciplinary fashion. Recent research pertaining to infectious diseases from a systems engineering perspective, however, skews towards a more narrow perspective. Research ranges from health system optimization [21] to risk assessment of pandemic influenza [22] to influenza modeling [23]. A similarly operations research-centric perspective can be used to minimize variability in a disease testing process, the layout for an infectious disease clinic, or logistics for shipping vaccines from one coast to another. Any of these problems fits within a current systems engineering problem set and yet each has deficiencies based on the adoption of a narrow perspective. None of these problems, on the surface, considers such questions as the efficacy of a given vaccine or delivery method (the purview of medicine) or necessary outreach programs for encouraging the public to vaccinate (the purview of education).

Certainly an engineering-centric perspective has room to improve. From a systems engineering perspective, a transdisciplinary approach makes sense.

Structural engineering is necessary to understand the planning and development of new hospitals, treatment facilities, and vaccine manufacturing plants. Environmental engineering is necessary to address concerns related to climate change and waste management and their potential role in exacerbating emerging infectious diseases. Computer engineering will be required to develop the next generation of biomedical systems to monitor patients' vital signs and to suggest treatment options based on advanced diagnostics. Systems engineering will need to be the integrating function to bring each of these perspectives together, as well as those from beyond engineering. Each of these perspectives provides an element of the solution, but only together can they solve the problem.

5 Viewing the Outbreak of Emerging Diseases Through a Global Health Lens

The current status of Global Health has gained momentum and now encompasses a broader understanding of health to include emerging diseases. There is also an integrated social, epidemiological and economic framework for understanding health from different perspectives to encompass the sustainability of the development among the communities. Fried et al. [24] defined global health as addressing the causes of poor health through scientific, social, cultural, and economic strategies and recognizing a multisectoral systems-based approach for sustainable community health. Based on the complex thinking of health that goes beyond humans and animals, these approaches are considered to be inextricable linkages beyond the human, animal and environmental interface. Collaboration between and amongst various professionals to understand the interactions and reactions to flux in a system can facilitate a clearer understanding of the health impacts on environmental, animal, and human health.

One of the important emerging health issues is infectious diseases, which pose a significant threat to human health, economic stability, and biodiversity. Mostly, the mechanisms emerging from infectious diseases are not fully understood and control measures depend on mitigating the impact of infectious diseases after they have emerged.

Despite efforts to control infectious diseases through improving sanitation, medical technology and pharmaceuticals, infectious diseases have continued to spread and are still a major cause of mortality around the world [25]. Widespread poverty, massive migration, urbanization, and globalization further facilitates the emergence and re-emergence of infectious diseases such as tuberculosis, HIV/AIDS, cholera, yellow fever, dengue, leptospirosis, leishmaniasis, malaria, and hantavirus [25–27]. In addition, major issues such as microbial resistance, inadequate antiviral medications, and the rise of unsafe farming and food-processing makes emergence and re-emergence of infectious diseases more likely [28]. Emerging infectious diseases create a critical threat to health, economic stability and ecological balance. The usefulness of being able to identify and recognize different perspectives in regards to emerging infectious diseases can be seen in regards to the Nipah virus case where the true cause of the virus was unknown. By obtaining different perspectives on the possible causes of the Nipah virus from various experts in different fields, it was found that the cause of the disease was due to complex social-ecological dynamics within the country [28]. Daszak et al. [29] concluded that many disciplines should work together to discover the underlying causes of infectious disease emergence such as the Nipah virus, which usually represent temporal shifts in the underlying environmental and demographic state. The current outbreak of the Ebola virus disease is another perfect example for this link. The Ebola virus disease outbreak poses many challenging questions around human rights, globalization, poverty, socio-economic issues, health systems, public health, engineering control, ethical issues, occupational health, "one health, one medicine" approach, climate change and national security. We should create a transdisciplinary approach to address these health challenges that occur across transnational boundaries and across different disciplines.

A transdisciplinary perspective is needed to fully understand the various challenges and the complexity of emerging and re-emerging infectious diseases in both developing and developed countries around the world. These systemic approaches may directly inform management and policy makers aiming to reduce emergence risk. The definition of transdisciplinary refers to different disciplines working together to create new theoretical, methodological, and translational innovations that move beyond discipline-specific approaches in emerging disease issues [30]. This type of approach can be used in many ways to help further the understanding of emerging infectious diseases around the world. For one, the transdisciplinary approach can aid in the formulation of infectious disease related research questions and in the design and analysis of an infectious diseases study due to a quick transfer of knowledge and shared interrelationships among scholars from various fields. Symposiums are a great tool to facilitate transdisciplinary sharing of knowledge and the creation of interrelationships. The Emerging Infectious Diseases and Social-Ecological Systems Symposium was held to quickly obtain ideas and experiences on dealing with emerging infectious diseases globally from a wide range of scholars in social science, international development, ecology, human medicine and veterinary [28]. Similarly, The First Symposium on Emerging Infectious Diseases, Biodiversity and the Environment gathered over 200 participants around the world to discuss the emergence of infectious diseases globally, emerging diseases and the environment, tools and techniques for health surveillance, research from a macro to micro scale, and the public health care networks around the world. The transdisciplinary approach to addressing emerging global infectious diseases at the symposium ultimately led to informative scientific and political debates on the issue of emerging infectious diseases and the creation of meaningful professional partnerships to improve research and health surveillance initiatives on emerging global infectious diseases [31]. Another way of using a systemic approach to addressing complex health issues can be helpful to the understanding of emerging and re-emerging infectious diseases is that a transdisciplinary view can help researchers and health professionals identify and recognize different perspectives on the same issue.

Overall, a transdisciplinary approach to emerging and re-emerging global infectious diseases can be a great solution to decreasing the gap in infectious disease knowledge around the world, increasing partnerships between various community and governmental organizations involved with global health and in generating cutting edge research on emerging infectious diseases that can bring researchers closer to preventing the spread of infectious disease globally. Health is at a crossroads as there is a need to more clearly work beyond the boundaries for the benefits of all populations as there is the continuing need for further efforts towards integration with the global community serving as the unit of health.

6 Viewing the Outbreak of Emerging Diseases Through an Education Lens

Big problems or messes as described by Ackoff [32] are not one dimensional, but are multi-dimensional and are the result of many systems and concentrating on addressing only one dimension of the problem will not solve the problem as long as other systems are left unaddressed. Only when we look at the full nature of the cause of the problem do we begin to see long-term results. This clearly is the case in dealing with infectious diseases as well as any major problem facing humanity.

One way of preventing and/or curbing the spread of infectious diseases is providing proper education to individuals and to whole communities. One of the educational goals is to assist with an accurate understanding of the disease and how to prevent the spread of the disease.

From an educational lens, solutions to complex "messes" are best dealt with from a systemic approach and not through an isolated impact approach [33]. The problem with the isolated approach is that even though the educational intervention may be well designed and be well meaning, it fails to address all factors of the problem or the unintended consequences of only looking at the issue from a single perspective. As McNall et al. [33] describe it, "the dynamics of the larger system that generated the problem remain unchanged, meaning that the problem reemerges when the intervention is scaled back. In addition, isolated impact efforts are frequently conducted in isolation from other related efforts, thereby failing to realize the synergies possible with more coordinated strategies" (p. 2).

An emerging example of systemic approaches to addressing complex issues can be found in the efforts of universities and colleges to become more engaged with their communities. In higher education university/community engagement work, we've found that successful university/community relationships embrace all the systems impacted with the key features being: (a) uses systems thinking to conceptualize complex problems and to organize responses to them, (b) recognizes the degree of uncertainty and unpredictably inherent to solving complex problems, and (c) employs transdisciplinary approaches to community-engaged social problem solving [33].

To effectively engage in university/community education efforts requires the preparation of professionals who understand and embrace transdisciplinary thinking and problem solving. Higher education institutions have had to take steps in modifying curriculum to create better problem solvers within a complex world that requires graduates to address challenges from a systems' perspective. The development of expertise in any discipline requires the acquisition of knowledge that will support the solving of complex problems. Exactly how that expertise develops has been the focus of much research on teaching and learning.

One perspective is that developing expertise is a vertical concept that occurs through the transformation of facts or the construction of knowledge through social constructs. Alternatively, the development of expertise may be viewed as a horizontal construct, involving complex contexts that require crossing barriers of knowledge that typically separate disciplines. This horizontal conceptualization of developing expertise is a multidimensional way of looking at learning and requires polycontextuality and boundary crossing. It has been suggested that polycontextuality occurs not in a single activity event but instead within activity systems that don't have a single beginning and end, but instead are cyclical and create permutations of responses, based on the contexts that are presented. As described by Engestrom et al. [34]:

> Polycontextuality at the level of activity systems means that experts are engaged not only in multiple simultaneous tasks and task-specific participation frameworks within one and the same activity. They are also increasingly involved in multiple communities of practice. (p. 320)

Boundary crossing, a second key dimension of the horizontal learning, calls upon experts to consider for a moment the expertise of others in relation to their own understandings. As Engestrom et al. [34] suggest, "boundary crossing calls for the formation of new mediating concepts. In this sense, boundary crossing may be analyzed as a process of collective concept formation." (p. 321).

Understanding the impact of these concepts on educating the community about outbreak of infectious diseases is critical. If we are to solve complex problems—either in medicine or in engineering—we must provide for learning experiences for professionals to develop this new level of expertise.

Specific examples of these changes in the curriculum can be found with health, early childhood education, and engineering, where the common thread is the concept of transdisciplinary collaboration. The concept of transdisciplinary collaboration in the health professions is an emerging concept that has taken hold in the UK and other countries as an approach to preparing health professionals to better serve patients. Recognizing that the increasing complexity and specialization in providing quality health care required the efforts of a broad range of medical and social care professionals, interprofessional education has provided a way of bringing professionals together in a teaching, learning and service arena. Fook et al. [35]

use the Centre for the Advancement of Interprofessional Education's definition that interprofessional education is "when two or more professions learn with, from and about each other to improve collaboration and the quality of care." "The imperatives for IPE and workforce transformation have become more urgent as health services struggle under the strain of global inequalities, population ageing, poverty and escalating health care costs with the increasing incidence of chronic illness and life-style diseases" ([35], p. 286).

Another example of interprofessional delivery of services is early intervention services to children and families have evolved over the past decades to reflect the philosophy that the child is part of a complex system of influences. To assume that addressing a single aspect of the child—such as health, motor development, or language development—without consideration for the ecosystem in which the child lives would result in short term, isolated development if, indeed, any developmental changes at all. Instead, the child is considered as an integral piece of a complex, ever-evolving dynamic interchange of energy, contexts, and influences.

The intersection of the transdisciplinary model in education and the interprofessional care philosophy in the health and social services sector results in a comprehensive delivery of services to children and their families that has mandated that the preparation of professionals be reconsidered and redesigned. Efforts at this are in place across the country with differing level of success and sustainability.

To address messes like infectious diseases requires sustainable education programs to prepare professions in interprofessional practices. There will certainly be some resistance to this type of preparation of students. This approach can create tension in defining the interprofessional education means feelings of exclusion for decision making, limited resources for implementation, and perceptual differences in understanding [35]. This new way of preparing professionals requires a change in how we have always done things. It demands that we consider the entire "system" of education and service delivery. Systems thinking assumes that it we must first unlearn what we know to better address new problems, that solutions to problems are not one dimensional and are often fluid, meaning the circumstances leading to the problems are changing and, thus, any solutions must recognize that the problems are changing, problems are based within systems and finding a solution for only one piece of the system will not lead to an overall solution unless a solution addresses multiple systems impacting the problem.

Whether from an educational, global health, or engineering lens, in order for systemic change that draws from the expertise of multiple disciplines in solving complex problems to be successful, several elements need to be evident. Recognition of the importance of shared values and goals is paramount. It is one thing to have the vision, and it is another to develop a system that will provide continues guidance and support to sustain the plan [35].

Regardless of the disciplines, fields of study, or the constituents that are involved in implementing the systems approach to problem solving and delivering services, there are several characteristics that must be present for the work to be successful. These are highlighted in the research of McNall et al. [33], which describes among others the importance of relationships, multiple lines of inquiry, the importance of centralized support and leadership and the appreciation that all parties are considered equal partners in the endeavor.

7 A Framework for Uniting Disparate Perspectives Through a Transdisciplinary Perspective

Complementarity is a necessity for understanding complex systems. Adopting a single discipline perspective in reasoning about complex messes leaves an individual open to committing a number of errors [36]. Thus, it is important to mutually consider multiple world views in attempting to understand a mess. Further, resolution of potentially disparate perspectives (i.e., disciplines) is neither trivial nor always achievable by means such as consensus. In lieu of such an approach, the authors advocate a method based on cognitive mapping. The focus of this proposed framework is to provide an approach for evaluating multiple stakeholders' (each potentially representing a differing discipline) perspectives on a given mess using cognitive mapping. In doing so, this approach invokes complementarity in order to understand a mess in a transdisciplinary manner.

Cognitive mapping was introduced by Axelrod [37] as a way to visually and logically capture the relationships between elements in a problem. Cognitive maps (CMs) are network-based collections of concepts (represented as nodes) and causal relationships (represented as arcs between the concepts). Arcs have signs associated with them that indicate the direction of a causal relationship; thus, a given relationship can be increasing or decreasing (i.e., A increases B or A decreases B) or unknown.

Additional complexities such as fuzzification of relationships may be included (and are useful for the extensibility of a CM), as outlined by Kosko [38], but are beyond the scope of what is proposed here. Rather, the authors will use CM construction as a mechanism to increase understanding of our mess by exploring the disparities between the unique disciplines representing this problem, following the proposed framework outlined below:

1. The discipline experts agree on a common problem to explore to provide a singular anchor for their individual CM;
2. Each discipline expert independently develops a CM of the problem landscape from their perspective, including directionality (positive, negative, or unknown) between concepts;
3. The discipline experts then have a collective discussion highlighting their individual perspective on the problem, using CM as the common language of comparison;
4. Experts are provided the opportunity to revise their CM based on the result of the conversation in Step 3. Thus, they can return to Step 2 (or 1 if needed), or proceed to Step 5;

5. A narrative is constructed to illustrate the similarities and difference between the unique perspectives;
6. Although it will not be undertaken in this analysis, further analysis including fuzzification can be invoked using CM maps to determine dynamic behaviors (see, e.g., Kosko [38]). This requires a consensus CM be generated. Given that this chapter is an initial assessment of this approach, a narrative discussion was determined to be a sufficient stopping point.

The following section will illustrate this approach on a sample emerging health problem.

8 Illustrative Example

In order to demonstrate the framework proposed in the previous section, the authors agreed, per Step 1, to explore the problem of the incidence rate of the Ebola virus disease. The Ebola virus disease was chosen as it is an illustrative example of an emerging health problem with a significant amount of recent media attention. This was divided into concepts of mortality and morbidity. From here, the authors representing each of the three disciplines discussed in this paper (engineering, education, and global health) each independently constructed an individual cognitive map of their conceptualization of these concepts (Step 2).

They then got together to discuss their individual perspectives (Step 3). The results were rather interesting. The three perspectives each generated unique results and yet, a surprising number of commonalities. After discussing their original maps, the authors modified their individual maps to reflect their updated perspectives (Step 4). While iteration could have continued until consensus was achieved on a singular representation (and further assessed in Step 6), it was more useful for the authors to see how their unique perspectives could be compared and contrasted using a common language, that of the cognitive map. Each of the three perspectives is discussed in the following subsections. Following the individual discussions, common themes presented across individual conceptualizations are discussed (as a result of Step 5 discussions).

8.1 Examining the Ebola Virus Disease through an Engineering Lens

The first articulation of the Ebola virus disease was examined from a systems engineering perspective. In doing so, the problem was thought of as a system. Thus, elements such as relevant stakeholders, policies, population attributes, etc., were generated in terms of their relationships to one another. This map is shown in Fig. 2. Several elements emerge as prominent. Health care availability, for one, is a

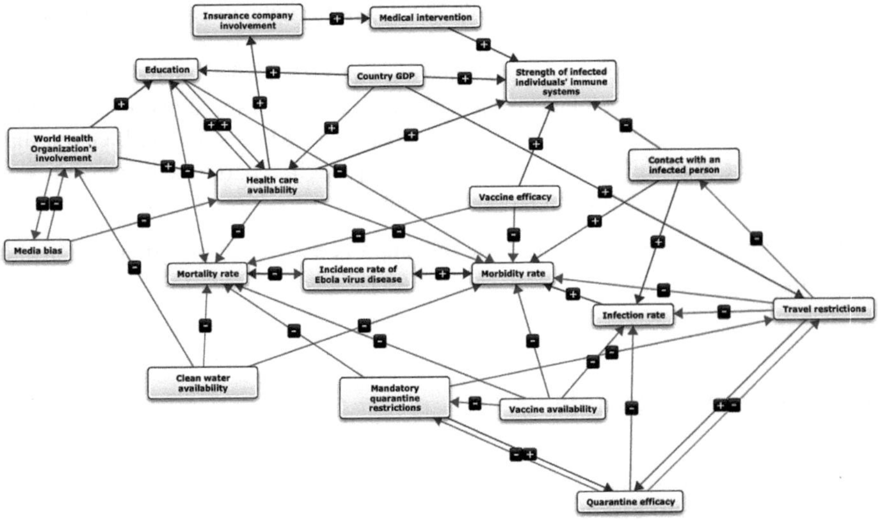

Fig. 2 Depiction of Ebola virus disease cognitive map through an engineering lens

large driver in an individual's ability to stay healthy and avoid becoming sick or dying from a given disease. Similarly, the strength of an individual's immune system is influenced by a number of factors, including contact with an infected person, medical intervention, the Gross Domestic Product (GDP) of the individual's resident country (a general indicator of the conditions in which an individual is likely to live), health care availability, and potential vaccine efficacy. Clearly, the problem is very complex and even though this mapping was crafted by an engineer, engineering elements played a very small role in that author's actual articulation of the problem.

8.2 Examining the Ebola Virus Disease through a Global Health Lens

From a global health perspective, Ebola virus disease resembles the cradle of social determinants of health. The natural history of Ebola virus disease includes various factors such as socioeconomic, health education needs, poor public health systems, etc. Lacking a transdisciplinary approach was one of the important issues in Ebola virus disease outbreak. Figure 3 shows how everything is connected with each other and how a lack of recognizing the critical connection between these factors leads to serious consequences as we experienced in this recent multi-country epidemic. Note the presence of multiple arcs marked with "?'s" indicating an unwillingness to assign directionality to causal relationships, derived by uncertainty surrounding the problem.

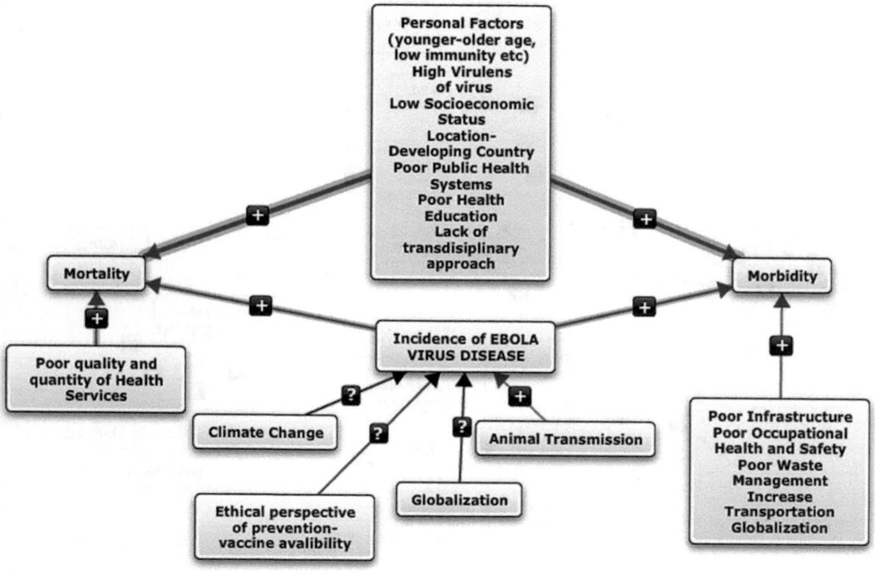

Fig. 3 Depiction of Ebola virus disease cognitive map through a global health lens

8.3 *Examining the Ebola Virus Disease through an Education Lens*

The final cognitive map (shown in Fig. 4) begins with the assumption that that the educational level impacts the mortality and morbidity rates of the incidence of Ebola virus disease. The underlying assumption is that higher educational levels of the population will result in lower the mortality and morbidity while also assuming that the lower the educational level of the population will result in increasing chances for mortality and morbidity rates of Ebola virus disease.

The CM also draws links to situations and contexts that potentially impact the educational level of the population. These include culture, the quality of the countries infrastructure, demographics, GDP level of the economy, and finally the presence and/or absence of the use of professionals who have been trained in transdisciplinary models of service delivery.

From an educational lens, one of the major changes to be addressed is moving from looking at the outbreak of emerging disease from a single content viewpoint to a perspective that multiple disciplines are involved and therefore need to be involved in the approach to control and eradication of the disease.

The transdisciplinary approach to preparing health professionals recognized the need to ensure that professional languages across disciplines are understood for their points of intersection and meaning. It requires that the professional identify knowledge each team member brings to the context is recognized and celebrated, building upon the strengths and experience of each. Recognizing that "messes"

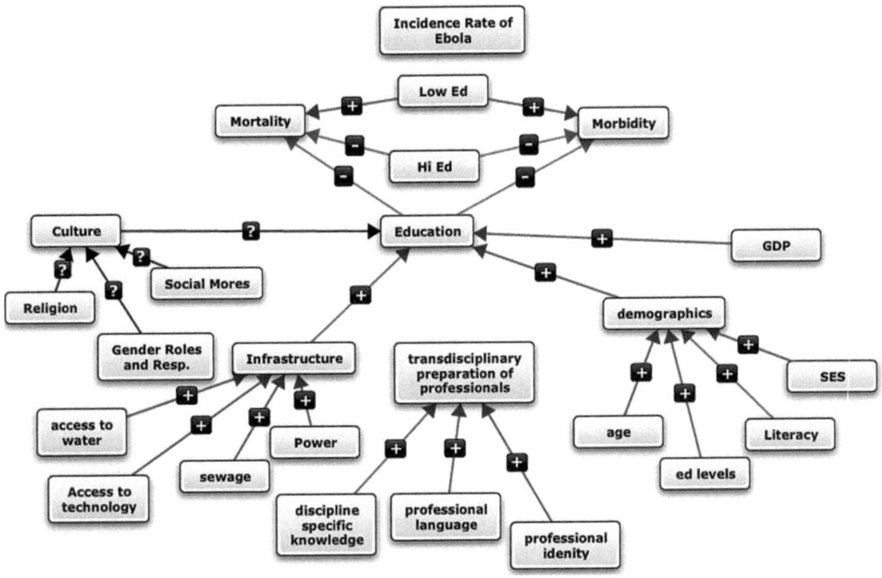

Fig. 4 Depiction of Ebola virus disease cognitive map through an educational lens

require a transdisciplinary approach in order to best serve all involved is a major step in addressing major problems. The CM as seen through an educational lens recognizes that the varying contexts of multiple systems must be addressed, in the solution of complex problems

The exercise of developing the CM through the education lens and then sharing that perspective through the other two lenses, engineering and global health, provided an opportunity to better understand how the complexity of our own personal and professional backgrounds influences how we see a "problem" and how we develop plans to solve complex issues.

Developing the CM forced the authors to see that complex problems, such as the crisis faced with the outbreak and spread of Ebola virus disease, must be addressed through transdisciplinary approaches. When we examined the three CM's that were developed through the independent lens of engineering, medicine and education, we found points of overlap and also not surprising, differences. While not part of this exercise, the next step would be for the authors to develop steps, based on the FCMs, in addressing the incidence the rate of Ebola virus disease in mitigating mortality and morbidity.

8.4 Examining CMs Across Disciplines

In examining all three CMs in concert with one another, several similarities emerge. Infrastructure, for one, plays a substantial role in all three. This may include elements such as travel accessibility and restrictions, access to clean water and health care, technology, sewage, and electricity. Certainly, a stronger infrastructure places individuals at an advantage when combating an infectious disease such as the Ebola virus. Additionally, GDP as a proxy for a country's financial health, was prominent across CMs. No doubt correlated to infrastructure, greater financial health for a country means better access to health care, a higher probability of having health insurance, and generally better living conditions leading to longer life expectancy. Demographics were also seen as key factors, including family history, age, race, sex, and education and literacy levels. Certainly these factors play into how susceptible an individual is to contracting, and ultimately dying from, an infectious disease. Other factors were identified as important but did not necessarily have consensus among the subject matter experts, including World Health Organization involvement, global health concerns, and transdisciplinary preparation of professionals. Each of these elements, it could be argued, were heavily influenced by the domain-specific nature of the experts generating the CM. Nonetheless, the exercise was enlightening for the participants in that it allowed them all to speak a common language and to discuss a complex problem in a manner that was able to be compared and contrasted. Additional work could certainly be performed to rectify the still-disparate perspectives into a unified whole, an objective made possible by using CM. However, the intent of this chapter was exploratory in that we sought to determine if we could utilize a technique to unite disparate worldviews through a common lens in an effort to examine a problem in a transdisciplinary manner. In this manner, the use of CM succeeded.

9 Conclusions

As stated earlier, complex problems present us with messes that are not one-dimensional in origin, design, or solution. Despite the proliferation of knowledge and information gained daily through research and technology, it is impossible to know what we don't know. We are humans and, as such, have a world view that supports our construction of knowledge and understanding, coloring our approaches to solving problems.

The examples presented, looking through three very different lenses—engineering, global health, and education—make the case that the concept of complementary approaches to problem solving results in the identification of solutions that hold potential for moving beyond the initial "mess" to considering even larger, more complex issues.

Further, we presented an approach to compare and contrast these disparate discipline perspectives in order to adopt a transdisciplinary perspective in reasoning about problems. This approach was demonstrated on a problem related to emerging diseases. Emerging diseases, global health, climate change, and sea-level rise are only a few of the challenges facing us that by virtue of the enormity of the situations they present and the severe consequences resulting if they are not addressed, demand a transdisciplinary approach in the search for resolution. This chapter demonstrates that even when we explore how our specific disciplines examine an issue, we have had a tendency to do so with a somewhat myopic approach. At the same time we have shown that expanding the conversations beyond a single perspective and applying a transdisciplinary process in solving messes yields more holistic results.

Further research into measurable impacts of a transdisciplinary approach in dealing with complex problems is needed if we are to apply the concept in a systematic and comprehensive way as messes present themselves. Additionally, future efforts should examine whether or not it is feasible to generate a single, unitary transdisciplinary perspective for a complex mess. Or, for that matter, is it even necessary? Is having the conversation between disciplines using a common language, as was proposed in this chapter, enough to move a problem solution effort forward sufficiently? These questions remain open research questions and should fuel future research in transdisciplinary approaches to problem solving.

References

1. Funke, J. (1991). Solving complex problems: Exploration and control of complex systems. In R. J. Sternberg & P. A. Frensch (Eds.), *Complex problem solving: Principles and mechanisms* (pp. 185–222). Hillsdale, NJ: Lawrence Erlbaum Associates.
2. Jackson, M. C. (2009). Fifty years of systems thinking for management. *Journal of the Operational Research Society, 60*, S24–S32.
3. Boulding, K. (1956). General systems theory—The skeleton of science. *Management Science, 2*(3), 197–208.
4. Jackson, M. C. (2003). *Systems thinking: Creative holism for managers*. Chichester: Wiley.
5. Jackson, M. C., & Keys, P. (1984). Towards a system of systems methodologies. *Journal of the Operational Research Society, 35*(6), 473–486.
6. Jackson, M. C. (1990). Beyond a system of systems methodologies. *Journal of the Operational Research Society, 41*(8), 657–668.
7. Flood, R. L., & Jackson, M. C. (1991). *Creative problem solving: Total systems intervention*. New York: Wiley.
8. Hester, P. T., & Adams, K. M. (2014). *Systemic thinking: Fundamentals for understanding problems and messes*. Cham: Springer International.
9. Bohr, N. (1928). The quantum postulate and the recent development of atomic theory. *Nature, 121*(3050), 580–590.
10. Adams, K. M. (2011). Systems principles: Foundation for the SoSE methodology. *International Journal of System of Systems Engineering, 2*(2/3), 120–155.
11. Aerts, D., Apostel, L., De Moor, B., Hellemans, S., Maex, E., Van Belle, H., et al. (1994). *World views: From fragmentation to integration*. Brussels: VUB Press.
12. Checkland, P. B. (1981). *Systems thinking, systems practice*. New York: Wiley.
13. Adams, K. M., & Meyers, T. J. (2011). Perspective 1 of the SoSE methodology: Framing the system under study. *International Journal of System of Systems Engineering, 2*(2/3), 163–192.
14. Crowley, L. V. (2007). *An introduction to human disease: Pathology and pathophysiology correlations* (7th ed.). Sudbury, MA: Jones and Bartlett Publishers.
15. World Health Organization. (1948). *WHO definition of Health*. Preamble to the Constitution of the World Health Organization as adopted by the International Health Conference, New York,

19–22 June 1946; signed on 22 July 1946 by the representatives of 61 States (Official Records of the World Health Organization, no. 2, p. 100) and entered into force on 7 April 1948.
16. engineering. (n.d.). *Dictionary.com Unabridged*. Retrieved January 7, 2015, from Dictionary.com website: http://dictionary.reference.com/browse/engineering
17. IEEE ISO/IEC. (2010). *IEEE and ISO/IEC Standard 24765: Systems and software engineering—Vocabulary* (pp. 1–418). New York and Geneva: Institute of Electrical and Electronics Engineers and the International Organization for Standardization and the International Electrotechnical Commission.
18. Goode, H., & Machol, R. (1957). *Systems engineering: An introduction to the design of large-scale systems*. New York: McGraw-Hill Book Company.
19. Hall, A. D. (1962). *A methodology for systems engineering*. Princeton, NJ: D. Van Nostrand Company Inc.
20. Smuts, J. (1926). *Holism and evolution*. New York: Greenwood Press.
21. Barker, P. M., McCannon, C. J., Mehta, N., Green, C., Youngleson, M. S., Yarrow, J., et al. (2007). Strategies for the scale-up of antiretroviral therapy in South Africa through health system optimization. *The Journal of Infectious Diseases, 196*, S457–S463.
22. Chen, S.-C., & Liao, C.-M. (2008). Modelling control measures to reduce the impact of pandemic influenza among schoolchildren. *Epidemiology and Infection, 136*(8), 1035–1045.
23. Davey, V. J., Cannon, D. C., & Glass, R. J. (2010, December). *The effect of healthcare environments on a pandemic influenza outbreak*. Paper presented at Modeling for Public Health Action: From Epidemiology to Operations, Atlanta, GA.
24. Fried, L., Bentley, M., Buekens, P., Burke, D., Frenk, J., Klag, M., et al. (2010). Global health is public health. *The Lancet, 375*(9714), 535–537.
25. Possas, C. A. (2001). Social ecosystem health: Confronting the complexity and emergence of infectious diseases. *Cadernos De Saúde Pública, 17*(1), 31–41.
26. Levins, R., Awerbuch, T., Brinman, U., Eck-hard, I., Epstein, P., Ford, T., et al. (1996). Globalization, development and the spread of diseases. In J. Mander & E. Goldsmith (Eds.), *The case against global economy* (pp. 161–170). New York: Sierra Club Books.
27. Stoltzfus, J., Carter, J., Akpinar-Elci, M., Matu, M., Kimotho, V., Giganti, M., et al. (2014). Interaction between climatic, environmental, and demographic factors on cholera outbreaks in Kenya. *Infectious Diseases of Poverty, 3*, 37. doi:10.1186/2049-9957-3-37.
28. Margot, W., & Parkes, L. B. (2005). All hands on deck: Transdisciplinary approaches to emerging infectious disease. *EcoHealth, 2*(4), 258–272. doi:10.1007/s10393-005-8387-y.
29. Daszak, P., Zambrana-Torrelio, C., Bogich, T., Fernandez, M., Epstein, J. H., Murray, K. A., et al. (2013). Interdisciplinary approaches to understanding disease emergence: The past, present, and future drivers of Nipah virus emergence. *PNAS, 110*(1), 3681–3688.
30. Aboelela, S. W., Larson, E., Bakken, S., Carrasquillo, O., Formicola, A., Glied, S. A., et al. (2007). Defining interdisciplinary research: Conclusions from a critical review of the literature. *Health Services Research, 42*(1), 329–346. doi:10.1111/j.1475-6773.2006.00621.x.
31. Gonzalez, J.-P., Lambert, G., Legand, A., & Debré, P. (2011). Toward a transdisciplinary understanding and a global control of emerging infectious diseases. *Journal of Infection in Developing Countries, 5*(12), 903–905.
32. Ackoff, R. (1999). *Recreating the corporation*. New York: Oxford University Press.
33. McNall, M. A., Barnes-Najor, J. V., Brown, R. E., Doberneck, D., & Fitzgerald, H. E. (2015). Systemic engagement: Universities as partners in systemic approaches to community change. *Journal of Higher Education Outreach and Engagement, 19*(1), 1–26.
34. Engestrom, Y., Engestrom, R., & Karkkainen, M. (1995). Polycontextuality and boundary crossing in expert cognition: Learning and problem solving in complex work activities. *Learning and Intervention, 5*, 319–336.
35. Fook, J., d'Avray, L., Norrie, C., Psoinos, M., Lamb, B., & Ross, F. (2013). Taking the long view: Exploring the development of interprofessional education. *Journal of interprofessional Care, 27*(4), 286–291.

36. Adams, K. M., & Hester, P. T. (2013). Accounting for errors when using systems approaches. *Procedia Computer Science, 20*, 318–324.
37. Axelrod, R. (1976). *Structure of decision: The cognitive maps of political elites*. Princeton, NJ: Princeton University Press.
38. Kosko, B. (1986). Fuzzy cognitive maps. *International Journal of Man-Machine Studies, 24*, 65–75.

On the Perception of Complexity and Its Implications

J.E. Manuse and Bogdan Sniezek

1 Executive Summary

The primary purpose of our system design community is to produce affordable and successful systems, regardless of scale and scope. To that end, we value the assurance of our design processes and the correctness and completeness of the resultant solutions. The fundamental problem with complexity is that it impedes the ability of our community to be successful. The complexity literature describes a strengthening correlation between the *measures of system complexity* and indications our community is not meeting its objectives. By historical standards, this dissatisfaction with our effectiveness seems to be a recent phenomenon.

The efforts to address the complexity problem focus around a few key areas:

Understanding complexity in order to improve the constructs and paradigms we use to describe the system so we can better predict its behavior;

- Managing complexity by focusing on improving different aspects of the process we apply to designing systems, including the application of risk/opportunity management methods, and augmentations to the underlying models;
- Selecting and training people to improve application of known "best practices."
- The literature implies a prevalent assumption that we are so unavoidably faced with the *burden of complexity* that the focus of our efforts should be towards sufficient understanding in support of *consequence management.*

This chapter takes a distinctive look at the problem of complexity in system design by considering the possibility that the system design community has been struggling with the consequences of a *deep-rooted assumption* concerning our core definition of a system. The chapter offers an alternative treatment of complexity in

J.E. Manuse, Ph.D. (✉) • B. Sniezek
The Charles Stark Draper Laboratory, Inc., Cambridge, MA, USA
e-mail: jmanuse@draper.com; bsniezek@draper.com

F.-J. Kahlen et al. (eds.), *Transdisciplinary Perspectives on Complex Systems*,
DOI 10.1007/978-3-319-38756-7_9

system design by extracting our community's viewpoint—the mental construction we perceive the world *through*—such that we are able to look *at it* less subjectively. In the process, the chapter lays out how the tendency to center system design on the core principle that a system is *a collection of interconnected parts that interact to produce some behavior* might manifest the historical hallmarks of an ungrounded assumption.

The chapter frames a normative comparative study intended to assess the ability of any viewpoint to successfully facilitate system design. The evaluation criteria are derived from a set of definitions expressed in a minimally sufficient patterned form conducive to generalized comparison. These evaluation criteria seem consistent with the values of the system design community: the assurance of our design processes and the correctness and completeness of the resultant solution to resolve the need. To establish the basis of similarity for the comparison, an abstract design viewpoint pattern is defined such that it captures the principles, beliefs and facilities that are necessary to support successful system design. The normative viewpoint is generated by applying our *deep-rooted assumption* to the abstract viewpoint pattern and querying the prevailing system design literature to fill in any open questions. This prevailing viewpoint is then applied to the minimally sufficient patterned forms in order to derive the general patterns of the design process and its solution as a consequence of the prevailing viewpoint. These general patterned forms appear to be consistent with empirical observations of the current design processes and resultant systems. The study comparative is similarly defined and derived around an alternative view of systems that intentionally contradicts our *deep-rooted assumption*. Finally, the prevailing and alternative viewpoints and their general patterned forms are compared against the evaluation criteria to assess their overall quality and analyze the conditions for their sufficiency.

The results show that the prevailing viewpoint becomes increasingly unsatisfactory *as systems exceed the capability and capacity of the practitioner to comprehend*, and therefore **brings into question the validity of our *deep-rooted assumption* for the design of complex systems**. The analysis further suggests that the system design community has been actively compensating *for the insufficiency of its models*. On the other hand, the alternative viewpoint demonstrates theoretical agreement with the evaluation criteria and shows the potential for systematic solution derivation, suggesting that the *alternate definition of systems might form a more suitable basis for the design of "complex" systems*. Unexpectedly, the analysis also uncovers the probable existence of at least two other viewpoints. *The four viewpoints appear to work collaboratively to affect adaptive change within and among systems*.

The system design community appears to be designing systems from within a single viewpoint that *views systems as constructions of things*. While this viewpoint has been successfully employed over thousands of years, it has recently exhibited the symptoms of exceeding its limitations: active *compensations for its insufficiency* that encourage the ***perception of complexity as significant to design***.

2 Chapter Roadmap

- **Introduction**: On Complexity and the Deep-Rooted Assumption
  - **Background and Motivation**: On Complexity in System Design
  - **Hypothesis**: On the Change-Prevention System and the "Big Assumption"
  - **Methodology:** On a Comparative Study Approach to Assess Validity
- **Framework**: Framework for the Normative Comparative Study
  - **Defining the Evaluation Criteria**: System Design Facilitation Criteria
  - **Establishing a Basis of Similarity**: Abstract Design Viewpoint Pattern
  - **Defining the Normative Viewpoint:** The Prevailing Viewpoint
  - **Defining the Comparative Viewpoint:** The Alternative Viewpoint
- **Comparative Study:** Comparative Study of Design Facilitation Alternatives
  - **Analysis and Results:** Comparative Analysis of the Forms and Qualities
  - **Evaluation of Hypothesis**: Assessment of Validity of the Assumption
- **Explanation of Results:** Suppositions on the Evolution of Misalignment
- **Conclusions:** The Path Forward

3 On Complexity and the Deep-Rooted Assumption

> Your assumptions are your windows on the world. Scrub them off every once in awhile, or the light won't come in.
>
> —Isaac Asimov

In systems literature, complexity is a term used to describe something with many parts *such that those parts interact with one another in multiple, often unexpected ways* [1, 2]. Complexity is viewed as both a property of systems and a metric for the amount of effort and resources that will be required for their design [3, 4]. A brief exploration of the term in the American Heritage dictionary brings in slight nuances such as "complicated" and "intricate" [5], which inject a notion of an inherent, painstaking effort necessary to solve or comprehend the parts of a whole [5]

In its simplest form, "complexity" merely refers to something with more than one part. The human experience begins to augment and alter its meaning. For instance, complexity in the current vernacular is often a surrogate for "it's difficult and I don't sufficiently understand it." When complexity is used in normal conversation, it is rarely meant in a positive connotation.

Perhaps the systems community has been using the term "complexity" in system design to represent something similar? Is it possible that unsatisfactory experiences working with systems containing many interacting parts has created and sustained a negative association with the word "complexity," perhaps encouraging us to *view it as a measure of something significant*?

3.1 On Complexity in System Design

> Making the simple complicated is commonplace; making the complicated simple—awesomely simple—that's creativity.
> —Charles Mingus, influential American Jazz musician

The primary purpose of our system design community is to produce affordable and successful systems, regardless of scale and scope. To that end, the community values the assurance of our design processes and the correctness and completeness of the resultant solutions. However, the literature is replete with examples of issues the community has encountered designing, building and operating systems as the complexity of those systems increases. The challenges are so great that new fields of study have been spawned—*complexity science* and *complex systems engineering*, for instance—in an effort to understand and manage the growing concerns.

The fundamental problem with complexity is that it impedes the ability of our system design community to be successful [6].[1] The greater the complexity, the larger the obstacle we must overcome.

These trends emerge as illustrated representatively in Fig. 1: An increase in complexity correlates with *increases in resource expenditure* and with *decreases in perceived project success* [7, 8]. The notion of what constitutes project success is subjective, spanning the range of timeliness, cost and requirement satisfaction considerations. For the purposes of this chapter, a successful system is one that sufficiently resolves a specified need, while resource expenditure keeps focus on cost, schedule and other design-related burdens.

Few studies have attempted to place hard numbers on these trends. In the absence of better information, it seems reasonable to assume the scale, slope and curve characteristics of the representative trends in Fig. 1 can vary between industries and system type. In any case, the overall sense of the situation should be clear: if the trend continues, the community faces an increasingly existential challenge with its sustainability.

Some industries have demonstrated a level of success with mitigating their complexity. The integrated circuit industry, for instance, has historically applied numerous abstractions, tools and methodologies to improve their design productivity in the face of increasing complexity. Despite the advantages of working with single-physics models, even this industry is beginning to struggle with the scaling issues of very large scale integration (VLSI) systems [9, 10].

In essence, the complexity literature describes a strengthening correlation between the *measures of system complexity* and indications our community is not meetings its objectives. By historical standards, this dissatisfaction with our effectiveness seems to be a recent phenomenon. It appears the systems we are trying to design have metaphorically applied a brake to our ability to design them.

[1]Calvano and John assert that "complexity poses a major challenge to our ability to achieve successful systems."

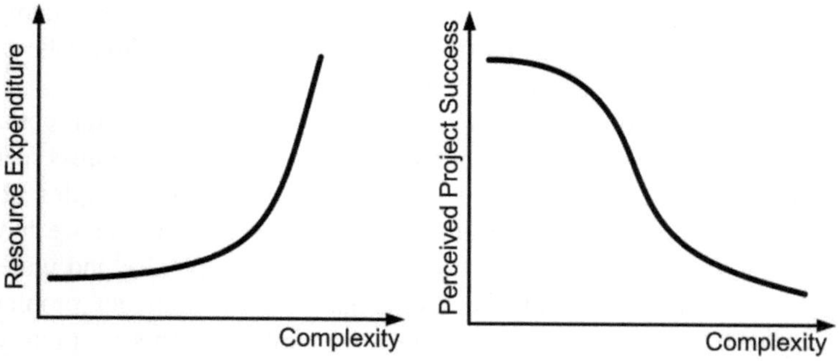

Fig. 1 Representative trends of (**a**) increasing resource expenditure and (**b**) decreasing project success (as perceived), correlated with increasing complexity

Perhaps this growing sense of resistance to success suggests the system design community should ask itself a fundamental question: are these circumstances truly how things are, or is there another explanation—perhaps one with a potential resolution?

The current efforts to address the complexity problem—as demonstrated by the available literature—are focused on understanding and managing complexity. A large portion of the research emphasizes the need to understand complexity in order to manage its effects, such as emergence and uncertainty, which create *tractability* and *predictability* issues. The general idea seems to be that by improving the constructs and paradigms we use to describe the system, we can better predict its behavior. Numerous attempts have been made at defining [11, 12], comprehending [13] and explaining [14] emergence, and yet, no theory or method seems to resonate and persist. Meanwhile, Stepney, Pollack and Turner have called for studying how to safely direct emergence to our own ends [15].

Other areas of research emphasize complexity management by focusing on improving different characteristics of the process we apply to designing systems and their underlying models. For instance, Fricke and Shulz argue for improving processes and architectures to support easier changes to the system so we can better handle rapid changes in operational context [16]. De Neufville focuses on taking an active approach to risk and opportunity management by developing principles, architectures and processes for managing the uncertainty in systems [17]. Hastings et al. consider quantifying and modeling uncertainty in the conceptual design of space architectures [18]. De Weck and Jones offer an alternative approach that visualizes performance tradeoffs using deterministic contours or stochastic "isoperformance" approach curves to support identifying acceptable solutions [19].

Some members of the community choose to focus their efforts on selecting and training people to use the best practices better. Numerous consultancy and training companies, including Project Performance International and the Applied Technology Institute, have emerged to train practitioners on how to apply the best practices more effectively [20, 21].

The literature implies a prevalent assumption that we are so unavoidably faced with the *burden of complexity* that the focus of our efforts should be towards sufficient understanding in support of *consequence management*.

Given the current trajectory, it would seem that the system design community would greatly benefit from a breakthrough [22]. One approach to address seemingly insurmountable problems is to step back and return to first principles [23]. An opening caution: correlation is not equivalent to causation. While we have perceived a strong correlation between our measures of complexity and undesirable outcomes, the correlation should not immediately imply that our problems are solvable by addressing the *technical* issues of complexity. This cautionary statement is not intended to imply that the community has confused correlation with causation; rather, it is meant to motivate introspection on whether the community has inadvertently chosen to treat a symptom (complexity) rather than to seek out the root cause (to be determined).

Therefore, this chapter postulates that the burden imposed by complexity—particularly its resultant resistance to success—is a symptom of a hidden malady; one that might be readily treatable once recognized.

To begin the exploration of what this hidden malady might represent, consider the historical record of people struggling to address the *technical* issues of apparent complexity—believing and accepting that there was nothing they could do to change the underlying reality, using their observations to confirm their beliefs and making accommodations to sustain the consequences—only to suddenly break through the barriers to their success.

- **The "Complex" Art of Alchemy**: The complex art of alchemy—based on an assumption of mythology and supported by ritualistic *practices* was rejected in Boyle's revolutionary treatise "The Sceptical Chymist [24]," marking the beginning of the transition to the systematic, modern science of chemistry. Boyle applied the empirical methods that had been recently developed by Sir Francis Bacon. The apparent complexity of alchemy ended up being due to an inaccurate comprehension of the world.
- **"Complex" and Unpredictable Chemical Elements:** The known set of elements appeared to be complex, incomplete and *unpredictable*, with no apparent pattern among them (assumption). By identifying and focusing on the underlying significance of the elements, Mendeleev was able to create an organizational construct (the periodic table) that greatly clarified the behaviors of the elements and enabled their prediction [25]. The apparent complexity of the elements ended up being due to an insufficiently complete comprehension of the world.
- **Filtering "Complex" and Dynamic Signals**: For many years, filtering complex, dynamic signals in the *space-time domain* (assumption) was a difficult and sometimes *intractable* problem. By extending the Fourier series that expresses a signal in terms of each of its constituent *frequencies* and identifying the conditions for a one-to-one correspondence between a time function (signal) and its transform in the frequency domain, Fourier enabled a readily tractable resolution to a seemingly "complex" problem [26]. The apparent complexity of

the dynamic signals ended up being due to an insufficiently complete comprehension of the world.

In these historical examples, the underlying systems never changed. The struggles experienced by the people of the time were not caused by the inherent complexity of the systems they were working with. Instead, their struggles were symptomatic of *ungrounded assumptions* based on an *inaccurate or insufficiently complete comprehension of the world*. It was the dramatic shift in their comprehension that gave them the clarity they needed to overcome their self-imposed barrier to success and thrive.

Therefore, this chapter will approach the problem of complexity from a new perspective: that our struggle with complexity may be symptomatic of an ungrounded assumption based on an *inaccurate or insufficiently complete comprehension of systems*.

3.2 *On the Change-Prevention System and the Deep-Rooted "Big Assumption" of the System Design Community*

The most reliable route to ultimately disrupting the immune system begins by identifying the core assumptions that sustain it.
—Kegan and Lahey, *Immunity to Change* [27]

The system design community is highly motivated to improve outcomes and has invested significant capital and manpower towards addressing the *technical* challenges associated with complexity and improving our understanding of the nature of the underlying systems. Given the extent of this commitment, if our community is indeed struggling with the symptoms of inaccurate or insufficiently complete comprehension, what has been preventing a big breakthrough?

3.2.1 On the Change-Prevention System

In their groundbreaking book, "Immunity to Change," Harvard Professor Robert Kegan and his colleague Miriam Lahey propose the existence of a psychological "immune" system relevant to human interactions and organizations [27]. This other "immune" system—a change-prevention system—acts by producing resistive behaviors to help us manage anxiety and prevent our worst fears from taking place. However, this same mechanism produces behaviors that can impede our ability to meet improvement goals. While Kegan and Lahey focus on an individual's resistance to change, this chapter postulates that similar protective systems can also exist within larger, social organizations of people.

Kegan and Lahey note instance after instance in which the individuals they worked with were *fully unaware* of the existence of these internal protective systems. These individuals were completely invested in meeting their improvement goals and yet they failed time and again. The solutions to the problems confronting these individuals were ***not technical*** in nature (e.g., they could lose weight by exercising more), but *adaptive* (e.g., they had to recognize that their need to eat more was a *compensation* for their lack of confidence).

According to Kegan and Lahey, "big assumptions" get in the way of a person's ability to adapt appropriately to new situations. Sometimes, we do not realize the manner in which we comprehend ourselves, the world and our sense of self in relationship to the world is merely a "mental construction." While we might view all of these comprehensions as accurate representations of the way things are—perhaps even as solidly as truths and facts—they may *not* be. Sometimes, these assumptions exist despite countervailing evidence, creating a distortion or blind spot. All mindsets suffer from blind spots.

To remove these impediments to success, each individual mapped out their immunity to change and identified the core assumptions that sustained it. By altering their perception of the assumption—its truth or fallacy—the individual could begin to adapt from the inside out. In other words, by adjusting their viewpoint to improve alignment with reality—a clarification—the individuals naturally adjusted their behaviors to match.

Conjecture: What if there is a similar change-prevention system acting within the system design community? What if despite everyone's best intention to address the issues of complexity and adapt as needed, there are built-in resistances, perceptions and behaviors that are actively preventing us from meeting our goals?

As supported by the complexity and system design literature, our understanding of systems and their inherent complexity seems to be built on a common, widely accepted mental construction—a form of *viewpoint*. This prevailing viewpoint has supported and sustained the belief that the *burden of complexity reflects the reality of the circumstances*. The undesirable consequences of complexity in system design are a relatively recent phenomenon, suggesting that this viewpoint has been a sufficiently valid mindset for a long time.

While we can easily see *through* our own viewpoint—often without recognition that we have one—to understand its limitations and biases, we have to be able to *objectively examine* it [27]. And when we observe evidence that our viewpoint has exceeded its limitations, such as a *consistent demonstration of increasing ineffectiveness*, it becomes even more critical to objectively examine it. Perhaps the increasing burden of complexity is a consequence of *not fully appreciating the limits and constraints of our current viewpoint*. Perhaps what we are experiencing is evidence of a growing misalignment between what we need our viewpoint to be and what it is.

3.2.2 Historical Example: The Geocentric Solar System Model

To understand what can happen when a viewpoint becomes misaligned with the larger world, consider a well-known historical example from astronomy: the Ptolemaic model of celestial motion.

As people could only make observations from the surface of the earth, they naturally favored an earth-centric viewpoint in their quest to understand the natural world. When it became useful to predict the position of certain celestial bodies, they made models. A standard example of these early mathematical representations is the Ptolemaic model, which rejected Aristarchus' version of heliocentrism in favor of a geocentric theory that described the motion of the celestial bodies as a combination of eccentric circles, the flexible and extensible Apollonian "deferents and epicycles" and an equant point [28]. Ptolemy's model provided "the first *comprehensive and systematic quantitative* account of celestial motions," and for many years it demonstrated itself useful for accurate time-keeping [28]. New observations were accounted for by adjusting the basic components of the model to fit the observed data.

By the time the model began showing signs of distress—over a thousand years after the time of Ptolemy—the theory was "as entrenched as a scientific paradigm could possibly be [28]." Despite the countervailing evidence indicating an increasingly questionable model, it was accepted and vigorously defended as an accurate representation.

The earliest indicators of a systematic problem must have been easy to ignore. Any issues with the *tractability* of the computations associated with "adding epicycles" to accommodate the improving observations would have been compensated for with a simple increase in manpower. By the sixteenth century, however, the models began to demonstrate increasingly serious inaccuracies [28]. In general, these difficulties with *predictability* were seemingly accepted as the *reality of the circumstance*. Eventually, the deviations grew so severe that people began to question the validity of the model. A storied line of inquiry into the validity of the model began with Nicholaus Copernicus, who initially intended to rectify the inaccuracies; he ended up questioning the internal consistency and coherence of the theory as a whole [28].

As we eventually confirmed with the theory of Newtonian mechanics, our society's viewpoint contained a deep-rooted "*big assumption.*" It was assumed that the Earth is at the orbital center of all the celestial bodies. This "big assumption" drove the geocentric solution to addressing the problem of celestial body motion prediction and sustained our society's "immunity to change" in spite of the growing countervailing evidence. The solution began in alignment with the intent to make predictions of celestial motion and it performed sufficiently well for a long time. However, the viewpoint had a

(continued)

fundamental misalignment with the nature of the solar system; namely, that the sun—not the Earth—is at the orbital center.[2] *Over time, this misalignment led astronomers to add more and more epicycles to compensate for the discrepancies between the model and their observations until the compensations could no longer be justified.*

3.2.3 Synthesis: The System Design Community's Deep-Rooted "Big Assumption"

Given the historical lessons and the general criteria that a "big assumption" is a mental construction of the world that must be capable of sustaining an immunity to change, consider the potential candidates for the system design community's deep-rooted assumption. Recall that the prevailing viewpoint—as indicated in the complexity and system design literature—supports the belief that the *burden of complexity reflects the reality of the circumstances*. This assertion is precisely the kind of assumption that sustains the immunity to change because it states that this perceived reality "is just how things are, and there is nothing that can be done about it."

What does the *burden of complexity* represent? In a sense, a burden describes anything that must be supported that does not directly support the objective. We might expect (and have observed) the design burden imposed by complexity to manifest along the lines of:

- The resource expenditure applied to predict the behavior of an increasingly massive number of interconnecting parts in order to better inform design decisions;
- The compensations built into the design intended to control, manage or otherwise mitigate the consequences of unexpected or undesirable behaviors that seem to scale with the number of interconnections.

What tenet in our mental model of systems and their design might sustain the belief that these compensatory burdens reflect the reality of the circumstance? Is it possible that our core definition of a system sustains this belief? If a system is defined around the notion that it represents *the collection of interconnected parts that interact to produce some behavior*, then the scale of interconnected parts

[2]Naturally, the heliocentric viewpoint has since expanded with our improved understanding of gravity and the effects of extra-solar celestial bodies on the orbital motions within our solar system.

becomes significant and behavior can only known as a consequence of the composition of those parts.

For these reasons, this chapter hypothesizes that the system design community's deep-rooted "big assumption" is the core definition of a system we apply to design, generally characterized as: **the collection of interconnected parts that interact to produce some behavior.**

The system design community may insist: "But we can observe and measure the parts and their behavior. We can observe their interactions. We observe systems to have these characteristics." There is no denying we can observe parts, their connections and their resultant behavior, and we can measure these things, thus making it significant to us, the technology maker. Yet, when the system design community asserts that we *must manage complexity*, this perceived significance might be *pursued to the detriment of our community*.

> What gets measured gets managed—even when it's pointless to measure and manage it, and even if it harms the purpose of the organization to do so.
>
> —Peter Drucker, major contributor to the philosophical and practical foundations of modern business [29]

The hypothesis does not contend the parts and their behavior cannot be measured or observed. *The question is whether we are applying a definition of a system that is valid when facilitating its design.*

After all, the *tractability* and *predictability* issues expressed in the complexity literature appear to meet the historical criteria for an ungrounded assumption.

Perhaps any discomfort with the proposed deep-rooted "big assumption" harkens back to the "whole before its parts" argument that originated in Aristotle's debate with Zeno. What is the system—is it the "whole before its parts" or do the "parts give rise to the whole?" *Why is there an implicit assumption in the framing of the question that the answer must be either/or?*

We often argue for a holistic view of systems and yet we seem caught within a *parts-centric* viewpoint [30, 31]. Is it because we struggle to work with and comprehend the abstraction of the whole? Is it possible we looked *through* a viewpoint that was *already grounded in the parts* and chose accordingly? Perhaps we could not comprehend the whole, perceiving it lacked the necessary "dynamicism of the parts" and implied "a pre-given coherent entity [13]." Perhaps we simply assumed the interconnected parts that interacted to produce some behavior were what were of significance to the design of systems?

Perhaps the proper answer to the Aristotelian question is that *a system is both* the "whole before its parts" and "the parts give rise to the whole," depending on the intent of the viewpoint. Consider the optical illusion illustrated in Fig. 2. There is

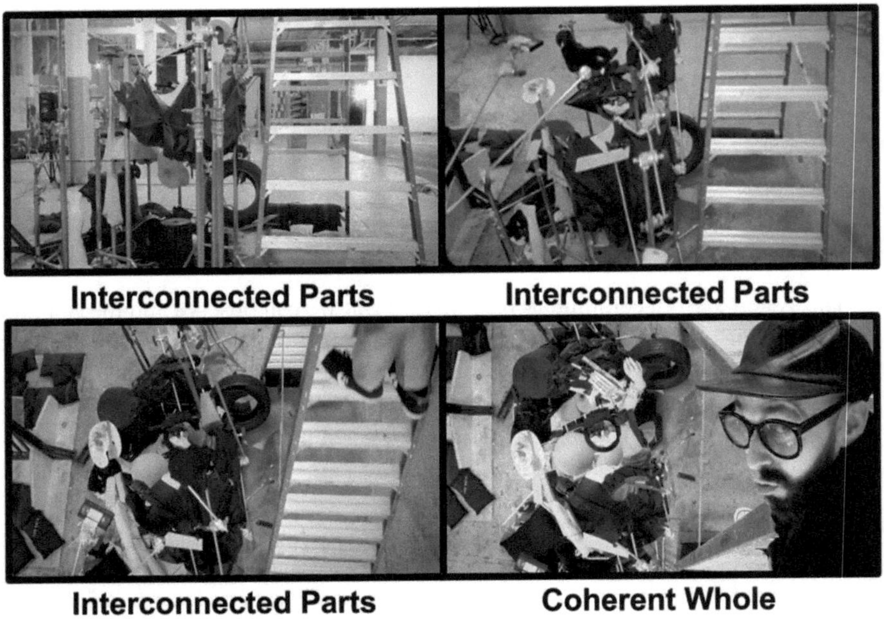

Fig. 2 Viewpoint illusion (Images used with permission. OK Go [32])

only one vantage point (at the peak of the ladder)[3] that provides awareness of the satisfaction of the *construction of interconnected parts* as a *coherent whole* (the image of a face). Any misalignment of viewpoint and the result is the perception of an unrecognizable collection of parts connected together in some seemingly arbitrary arrangement—rather than the intended image. When might we want to take the viewpoint of the coherent whole versus that of the interconnected parts?

3.3 On a Comparative Study Approach to Assess the Validity of the Deep-Rooted Assumption

> We have been focusing on improving parts of the system rather than focusing on the system as a whole. As a result, we have been improving the parts but not the whole. We have got to restart by focusing on designing the whole and then designing parts that fit it rather than vice versa.
>
> —VP of AT&T, in 1951, as related by Russ Ackoff in Idealized Design [30]

This chapter has proposed the existence of a deep-rooted "big assumption" in the system design community. This assumption meets the criteria for sustaining the community's immunity to change ***and*** may be connected to the observed

[3]Does this suggest the importance of being cautious whose vantage point we construct around?

manifestation of *tractability* and *predictability* issues (qualities consistent with the historical hallmarks of an ungrounded assumption). To what extent can the validity of our deep-rooted assumption be substantiated? In other words, to what extent can this assumption demonstrate that it is well-grounded and produces the desired results? What might the results mean for the systems community?

3.3.1 A Normative Comparative Study Approach

Tools can be assessed for their utility and applicability to accomplish some purpose. Based on this assessment, we can choose the tool that best resolves our needs. Perhaps we can apply similar methods to assess our *cognitive tools*; in a sense, viewpoints are *cognitive tools* we apply to comprehend our world.

One approach for assessing tools or products is a comparative study. A comparative study is also a valid method for "the early stages of the development of a branch of science [33]." The study compares two or more objects—in this case, viewpoints—that are different in some respects, but share similarities that support the evaluation of the differences. A normative comparison is a special form of the method that applies evaluation criteria like "satisfaction' in order to assess alternatives. This approach is effective at identifying cases that agree with the given criteria [33].

In order to assess the validity of the deep-rooted assumption, this chapter frames a *normative comparative study* intended to assess the utility and applicability of any viewpoint *to successfully facilitate system design*.

- **The Evaluation Criteria**. The evaluation criteria are derived from a proposed set of minimally sufficient definitions significant to system design. First, the intent of system design is offered in a *minimally sufficient form conducive for generalized comparison*. The minimally sufficient patterns of the design process and its resultant solution are derived from this statement of intent. Based on these patterns, success criteria are identified that describe the values and concerns that must be fully satisfied in order to meet the intent.
- **The Basis of Similarity.** To establish the basis of similarity for the comparison, an abstract viewpoint pattern is defined such that it captures the principles, beliefs and facilities that are necessary to support successful system design. Therefore, this design viewpoint pattern is directed toward meeting the intent of system design.
- **The Normative Viewpoint and its Facilitation of System Design.** The ***prevailing*** viewpoint is generated by *applying our deep-rooted assumption* to the abstract design viewpoint pattern and querying the prevailing system design literature to fill in any open questions. This prevailing viewpoint is then applied to the design process and resultant solution patterns in order to derive the general forms they *must take* as a direct consequence of the viewpoint and its core tenets. These derived general forms enable the evaluation of the prevailing viewpoint against the evaluation criteria.

- **The Comparative Viewpoint and its Facilitation of System Design.** The study comparative is similarly defined and derived, *except that the basis of this* ***alternative*** *viewpoint is an alternative view of system that intentionally contradicts our deep-rooted assumption.*

Finally, the prevailing and alternative general forms of the design process and resultant solution are compared against the evaluation criteria. The analysis assesses the overall quality of the viewpoints, identifying the conditions for their sufficiency and insufficiency to meet the intent of system design. In doing so, the analysis highlights the biases, limitations, dependencies and capabilities of each viewpoint.

3.3.2 On the Consequences of the Assessment of the Deep-Rooted Assumption

What does it mean for the system design community *if the prevailing viewpoint demonstrates agreement* with the evaluation criteria? This result will imply that our core definition of a system successfully supports the intent of system design: it will have shown itself to be sufficiently well-grounded to produce the desired results. Therefore, the results will have substantiated the assumption that the burden of complexity is just how it is. From this, we can infer that the best we can do is to continue our efforts to understand complexity and improve our management of its consequences. Since the evidence around us suggests that our comprehension of systems remains incomplete, it might behoove the community to seek an organizational construct for systems analogous to the periodic table of elements.

What does it mean for the system design community *if the prevailing viewpoint does not satisfactorily agree* with the evaluation criteria under certain conditions? This result will imply that our core definition of a system is *of questionable validity* when facilitating system design under those conditions. The burden of complexity may be the consequence of protective behaviors that have actively prevented us from accurately and sufficiently comprehending the systems we work with. As the system design community alters their perception of how to view systems, they can adapt from the inside out and adopt behaviors that improve alignment with what is needed to successfully facilitate system design under the aforementioned conditions. Therefore, the potential exists to release the system design community from the burden of complexity.

What does it mean for the system design community *if the alternative viewpoint demonstrates more agreement* with the evaluation criteria? This result will imply that the definition of a system proposed to form the study comparative may be *a more suitable basis for our comprehension of "complex" systems*. In this scenario, not only is there a potential to release the system design community from the burden of complexity, *a potential path to its resolution will have been framed and characterized to guide future exploration.*

4 Framework for the Normative Comparative Study to Assess the Validity of the Deep-Rooted Assumption

Science is nothing but perception.
—Plato

The framework for the comparative study consists of four parts:

- **The Evaluation Criteria.** This part offers the intent of system design in a *minimally sufficient form conducive for generalized comparison*, derives the minimally sufficient patterns of the design process and its resultant solution, and identifies their success criteria.
- **The Basis of Similarity.** This part defines a viewpoint query directed toward meeting the intent of system design.
- **The Normative Viewpoint and its Facilitation of System Design.** This part describes the prevailing viewpoint based on our deep-rooted assumption and applies this viewpoint to derive its general forms for comparison to the evaluation criteria.
- **The Comparative Viewpoint and its Facilitation of System Design.** This part describes the alternative viewpoint based a proposed alternative view of systems, and applies this viewpoint to derive its general forms for comparison to the evaluation criteria.

4.1 The Evaluation Criteria

This section establishes the evaluation criteria for the normative comparison study. First, the intent of system design is offered in a minimally sufficient form conducive for generalized comparison. The minimally sufficient patterns of the design process and its resultant solution are then derived from this general form of system design, enabling identification of the criteria for their success.

4.1.1 Generalizing the Intent of Systems Design

As stated in the introductory section, the purpose of the system design community is to produce affordable and successful systems regardless of scale and scope. If a well-formed need encompasses success criteria, a sense of the necessary and sufficient scale and scope of the solution to resolve that need, and a notion of the acceptable resource expenditure to provision that solution, then the minimally sufficient general form of the intent of systems design can be described as *to provision a solution to resolve a need*. The addition of any other qualifiers would arguably bias the statement and reduce its generality.

4.1.2 Understanding Facilitation to Meet the Intent of Systems Design

The statement *to provision a solution to resolve a need* suggests three concepts are required to facilitate its satisfaction: an expression of need, an approach to provision a solution to resolve the need, and the resultant solution.

A need has been asserted to encompass success criteria, a sense of the necessary and sufficient scale and scope of the solution to resolve that need, and a notion of the acceptable resource expenditure to provision that solution. Each of these elements seems to describe some aspect of what is required to sufficiently resolve some dissatisfaction. Thus, the minimally sufficient general form of need might be stated as *a dissatisfaction that must be sufficiently resolved.* Consider that even if a designer speculates on the benefits of building a system opportunistically, it is because of the perception that the future system will be insufficient or unsatisfactory in some manner. It is strongly implied that the need is pre-given to any design viewpoint, and that *the need represents the ultimate significance to the design of the system*: it is the driver for producing that system in the first place.

On the other hand, the form of the approach to provision a solution to resolve a need and the quality of the resultant solution it provides are fully within the purview of the facilitator of satisfaction: the designer.

4.1.3 Deriving the Minimally Sufficient Form of Design Approach and Its Success Criteria

The minimally sufficient general form of an approach might be described as *a process of facilitation to meet intent.* If the core intent of system design is to provision a solution to resolve a need, then a design approach may be described as *a process of facilitation to provision a solution to resolve the need.*

How might this process of facilitation be assessed? In a sense, the process represents a mechanism to close a gap between what is expected (the need) and what is (the current solution to the need). Perhaps it is reasonable to value how well the process of facilitation *assures* this gap closing. We might expect *assurance* to require the capability and capacity to assess the *fitness* of the solution to resolve the need. Otherwise, there is no mechanism to know how well the process of facilitation is being performed. What does this *fitness* represent? Perhaps it is reasonable to describe it as an *alignment* between the need and the solution such that the solution is relevant, significant and sufficient to satisfy the values, cares and concerns expressed by the need. The observation of sufficient *alignment* to some intent might be described as a *conformance*.

Therefore, we can expect any system design process to require the facility to assess the *fitness* of the solution to resolve the need. The success criterion is the *conformance* of the solution to resolve that need.

4.1.4 Deriving the Minimally Sufficient Form of Design Solution and It Success Criteria

The design process concludes successfully when there is *conformance* of the solution to resolve the need. What does it mean for the resultant design solution to be conformal to the resolution of the need? The optical illusion example in Fig. 2 provides guidance on the necessary qualities. In that example, a coherent whole only appears when the solution is sufficiently aligned with the intended vantage point. This phenomenon seems to occur because the need (to see an image from a particular vantage point) has an intrinsic cohesion with the solution that resolves the need (the construction of the parts). The example implies that to successfully design the solution, the facility to observe the coherent whole is required; otherwise there is no assured mechanism to know how well the need has been resolved.

For these reasons, the minimally sufficient form of the design solution appears to be that of *the coherent whole that conforms to the resolution of the need*. What does the conformal coherent whole represent? In order to arrive at a conformal coherent whole, the solution must be ***correct*** (it aligns with the interests and concerns of the need) and ***complete*** (it aligns with the scale and scope of the need).

Therefore, we can expect that the assessment of any resultant design solution requires the facility to observe the *coherent whole that conforms* to the resolution of the need. The success criteria are *correctness* and *completeness* of the solution to resolve the need.

4.2 The Basis of Similarity: A Viewpoint Pattern Directed Toward Meeting the Intent of System Design

To recap, a viewpoint can be usefully thought of as a cognitive tool for comprehension. A viewpoint describes a set of underlying principles and patterns that affect how we view and assess ourselves, the world and our sense of self in relationship to the world. It can include characteristics, values, concerns and beliefs. Viewpoints do not describe how—or in what manner—aspects like values or concerns are satisfied. They can, however, limit the language we use to describe the world around us [34].

Viewpoints are abstract entities; therefore, to capture its principles and patterns for understanding and analysis, it must be queried and its responses approximated. Fig. 3 illustrates the viewpoint query applied by this chapter to assess the facilitation of system design.

Since a viewpoint can be considered a cognitive *tool*, it is always in service to some end. The specific instance of viewpoint we apply can be in or out of alignment with this intent. As evidenced in the historical examples mentioned in the introduction of this chapter, a well-aligned viewpoint primes us for success to meet the

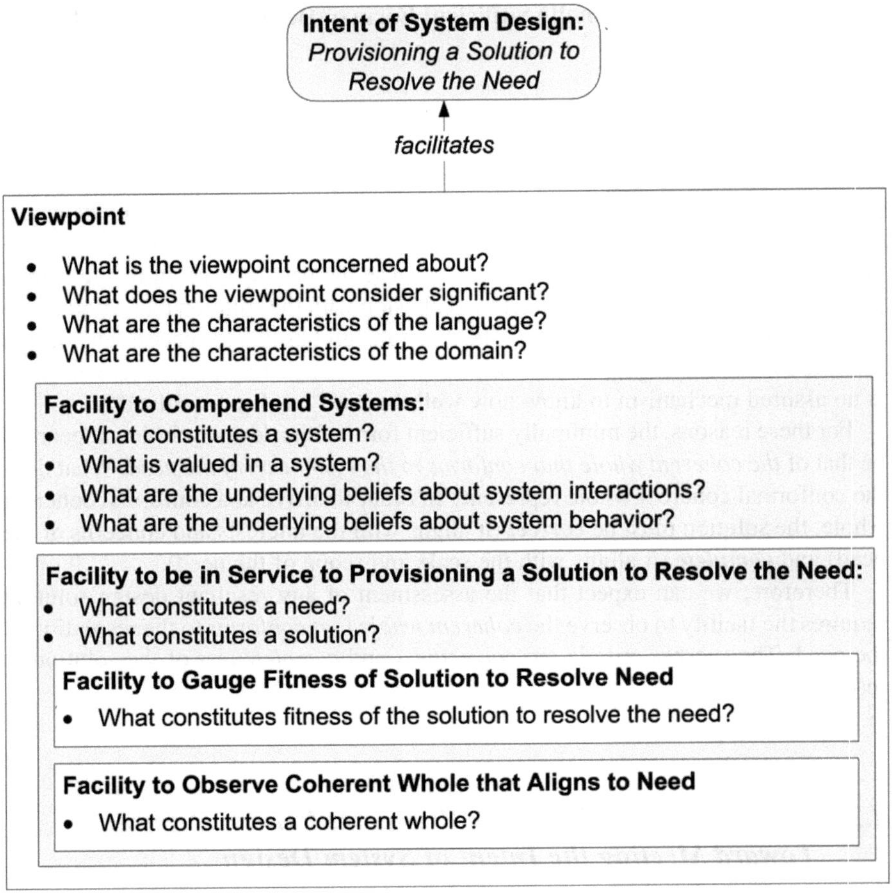

Fig. 3 A viewpoint pattern directed toward meeting the intent of system design

intent. However, if our viewpoint is out of alignment, we may struggle with the consequences of a lack of clarity.

For any given viewpoint to be aligned with its intent, it must demonstrate that it has the capability and capacity to meet the success criteria of that intent. Here, the intent of the viewpoint is to be in service *to provisioning a solution to resolve a need*. The two supporting elements are: the design process that acts to provision the solution such that its fitness to the resolution of the need is *assured*, and the design solution that must be provided as a coherent whole in order to resolve the need *correctly* and *completely*.

4.3 Defining and Applying the Prevailing Viewpoint

This section queries the prevailing viewpoint and provides brief responses that attempt to capture a general approximation of its underlying principles, values, and beliefs. The prevailing viewpoint facilitates provisioning a solution to resolve a need such that our deep-rooted assumption is one of its core tenets.

4.3.1 Querying the Prevailing Viewpoint

Fig. 4 captures the responses to the query given the prevailing viewpoint. These responses are by no means intended to represent all variations of the prevailing viewpoint; rather, they are intended to represent an observable pattern of thought within the world of systems thinking based on the literature thus far discussed.

4.3.2 On the Form of the Design Approach Within the Prevailing Viewpoint

Recall that the minimally sufficient form of a design approach is a *process to facilitate provisioning a solution to resolve a need*. In a sense, the design process represents a mechanism to close a gap between what is expected (the need) and what is (the current solution), and therefore requires the facility to assess the *fitness* of a solution to resolve the need. What is the form of the process that manifests when the prevailing viewpoint is applied to the minimally sufficient form?

> The prevailing viewpoint asserts the belief that systems form as a consequence of interactions of smaller, less complex systems and that their behavior is a consequence of the interacting parts. For these reasons, the only way to assess the fitness of a solution to resolve the need is *to construct a system and observe its behavior within some context*, comparing its characteristics and performance against some desired standard. *Any other approach would assert a contradiction to the underlying belief.*

Through the prevailing viewpoint, the need *for a system* can only be understood *as equivalent to the manner in which a thing is constructed* (i.e., "a construction is good if it looks like this and does these things"). Therefore, the expression of the perceived need (no matter its true form) must take the form of a ***need surrogate***: a base construction that is intended to represent the need. To assess the fitness of any solution to resolve the perceived need, some outcome of the construction process must be compared against this need surrogate. The result of this comparison informs the specification of the system to be built.

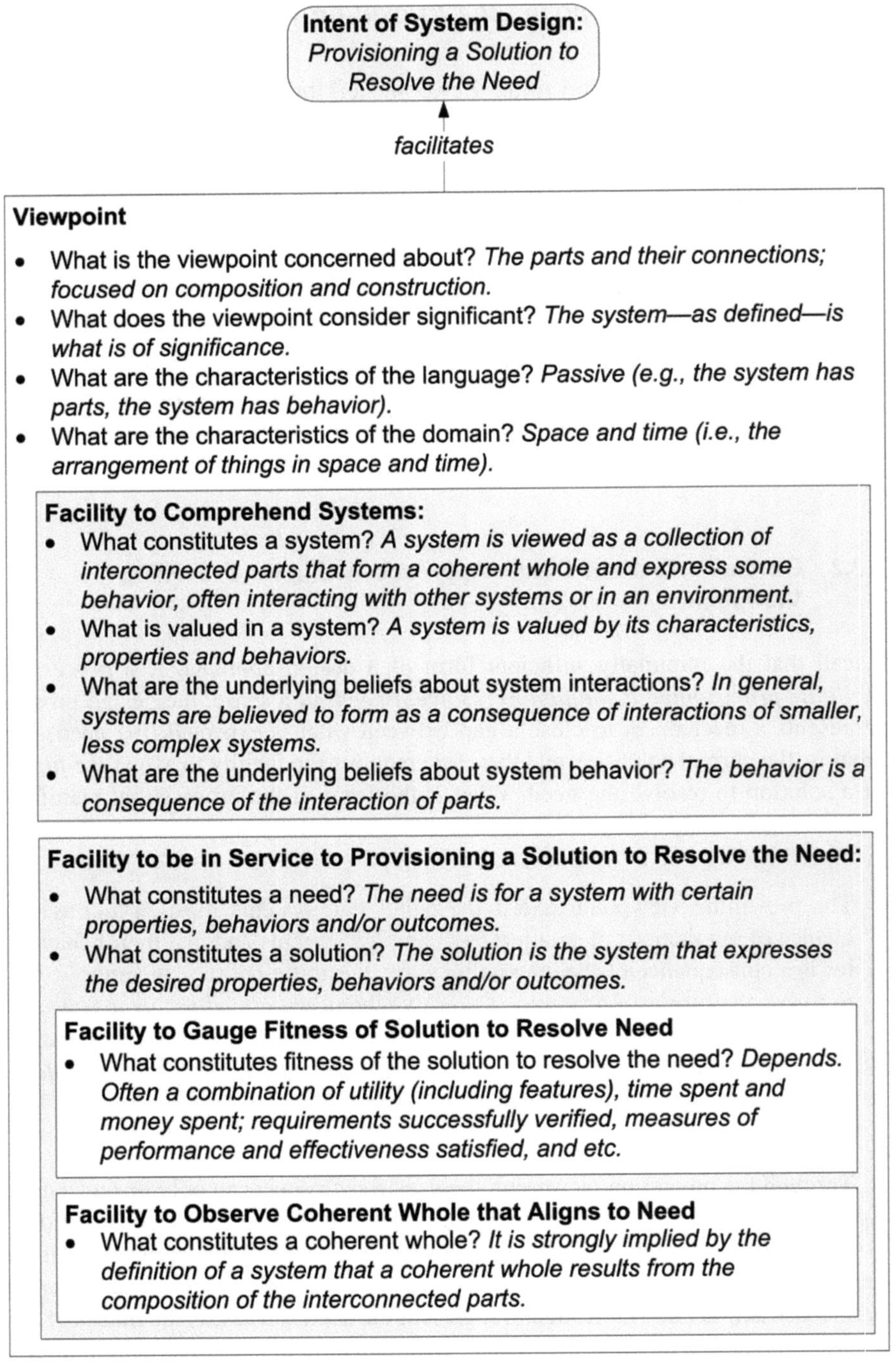

Fig. 4 The prevailing viewpoint query with responses [35, 36]

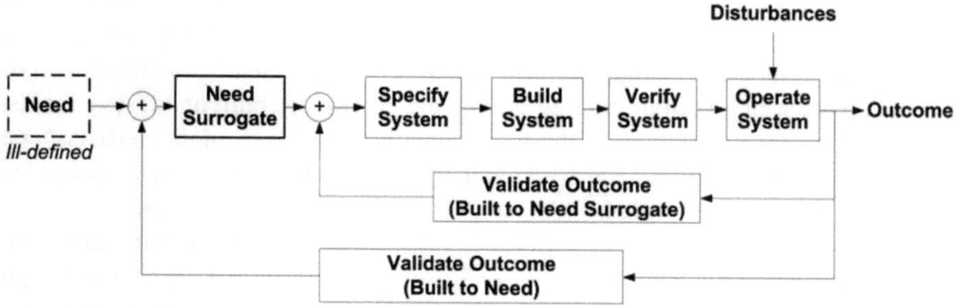

Fig. 5 The general form of the design approach that must be taken as a consequence of applying the prevailing viewpoint: a feedback control system around the need surrogate

Prior to the completion of the first construction cycle, the need surrogate has no system construction to compare against. Instead, the initial specification must be informed by experiences designing systems for some combination of need surrogates that appear most likely to match the overall surrogate. Thus, the specification process takes the form of experientially driven "pattern-matching" of similar systems to apparent needs.

Once the initial specification is completed, the system can be built and verified against the specification. Then, it can be placed in some form of operation in order to measure its outcomes to inform the next comparison against the need surrogate.

Thus, the generalized pattern for the design approach that forms as a direct consequence of applying the prevailing viewpoint is illustrated in Fig. 5.

From within the prevailing viewpoint, any mechanism that closes a gap between a need and a solution will take the *form of a feedback control system operating around a need* ***surrogate***.

An initial conceptualization exercise, an early first-step in system design, might consist of one or more *cognitive* construction feedback loops taking place within someone's head. A concept study might consist of *multiple on-paper or virtual* construction feedback loops through a *low-fidelity* model of the built system, varying inputs to observe outcomes. A trade study might consist of *many virtual* construction feedback loops through a *higher-fidelity* model of the built system, likewise varying inputs to observe outcomes.

In both cases, the models are checked to verify they meet the expectations for their initial specification, and are consistent with similar systems in terms of their behavior. This process is known as benchmarking. The models are then run within some form of simulation environment to assess their expected performance.

The concept or trade study can be used *to generate a need surrogate* in the form of an "optimized" set of design variables for a built system. In these cases, the seed for the need surrogate takes the form of an objective or utility function, or some similar representation of the expectations *for the system* (all lightweight forms of need surrogates). Trade studies and other forms of modeling and simulation can also be used to guide decision-making throughout the specification process.

A prototyping process might consist of one or more *physical* construction feedback loops in which various configurations of a system are built and tested against the need surrogate. Hardware-in-the-loop emulation processes might consist of one or more *physical* and *virtual* construction feedback loops in which various configurations of virtual models and physical manifestations are built and tested.

In all cases, knowledge and performance gaps are filled by instantiating additional construction feedback cycles of one form or another. In a sense, the design approach driven by the prevailing viewpoint is a process of successive approximations *around the construction of a system.*

4.3.3 On the Form of the Design Solution Within the Prevailing Viewpoint

The design solution within the prevailing viewpoint must take the form of the construction of parts connected to other parts, interacting with each other in proscribed ways in order to produce the desired behavior. This form appears to be true regardless of the fidelity of the instantiation.

4.4 Defining and Applying the Alternative Viewpoint

> Two roads diverged in a wood and I—I took the one less traveled by, and that has made all the difference.
>
> —Robert Frost, from "The Road Not Taken"

This section defines an alternative viewpoint to facilitate system design by providing brief responses to the viewpoint query that intentionally contradict our deep-rooted assumption.

4.4.1 Defining the Alternative Viewpoint

For our comparative study, an alternative viewpoint to facilitate system design must be defined such that it contradicts the assertion that a system is expressed in terms of the *collection of interconnected parts that interact to produce some behavior*. The "whole before the parts" argument suggests that in order to contradict this assertion, the viewpoint must be focused around the whole in an actionable way.

The chapter authors assert that to contradict the *parts-centric* view of systems, this alternative perspective must *take the viewpoint of a single, unified system.* To do so will require understanding what the *system* finds significant to its own design, and in forewarning, this exposition may appear both familiar and foreign. The *system-centric* responses defined for this comparative viewpoint are captured and summarized in Fig. 6.

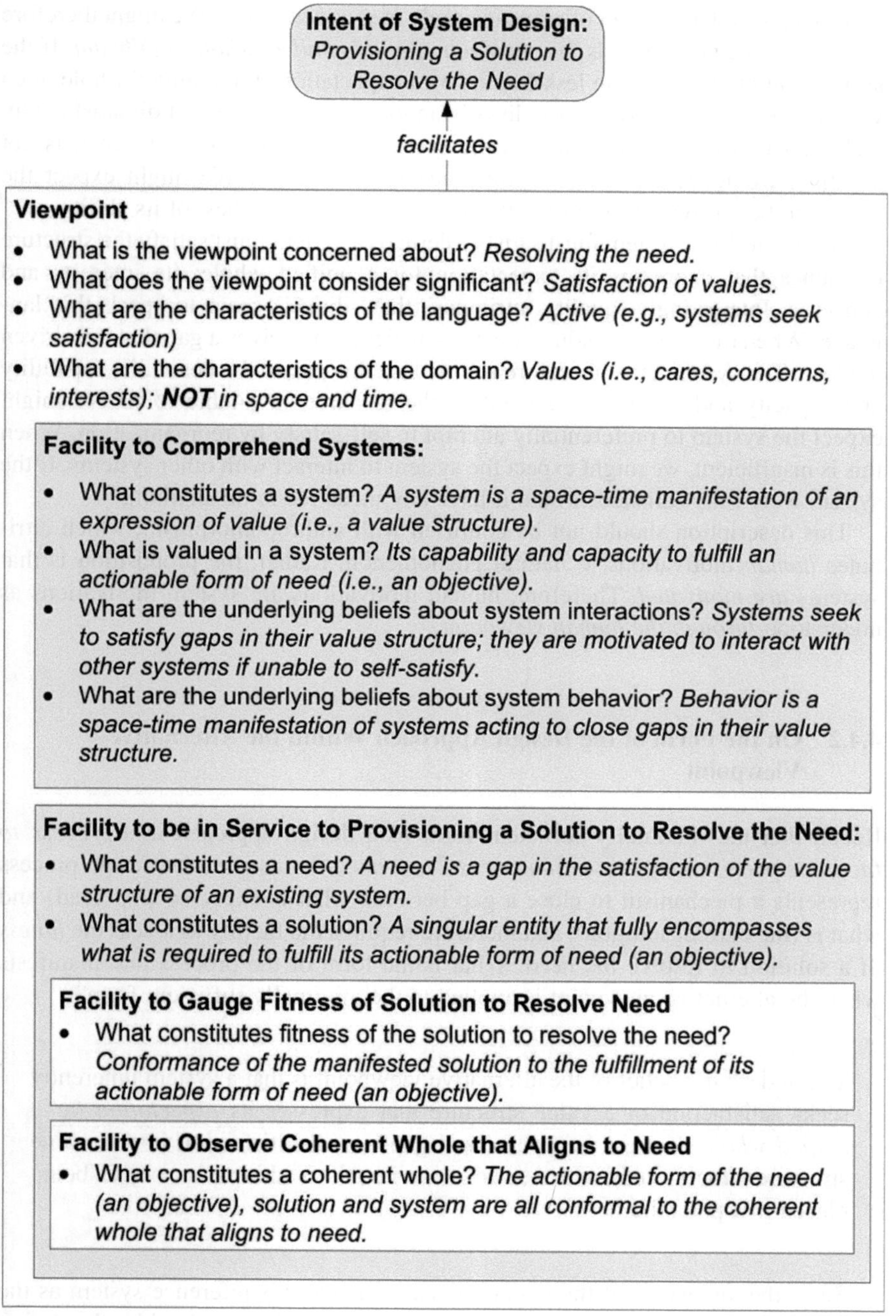

Fig. 6 The alternative viewpoint pattern with responses

From the system's point of view, it simply ***is*** *a unified whole*. We might therefore expect that the system finds the *satisfaction of this unified whole significant*. If the system were to be anything less than its own expectations for a unified whole, then we might expect it to experience dissatisfaction. This perception of dissatisfaction, including its overall importance to the system and the degree to which it is not satisfied, would be of utmost significance to the system. We might expect the system to be motivated to resolve this dissatisfaction to the best of its ability.

What is it that a system must satisfy? Perhaps a system must satisfy the structure of values that expresses its expectation for a unified whole; its interests and concerns. Interestingly, quality attributes (the '-ilities') seem to speak this language. When the systems' values are not satisfied, it perceives a gap that it is driven to close. The system must therefore act to close this gap to the best of its capability and capacity and within the context of the resources available to it. We might expect the system to preferentially attempt to self-satisfy by reorganization. When this is insufficient, we might expect the system to interact with other systems. If the system were fully satisfied, would it have any incentive to do anything?

This description should not be confused with anthropomorphism, which attributes *human* motivations to natural phenomenon. Rather, the proposition is that systems ***are*** *motivated*. Therefore, human motivations are system motivations as understood *through the human viewpoint*.

4.4.2 On the Form of the Design Approach Within the Alternative Viewpoint

Recall that the minimally sufficient form of a design approach is a *process to facilitate provisioning a solution to resolve a need*. In a sense, the design process represents a mechanism to close a gap between what is expected (the need) and what is (the current solution), and therefore requires the facility to assess the *fitness* of a solution to resolve the need. What is the form of the process that manifests when the alternative viewpoint is applied to the minimally sufficient form?

> The underlying belief of the alternative viewpoint is that a system inherently seeks satisfaction of a value structure that expresses its *expectation for a unified whole*. If there is a perceived gap—or dissatisfaction in this value structure—the system is driven to close it; this is akin to an object being placed in a potential field.

Since the viewpoint of the system is taken, define this reference system as the ***supported system***: the system that has a need that is being resolved by the design process and its resultant solution. If the supported system is driven to resolve its need—to close a gap in its satisfaction—the implication is that the system inherently has the facility to do so. In order for this to be true, the system must have the ability to:

- Comprehend the need in its abstract value form;
- Translate the abstract need into an actionable form (a need objective);
- Systematically derive and validate a correct and complete solution that conforms to the need objective (the solution objectives); and
- Interact in space-time in order to *form a system* that acts to fulfill the solution objectives that resolves the need.

By itself, the need is not actionable. It simply states that there is a gap between what a value needs to be and what the value is perceived to be in its context.[4] Therefore, this deficiency must be translated into an *actionable form of values*; in this case, a need objective that will be sufficient to close the gap and thereby resolve the need by satisfying the value structure in its context.

However, a need objective is likewise insufficient on its own. It provides an actionable form of value and a sense as to what extent this actionable form must be satisfied, but this is not enough to specify the form of a system that will resolve the need. Therefore, the need objective must be translated into an *active, value-based form of organizational structure* that describes a correct and complete and well-balanced solution that will be sufficient to fulfill the need objective. This correct and complete and well-balanced solution can be viewed as *a composite set of aligned solution objectives that form a singular entity* which fully encompasses what is required to fulfill the need objective.

These aligned solution objectives must then be matched (bound) to systems in space-time in order to form the system that can act to resolve the need.

The facility to assess the fitness of the solution to the need is likewise built-in. The need is the original definition of correct and complete. So long as the integrity of correctness and completeness persists throughout the process, the alignment at each translation step can be *continuously assessed* for conformance and negotiated for overall satisfaction. The required negotiation would take the form of a request for conformance (i.e., *derive* a solution that fulfills this need objective) and a *validation* response that indicates how well that conformance is being met. However, for a solution to be systematically derived from the need, the need must be expressed in a particular form.

Thus, we arrive at the high-level form of design approach driven by the alternative viewpoint as illustrated in Fig. 7.

From within the comparative viewpoint, any mechanism that closes a gap between a need and its solution will take the form of *a continuously-assessed change*[5] *negotiated and balanced to resolve the dissatisfaction.*

[4]There is a distinction between an abstract value (something that is of significance) and the specific, measurable instance it takes on within a context (its set-point).

[5]Change in the sense of "an act to make something different."

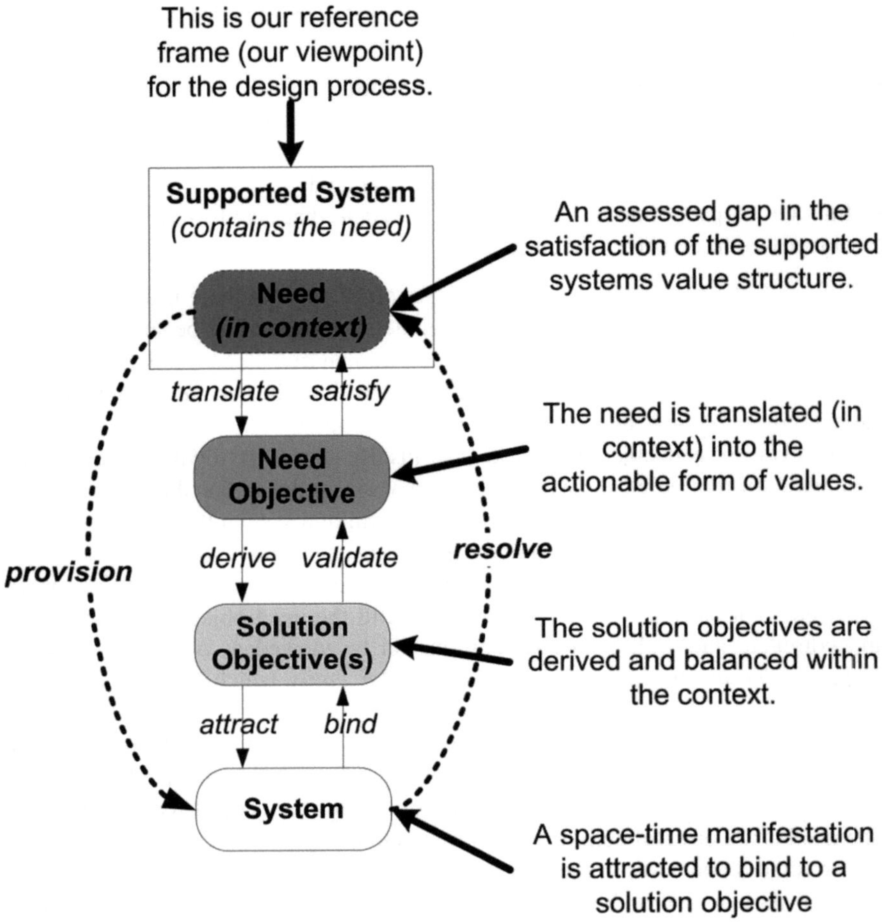

Fig. 7 The form of the design approach driven by the comparative viewpoint: a continuously-assessed change negotiated and balanced towards resolution of dissatisfaction

Yet, the high-level form seems unsatisfactory. What does the value structure and a well-formed expression of need look like? What does a well-formed actionable objective look like?

On the other hand, the conceptual form of the transform space—the derivation of the need objective into the solution objectives and their validation—can be summarized given the constructs asserted thus far, and is shown in Fig. 8.

The form of the transform space is that of a process of successive approximations *toward resolving the need*, similar in concept to the observable space-time manifestation of the cellular differentiation that occurs in the growth process of a developing organism.

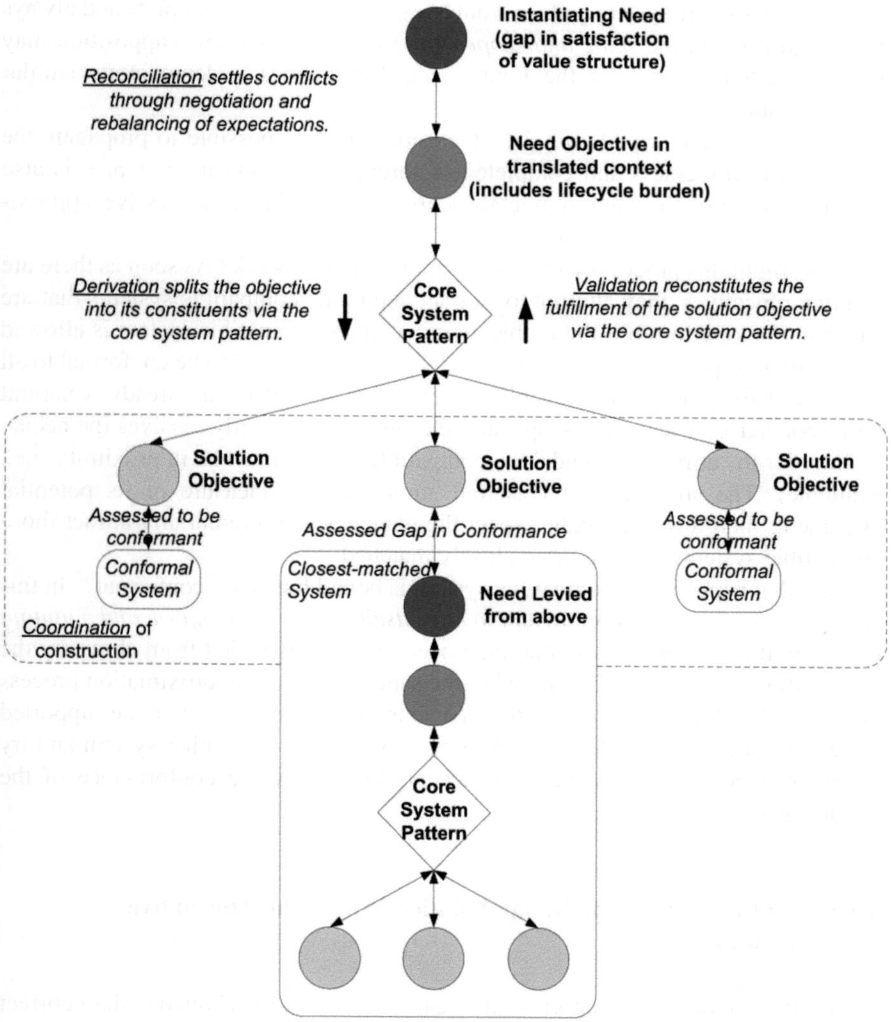

Fig. 8 Conceptual form of the transform between need objective and its conformal system

An immediate challenge arises with the required systematic translation of the need objective into an active, value-based form of organizational structure: A composite set of aligned solution objectives that form a singular entity which fully encompasses what is required to fulfill the need objective. How can the integrity of correctness and completeness persist through this systematic translation? The chapter authors posit that the integrity of correctness and completeness can persist, if and only if, a pattern exists that all systems must conform to; specifically, one that appropriately exposes (i.e., splits) the need objective into the minimally sufficient set of aligned constituent objectives and then reconstitutes the fulfillment of those objectives to validate alignment with the original need objective. This core system pattern would be—by definition—complete and always true, *regardless of the scale and scope of the system.* While this supposition may seem far-fetched, consider the Fourier transform: Does it demonstrate similar characteristics?

If this core system pattern exists, then not only is it possible to propagate the integrity of correctness and completeness through this translation step, it is also possible to apply the pattern fractal-recursively to achieve successive approximations *regardless of the system or its scale and scope.*

How might this process of successive approximation work? As soon as there are solution objectives, they attempt to "attract and bind" compatible systems that are capable of acting to fulfill these objectives. A single compatible system is allowed to bind to multiple solution objectives so long as it is assessed to be conformal to all of them. This situation is arguably preferred because there is already a natural correspondence in scale and scope, and the system sufficiently resolves the need.

In order to "attract and bind," the compatible system must be in proximity (i.e., available). The strength of the shared attraction is a measure of its potential alignment; therefore, we might expect the tendency to preferentially attract those compatible systems that are more closely matched.

But what happens if a compatible system is bound but is not conformal?[6] In this case, *the bound system must attempt to alter itself in order to conform and maintain the bond.* It does this by instantiating a need, one that is levied from above by the non-conformal bond, and by following the same successive approximation process to resolve it. If this *supporting system* is unable to fully conform, then the supported system has a choice: break the bond to attempt a match to another system and try again, or negotiate a rebalance that accepts the achievable conformance of the bound system.

4.4.3 On the Form of the Design Solution Within the Alternative Viewpoint

The design solution is in the form of a construction of systems bound to their correct and complete set of aligned objectives that specify the necessary and sufficient quality of their actions to achieve the expectations for the unified whole. This form appears to be true regardless of the fidelity of the instantiation.

[6]Conformal includes the notion of what is sufficient. Therefore, non-conformal implies insufficiency.

5 Comparative Study of the Viewpoints to Facilitate System Design

> It was the best of times, it was the worst of times, it was the age of wisdom, it was the age of foolishness, it was the epoch of belief, it was the epoch of incredulity, it was the season of Light, it was the season of Darkness, it was the spring of hope, it was the winter of despair, we had everything before us, we had nothing before us, we were all going direct to Heaven, we were all going direct the other way—in short, the period was so far unlike the present period, that some of its noisiest authorities insisted on its being received, for good or for evil, in the superlative degree of comparison only.
>
> —Charles Dickens, from "A Tale of Two Cities"

This section compares the viewpoints and their derived general forms to the evaluation criteria. The analysis includes an assessment of each viewpoint's disposition towards meeting intent, capabilities, limitations and dependencies. This information serves as a basis for assessing the validity of our deep-rooted assumption.

5.1 Comparative Analysis of the Viewpoints and Their General Forms

5.1.1 Agreement with Comparative Study Evaluation Criteria: Disposition Towards Meeting Intent

Recall, the generalized intent of systems design is *to provision a solution to resolve a need*. This section assesses the overall alignment—or disposition—of each of the viewpoints towards this intent.

The Prevailing Viewpoint

The prevailing viewpoint response demonstrates a strong focus towards *the system and its construction*—a form of solution provisioning—rather than towards resolving the need. Need is lightly addressed in the responses, in which the expression of need is *for a system* that has desirable properties, behaviors, and performance outcomes, which are then compared to some set of measures characterizing one or more realized *systems*. As a consequence of this disposition, we might expect that the viewpoint can only support an indirect comparison of a solution to a "solution standing in for need," rather than a direct solution-to-need comparison.

To support this supposition, consider the opening of the mission statement of the hypothetical FireSat mission provided as a worked example in Space Mission Analysis and Design, a book often referred to when training new space systems engineers:

> Because forest fires have an increasing impact on recreation and commerce and ever higher public visibility, the United States needs *a more effective system to identify and monitor them* [37].

Does the underlined phrase describe "a dissatisfaction that must be sufficiently resolved?" Or does it describe an assumption of the form of the solution—a surrogate for the need? In other words, has the required comprehension and translation of the need been short-circuited by an *assumption of form* based on the notion that a need *is for a system*?

The fundamental disposition of the prevailing viewpoint indicates an immediate disconnect with the need. If this observation holds true, then the viewpoint starts off (and can only ever be) *ungrounded* relative to the need—which is of ultimate significance to the design of the system.

When might comparison of a solution to another solution be applicable? Perhaps this facility is more appropriate when a solution needs to be constructed and replicated as it does during construction and manufacturing.Therefore, the prevailing viewpoint appears disposed towards *provisioning* the *construction design* and *guiding* the subsequent manufacturing of a system.

The Alternative Viewpoint

The alternative viewpoint response demonstrates a strong focus towards *resolving the need by forming a solution*. The construction of the system is described only as a manifestation in the space-time domain that binds to the solution objectives.

Within this viewpoint, the system is viewed as active, capable and possessing a set of values it is driven to satisfy. In a sense, a system's need guides the formation of a supporting system that resolves that need.

Consider how the alternative viewpoint might look at the forest fire problem presented in the previous discussion. Based on the mission statement, the alternative viewpoint might begin by viewing forest fires as *disruptive*—an impediment toward satisfaction of the values of society as a whole. It might note that the decision-makers have already assumed forest fires to be a *disruption of significance*. Here are a set of questions that might come up to help identify the true need:

- What is the specific *dissatisfaction* to recreation and commerce?
- Are there other relevant *dissatisfactions*?
- Who is dissatisfied by the "public visibility" and in what manner? Is this *relevant* to addressing the *disruption*?
- Given each specific *dissatisfaction*:
  - To what extent are we looking to make society *robust* to the disruption (through the prevention of, or resistance to, that disruption)?
  - To what extent are we looking to make society *resilient* to the disruption (through recovery from that disruption)?

The alternative viewpoint focuses on specifying "what" a correct and complete solution must do and "how well" it must do so in order to resolve the need. While the viewpoint has some visibility into a system's construction (through guiding the attraction of systems that *can act* to fulfill the set of objectives that resolve the need), it seems to have an intentional blind spot towards construction in terms of parts connecting to parts.

Therefore, the alternative viewpoint appears disposed towards *provisioning* the *solution design* and *guiding* the resultant generation of a system such that it resolves the need.

5.1.2 Agreement with Comparative Study Evaluation Criteria: Assurance

Recall, the generalized form of design approach for systems is *a process of facilitation to provision a solution to resolve a need.* This section assesses the overall agreement of each of the viewpoints relative to the evaluation criterion: *assurance* of the solution to resolve the need. In other words, how do we know we have *the right thing*?

General Form of the Prevailing Viewpoint

What is the form of assurance that can be supported by the general design process derived from the prevailing viewpoint? The viewpoint is well-disposed towards assuring that the construction of the realized system exhibits the same characteristics as the surrogate system. The immediate concern with this form of assurance is that there is no intrinsic capability to assure that the need itself can be resolved with the surrogate system.

> Because the prevailing design process can only comprehend the construction of a system, the assurance it supports *can only be as good as the assurance of the surrogate system.* At best, a surrogate represents an *assumed approximation* of the need that is *dependent on the circumstances of the assessment* and the *skills of the assessor.*

To *compensate* for this lack of intrinsic capability, methods are applied at the beginning of the design process in an attempt to improve the pattern-matching of the surrogate system to the need *for the system.* For instance, needs analysis and stakeholder analysis methods sort through a maze of conflicting expectations relative to *the system* that must be constructed [38, 39]. These methods work through the lens of the *system as significant to the design*, which immediately places the users and supporting stakeholders as external actors. These methods set up the user and customer requirements—in terms that are significant *to the system*—that are then passed down the chain of design. Other methods include system architecting and system engineering processes [31, 35, 36]. Regardless of the method, the resultant artifacts take the form of measures of *the system*: measures of effectiveness, measures of performance, an objective or utility statement, a document describing the concept of operations, and etc.

In all cases, the need for a system is *understood and accepted to be ill-defined or "fuzzy."* Any standards for what these represent are captured as "best practices." We struggle to *assure that the surrogate* sufficiently represents the true need.

Once we have maximized the assurance of our surrogate, we shift our attention to maximizing the assurance of our construction *to the surrogate*. On what basis are design decisions made? Without comprehension of need, the choices made by the designer might tend toward supporting what the designer finds significant, rather than what is significant to the need. We *compensate* for this designer drift by imposing standards that represent some lower-level construction surrogate for perceived needs that may or may not support the resolution of the overall need.

Once the construction has been sufficiently manifested, we can verify that the system conforms to the construction surrogates. We place the system in use and observe its behavior to validate that it matches the surrogate of the need *for the system*. And yet, when we field the system, we often find it unsatisfactory; we experience a sense of dissatisfaction towards the true need.

Depending on the level of assurance required, we add feedback cycles at various places within the process in an attempt to improve our assurance (see Fig. 5). All of this occurs subject to the availability of funding. Once funding runs out, we either accept the results or cancel the project and rationalize the outcomes as "that is just how things are" or "we need to do better next time."

Therefore, we might expect that the level of assurance correlates with the availability of resources, the skills of the design practitioners and the external need assessors, and the scale and scope of the system.

Conditions for Sufficiency or Insufficiency to Satisfy Criteria and Meet Intent

Under what conditions can the process driven by the prevailing viewpoint assure that a solution will resolve the need? The *burden of best-case assurance* is placed on the shoulders of *the practitioner who assesses the conformance of the surrogate construction to the perceived need.* Therefore, to sufficiently satisfy the assurance criteria to meet the intent of system design, the practitioner must be capable of:

- Empathizing with the true need such that they can sufficiently place themselves as a surrogate;
- Comprehending, in totality, the potential constructions such that they can be assessed and matched to the true need.

So long as these criteria are met, the prevailing viewpoint is capable of satisfying the criteria to meet the intent. If the practitioner is ***not*** capable of empathizing with the true need and can only assume an approximation based on the circumstances at the time of the assessment, then the prevailing viewpoint is capable of satisfying the criteria *only if the constructed system is operated in accordance with those assumed circumstances*. In all other cases, it is arguably insufficient.

Therefore, the prevailing viewpoint's capability to satisfy the criteria and meet intent is contingent on the skills of an *external agent* known as the "practitioner" who generates the *surrogate construction* for the need—the assumed entry into the design process *because the viewpoint can only comprehend construction*. The support provided to the "practitioner" consists of the provision of construction models that provide some level of prediction, experientially-based standards, "best practices" and construction processes.

What might we expect to observe in outcomes, as the system grows beyond the scale and scope comprehendible by the ill-supported practitioner? Given the above, we might expect a rapid drop in overall system success as the scale and scope of the system exceeds the capability and capacity of the practitioner. Given the propensity to add in feedback cycles in an attempt to *compensate* for the insufficient assurance provided by the practitioner, we can expect that there would be a corresponding rapid rise in resource expenditure (see Fig. 1).

> Examination of [current defense and space] programs shows that our systems engineering processes simply do not work. Requirements are set, but the artifacts do not meet the requirements. Multiple redesigns to meet the requirements, inevitably trading weight, performance, and reliability with cost, is the direct cause of most of the delays and overruns. Risk management systems are put in place and carefully executed, but fail to prevent major delays and requirement shortfalls on almost all programs... Decades of cutting edge methods, processes and tools have been injected into the systems engineering process to no avail. Despite major innovations enabled by desktop computing and networks, bottom line performance in this class of engineering development programs has worsened, not improved, since the late 1970's and early 1980's [7].

General Form of the Alternative Viewpoint

What is the form of assurance that can be supported by the general design process derived from the within the alternative viewpoint? The viewpoint is well-disposed towards providing assurance. It is strongly built to derive an aligned, complete and correct solution regardless of the scale and scope, and to continuously validate its progress towards resolving the need. There appears to be a persistent situational awareness of how well assurance is being met at any and all translation steps.

The need—which is embraced as an abstraction and is explicitly defined by its significance—forms the root of the systematic derivation process. In combination with the reconciliation of conflicting interests, the fitness of the emerging solution is continuously gauged, negotiated and assured.

The most challenging aspect of assurance takes place within the transform space between the abstract need objective and the set of solution objectives. Even here, assurance is enforced at every exposition through a mechanism of precise and rigorous application of the core system pattern in combination with continual negotiation, reconciliation and validation.

Finally, assurance of the alignment of the matched systems to the solution objectives is enforced through the ability to instantiate needs within the matched systems to close any conformance gaps, to rebalance expectations such that the need is sufficiently resolved, or to renege on the current matching in favor of another. This becomes a process of successive approximations *until the need is sufficiently resolved.* Convergence occurs when no new supporting needs are generated.

Because the comparative process is aligned toward assurance-seeking, we can expect a theoretical limit of 100 % assurance, practically constrained by the capability and capacity of the system performing the process, the level and quality of provisioning of the compatible systems, and (only when necessary) the capability and capacity of the supported system to accept and absorb the consequences of a negotiated rebalancing towards satisfaction of its values.

Conditions for Sufficiency or Insufficiency to Satisfy Criteria and Meet Intent

Under what conditions can the process driven by the alternative viewpoint assure that a solution will resolve the need? The primary burden of assurance is placed on *our comprehension of the system's value structure and the consistent expression of a well-formed need.* Once these have been systematically derived and validated, we will have the foundations necessary to drive toward the form's theoretical limit of 100 % assurance.

Unlike in the prevailing viewpoint where we have numerous examples to draw on, the alternative viewpoint is still within its formative stages, so a detailed comprehension of its practical limits is still out of reach. However, there are several reasons why we can expect practical limits on the assurance. First, the form must still be implemented and the manifestation of the process will impose its own limitation that can affect the overall assurance it provides (the capability and capacity of the system performing the process). Second, we must work with the level and quality of the compatible systems that form the solution. Since our component systems are built to conform to construction standards rather than toward collaborative alignment in order resolve a need, we might expect continued—though significantly reduced—issues with assuring their integration. We may also have to work with the varying levels of human capability and capacity to handle the very different kind of need-based negotiations expected from this approach.

The assurance of the comparative process is also *dependent on the existence of the core system pattern*. Even a rudimentary form of this core pattern would serve to move us away from practitioner-dependence and towards a systematic science.

What are the conditions for insufficiency? Within the theoretical form of the design process, all parties inherently seek to find the most mutually agreeable solution, given their capability and capacity. What happens if one or more systems that are brought into the solution are not aligned toward the common objective?

> We might expect that inefficiencies in resolving the need will occur if the objectives of the interacting systems that compose the overall solution are misaligned (i.e. antagonistic or parasitic) rather than aligned and synergistic.

5.1.3 On the Agreement with Comparative Study Evaluation Criteria: *Correct & Complete*

Recall, the general form of the design solution for a system appears to be *the coherent whole that conforms to the resolution of the need*. This section assesses the overall alignment of each of the viewpoints towards *conformance* of the solution to resolve the need. In other words, the solution must be *correct* (it aligns with the interests and concerns expressed by the need) and *complete* (it aligns with the scale and scope of the need). The design process assesses the alignment of the solution to the need, and concludes when the design solution is conformal.

General Form of the Prevailing Viewpoint

What is the form of *correct* and *complete* that can be supported by the resultant design solution generated from within the prevailing viewpoint? As in the design process discussion, the viewpoint seems well-disposed towards assuring that the construction of the realized system exhibits the same characteristics as the surrogate system. While the viewpoint can support assessing the *correctness* and *completeness* of the realized *construction* to the need surrogate, the immediate concern with this facility is that it lacks any notion of *correct* and *complete conformance* to *resolve the need.*

We can observe this phenomenon in the application of formal methods, such as "correct by construction," to assess the *correctness of the system construction.* Correctness *to the need* can only be assessed by an external observer, who, because of their arguably indistinguishable viewpoint, is likewise limited in his or her capability and capacity to assess true alignment to need. In general, the feedback we receive is a vague sense of discomfort or *dissatisfaction* on the part of the external observer, encouraging us to reassess yet another surrogate construction in an attempt to improve their satisfaction.

How do we assess whether the solution is *complete*? We might hear this question expressed as: "How do we know we're done?" Not only do we struggle to assess this relative to our vague notion of need, but we also seemingly struggle to answer this relative to the specification—an odd phenomenon considering the specification should be a standard of conformance to another construction. By observation, this uncertainty in completeness appears to be a consequence of *discovery*. As we progress through the design process we *discover* items that are not covered in the specification that must be addressed in order to continue to support the specification.

> Perhaps we have no mechanism for assessing *completeness* because we insist on placing elements critical to the success of the overall solution—including people—outside the bounds of the system we are constructing. If any part of the solution to the need exists outside the "bounds" of the system, to what extent is the *completeness* of the system to the need *knowable*?

On the other hand, the prevailing viewpoint supports the assessment of a physical, space-time sense of fit *completeness*; we can place a part in the context of other parts and assess whether it is conformal in scale and scope to its ecosystem of parts. This facility is quite necessary in the realm of *construction design* and manufacturing.

The General Form of the Alternative Viewpoint

What is the form of *correct* and *complete* that can be supported by the resultant design solution generated from within the alternative viewpoint? As in the design process discussion, the viewpoint is well-disposed towards providing assurance that a solution resolves the need. It is strongly built to derive an aligned, complete and correct solution regardless of the scale and scope, and to continuously validate its progress towards resolving the need. There appears to be a persistent situational awareness of how well assurance is being met at any and all translation steps.

> As the comparative process is a process of successive approximations derived systematically from the need, *the intermediate systems are always as* ***correct*** *and* ***complete*** *to resolve the need as possible* (i.e., given the available compatible systems). Any gaps in the assessed ***correctness*** and ***completeness*** drive additional approximations. Thus, the system at the termination of the

(continued)

design process is conformal to the supported system's notion of ***correct*** and ***complete***: it originates in alignment and correspondent in scale and scope to the need, and this definition *is enforced throughout the process.*

5.2 *Assessment of the Validity of the Deep-Rooted Assumption*

This section assesses the validity of our deep-rooted assumption and discusses what the study suggests about the nature of complexity in system design and the relationship between the viewpoints. Recall that the hypothesized deep-rooted "big assumption" is the core definition of a system applied to its design, generally characterized as: *the collection of interconnected parts that interact to produce some behavior.*

5.2.1 On the Validity of the Deep-Rooted Assumption

The prevailing viewpoint demonstrates agreement with the comparative study evaluation criteria *only so long as the practitioner who generates the surrogate for the need has sufficient capability and capacity* to empathize with the need *and* to comprehend, in totality, the construction that will conform to that need. Once the system exceeds the scale and scope comprehendible by the practitioner, the prevailing viewpoint increasingly shifts out of agreement with the evaluation criteria. Likewise, if the need exceeds the ability of the practitioner to empathize, the surrogate constructions begin to lose their grounding to the true need.

The prevailing viewpoint analysis also manifests many of the historical hallmarks of misalignment described in the introduction to this chapter, including *compensatory measures* to improve *tractability* and *predictability.*

These results **raise questions as to the validity of our deep-rooted assumption for the design of "complex" systems**: the prevailing viewpoint does *not* produce the desired results *as the system exceeds the capability and capacity of the practitioner to comprehend.*

On the other hand, the alternative viewpoint demonstrates theoretical agreement with the evaluation criteria.

The existence of an alternative viewpoint (*based on a distinctively different comprehension of systems*) that demonstrates *improved agreement* with the evaluation criteria suggests that **this alternate view of systems may be a more suitable basis for the design of "complex" systems.**

Based on historical precedent, this preliminary result is good news for the system design community: The problems we have with the *burden of complexity* can be connected to our mental construct, and mental constructs *are alterable*. Recognizing misalignment in our mental constructs enables us to adapt—from the inside out—toward behaviors that can improve our success.

5.2.2 On the Nature of Complexity in System Design

The prevailing viewpoint's implicit assumption that *the collection of interconnected parts that interact to produce some behavior* is what is of significance to system design necessitates the construction of an *instance* of a system (virtually and/or physically) in order to assess its fitness to a *perceived* need. As the number of the parts and their connections defining this construction increases, our ability to comprehend the coherent whole decreases *even with the cognitive assistance of our construction models*, encouraging us to implement feedback cycles to better inform our *construction design* decisions.

If we instead applied a viewpoint in which the coherent whole *is the reference standard for solution design* rather than an assumed construction, would we perceive complexity in system design in the same manner? Is it significant to this viewpoint to concern ourselves with the number of parts and their connections? Or is it *only significant* to a viewpoint that must concern itself with *construction design* and replication? After all, understanding the number and kinds of parts and how they are connected (as required to meet a solution specification) informs the scaling and scoping of the resources that must be available in order to meet a request for construction or replication.

5.2.3 On the Viewpoints and Their Relationships

The viewpoint illusion in Fig. 2 can provide an immediate conceptual sense of the relationships between the prevailing and alternative viewpoints; a useful analogy supported by the results of the comparative study. From the vantage point of the coherent whole, we can comprehend and *validate* the significance of the parts to the whole ('is it doing the right thing?'), but we have no comprehension of how the parts are connected and arranged in space and time. Therefore, the alternative viewpoint seems capable of supporting the *derivation of the conformance specifications* for the interconnected parts towards the whole, but it *cannot* comprehend its construction. From the vantage point of the interconnected parts, we can comprehend *the arrangement and connections of parts in space and time* to meet some specification ('is it doing the thing right?'), but we have no inherent comprehension of the significance of these parts to the whole. While the prevailing viewpoint can sense the existence of a coherent whole, arriving at the *intended* one without a specification generated from the coherent whole is—at best—an educated guess.

> The viewpoint of the coherent whole guides the *design of the solution* that forms the coherent whole to resolve the need. The viewpoint of the interconnected parts supports the *design of the system construction* that conforms to that solution. Therefore, we can infer that ***both viewpoints are required*** to *successfully provide a system to resolve a need.*

If the coherent whole represents the viewpoint of the system, then this suggests that the key to successful system design is *to take the viewpoint of the system*. As a corollary observation, what happens if we choose a viewpoint other than that of the system? How many are there? Is it a *countably infinite* set? Are they all valid? Are they all consistent?

Even though the viewpoint of the coherent whole does not comprehend *how* the parts are connected and arranged in space and time, it still *encompasses* the viewpoint of the interconnected parts. If the system manifested in space-time is out of alignment with its specification, the vantage point of the coherent whole has immediate feedback of this misalignment: the resultant image appears fuzzy. This phenomenon does not occur looking the other direction.

5.2.4 On the Potential for Additional Viewpoints

The observation of encompassment leads us to an interesting question: Are there *other viewpoints* encompassed within the coherent whole (system) viewpoint? Recall the discussion of the form of design approach within the comparative viewpoint. It was observed that the viewpoint required a proper and well-formed expression of need. Where does this come from? What forms this expression of need? Arguably, neither of the two viewpoints discussed thus far are capable of forming the proper expression of need. Therefore, the chapter authors *assert the existence of a third viewpoint*: one that *forms the expression of need* used by the alternative viewpoint to generate the solution to resolve it.

Three potential viewpoints have been identified—are there any others? Both the prevailing and alternative viewpoints acknowledge that systems interact, yet they arguably do not place those *system interactions* as their primary focus. This gap implies a *fourth viewpoint*: one that *performs the interactions among systems that resolve the need.*

Why might these *four viewpoints* exist? Based on the design process derived from the alternative viewpoint, assuring a solution to resolve a need requires the other viewpoints: to provide a *well-formed need*; to provide a *conformal system*; and to provide the *resultant interactions that perform the resolution of the need*. These viewpoints seem to take *distinctive, non-overlapping roles* that *collaboratively support each other* towards resolving the originating need as well as towards providing the facility to responsively identify and resolve new needs. This behavior suggests the ability *to affect adaptive change within and between systems.*

If these observations hold true, it begins to imply that *all four viewpoints must be sufficiently satisfied* in order to successfully affect adaptive change within and between systems.

Each viewpoint seems highly directed towards favoring and emphasizing certain aspects of a whole system. We might expect this specialized focus to result in a de-emphasis of the facilities required to fully support the *other roles*. If any viewpoint were to be applied to a role it did not favor, we might expect an immediate misalignment of emphasis and the application of ill-suited or de-emphasized facilities in an attempt to support the role. Overall, the effect would be a reduction in effectiveness. By extension, *if only one viewpoint were to be applied to all roles*, then we might expect the potential for serious problems to arise. In this case, three of the four roles could experience a reduction in effectiveness *and* we might expect the potential for blurring of the lines between the distinct roles, leading to confusion, uncertainty and distrust. Overall, the system could experience a *reduction in its ability to successfully affect adaptive change towards the resolution of need.*

6 Suppositions Concerning the Evolution of the Observed Misalignment

If this chapter's preliminary conclusion that we have been applying a misaligned viewpoint to system design holds up to further scrutiny, how is it that the system design community ended up here? The chapter authors suggest the observed misalignment may be a natural consequence of being human. The earlier discussion concerning the psychological change-prevention system explains how people are built to ignore countervailing evidence. This section will explore how such a misalignment could arise in the first place and evolve to the current situation.

6.1 *On the Trajectory of Growing Misalignment*

Based on historical evidence, the deep-rooted assumption at the core of our observably misaligned viewpoint seems to have been a valid approximation for a long time. Therefore, it seems reasonable to ask: At what point did our alignment start to shift so far and why? When and how did our *parts-centric* viewpoint originate, and how could it be so pervasive that we have been unable to recognize it as a mindset?

6.1.1 A Brief History of Technology Development

The history of technology development from the earliest fossilized records of our existence can be summarized as follows: humans as the consummate toolmakers. Why? Perhaps it is simply because we are inherently tool users. We seem to have a built-in need and ability to extend our capabilities and ourselves. The consequence of being a tool user is that we developed the need to provision better tools via tool-making. What is the significance of technological systems to us? Do they represent a mechanism for extending our capability and capacity [40]?[7]

The chapter authors propose that tools fall within a continuum of relational significance to the tool users and can be roughly approximated in terms of three categories: *passive*, *amplifying* and *collaborative* extensions. These categories acknowledge what motivates us to construct systems—the value satisfaction they represent to us. They also form the basis of a useful approximation of the historical paradigms of tool-making—what we now call technology development—reframed to acknowledge what motivates us to construct systems.

As humans have advanced through the levels of technology development, the motivations for the earlier stages have remained for many kinds of systems, necessitating the persistence of those paradigms in parallel with the emerging one. Therefore, the strategies and mental constructions that enabled successful crafting within those paradigms have persisted as well, *strongly influencing the evolution of our comprehension of systems and its resultant models.*

Level 1: Passive Extension Development

Passive Extensions are systems that passively extend the capability and capacity of the tool user (e.g., clothing, roadways, a bridge, and etc.) or that require external actualization to extend the capability and capacity of the tool user (e.g., a projectile, a pencil, a bicycle, and etc.).

These extensions harness the **passively-transformative** capability and capacity of natural systems. We value their *properties* and their alignment with the

[7]The reference describes the use of technology as something that "enhances" the human.

mechanics we need to effectively *transform the physical scale and scope of our structure directed towards adapting us to the conditions of the environment—or towards adapting the environment to us*. In this first level of technology development, we repurpose, modify and combine natural and passive systems that demonstrate these desirable characteristics (e.g., hardness, durability, leverage, force, thermal transference, and etc.). This paradigm demonstrates the highest level of predictability in system behavior.

Within this paradigm, we might expect it to be quite natural to apply a mental construction of parts fitting together with parts.

Level 2: Amplifying Extension Development

Amplifying Extensions are systems that extend the capability and capacity of the tool user using some degree of self-actualization, but are fully directed in their actions by the human (e.g., many cars, a gun firing a projectile, calculators, and a basic cell phone).

These extensions harness the ***actively-transformative*** *capability and capacity* of natural systems. We value their *behavior* and alignment with the *mechanics and active production* we need to effectively *amplify the functional scale and scope of our actions directed towards adapting us to desired interactions* [41]. In this second level of technology development, we repurpose, modify and combine natural, passive and amplifying systems that demonstrate these desirable characteristics and support these functions (e.g., transportation, disruption, communication, and etc.). A flour mill that harnesses hydropower in order to grind wheat would be an example of an early amplifying extension.

This paradigm requires enough knowledge of the natural world to be able to harness system behaviors for our own ends. In this paradigm, we have observed increasing difficulty with predicting overall behavior as the scale and scope of these systems have increased, but have found reasonable compensations by improving abstractions, methods and tools.

Within this paradigm, we might expect it to be natural to augment the *parts-centric* model of systems developed within the passive extension paradigm with an overlaid behavioral model.

Level 3: Collaborative Extension Development

Collaborative Extensions are systems that are self-motivating and work in concert with the tool user to extend his or her capability and capacity (any system that has any semblance of autonomy; a personal computer, for example).

These extensions harness the ***collaboratively-transformative capability and capacity*** of natural systems and therefore begin to emulate the highly interactive, *social behavior* of natural systems. We value their *quality of interaction* and our mutual alignment of *interests and concerns* we need to effectively *negotiate the* *collaborative interaction directed towards adapting to the scale and scope of our objectives*. In this third level of technology development, we repurpose, modify and combine natural, passive, amplifying and collaborative systems that demonstrate these desired characteristics, functions and interaction quality. We defer more and more directive authority to the systems we create (e.g., autonomous systems, aerospace systems, artificially intelligent services, and etc.). These systems tend to become highly inter-dependent and inter-networked with each other (socialized) in order to perform their duties (e.g., the Internet).

This paradigm requires significant comprehension of the natural world, with increasing significance placed on the quality of our collaborative interaction with the system rather than on its properties and behaviors. In this paradigm, we have observed extreme difficulty predicting behavior and quality of interaction, which only worsens as the scale and scope of the system increases. Furthermore, we have found typical compensations insufficient.

In the collaborative extension paradigm, we are *attempting to emulate the social behaviors of natural systems*. Although we have always worked with natural systems in this manner—livestock, agricultural cultivation and social engagements with other people—we have only recently attempted to *craft* such systems (e.g., the internet, future transportation with autonomous vehicles, smart energy grid, and etc.).

Within this paradigm, we might expect it to be natural to attempt application of the existing behavioral-augmented *parts-centric* model of systems and perhaps replicate its use in an attempt to model the observed collaborative behaviors of multiple interacting systems (e.g., system-of-systems). Does this form of comprehension support us in our negotiations of interactions with collaborative systems?

6.1.2 On the Emergence of the "Craftsman" Mentality

Early in the development of passive extensions, a "craftsman" mentality emerged, and was then sustained and encouraged throughout the development of the amplifying extension paradigm. Hughes' "Human-Built World" supports this notion of a historical craftsman and dates the emergence of this mentality to the first time that humans needed to "transform an uncultivated physical environment into a cultivated and human-built one with all of its artifacts and systems [42]."

In general, the *craftsman* mentality is the mindset of one skilled in a trade; particularly in relationship to the skill of combining existing systems to produce new systems that enable some new purpose within a well-defined domain (i.e., parts connecting to parts). This mindset often manifests as incremental improvement of something that has worked well in the past; ballistic weaponry, for instance. Improvements are valued along predictable dimensions: the degree of damage to the target relative to the desired threshold for disruption, the precision and accuracy of hitting the target with the intervening system—be it a rock, spear, round or missile—and the range of action required in order to avoid having our own effectiveness compromised.

The *craftsman* mentality can also manifest in a creative, disruptive sense. How often is an industry shaken when someone combines a unique set of technologies, usually a set that would seem counter-intuitive to combine in the first place, and produces something truly unexpected? It seems natural to consider this an altogether different, innovative view on the world; and yet, this is still viewing tools (human extensions) in terms of an assemblage of known parts and pieces.

The pervasiveness of this mindset seems to be supported in this quotation provided from a paper abstract:

> Technology—the collection of devices and methods available to human society—evolves by constructing new devices and methods from ones that previously exist, and in turn offering these as possible components—building blocks—for the construction of further new devices and elements. The collective of technology in this way forms a network of elements where novel elements are created from existing ones and where more complicated elements evolve from simpler ones.
>
> —The evolution of technology within a simple computer model [43].

It might therefore be inferred that given its *human-scale* historical pervasiveness and longevity, the system design community has inherited the *craftsman* mentality from the first two stages of technology development—a mindset that values parts connecting to parts. And with the well-practiced—and thus far successful—*craftsman* mentality at our disposal, we have begun our transition into the collaborative extension stage, in which we seek to *craft our systems in full social emulation of the natural world.*

6.1.3 On Comprehension of Natural Systems and the Perception of Complexity

What might we expect to observe if our comprehension lags behind what our appetite for technology demands of our emulation of natural systems? Fig. 9 depicts a notional curve of the level of comprehension of natural system interactions that are required for their successful emulation. As the emulation fidelity required for the design of a system increases, then we might likewise expect an increase in the level of comprehension required to do so successfully. Given our experience with comprehending interactions, we might expect the curve to be non-linear as shown.

If the level required for successful emulation falls within our level of actual comprehension, then we might expect to be in a "grasp zone," in which we perceive clarity, tractability, certainty and predictability. However, if the level required

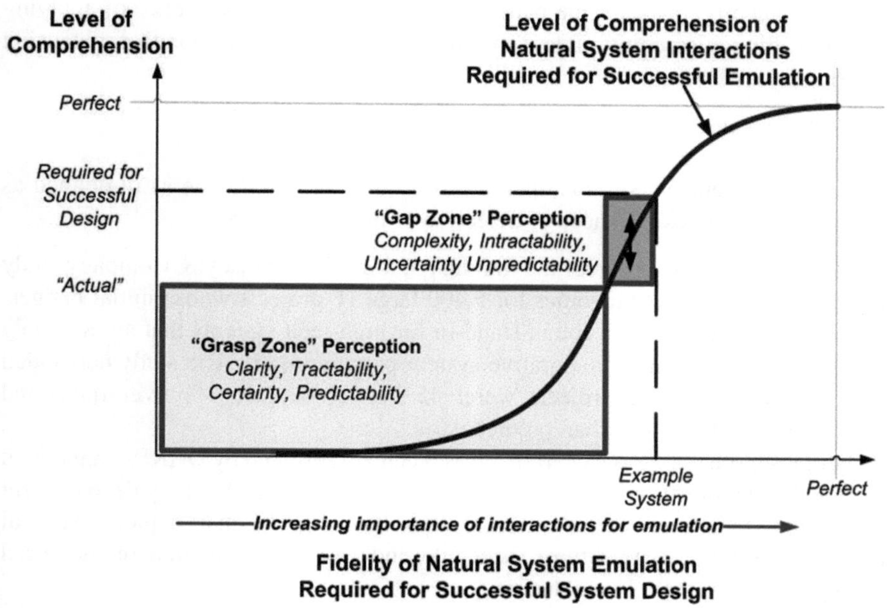

Fig. 9 The level of comprehension of natural system interactions required for successful emulation and its implications for the perception of complexity

exceeds our level of comprehension, we might expect to be in a "gap zone," in which we perceive complexity,[8] intractability, uncertainty and unpredictability.

The historical examples provided in the first section of this chapter support the aforementioned perceptions we tend to experience from within the "gap" and "grasp" zones. The examples also support the notion that a *shift in comprehension* was required to *transition* a "gap" zone *into an enlarged* "grasp" zone.

6.1.4 Brief Summary of Evidence in Support of Suppositions

Is there any other evidence to support these suppositions concerning the evolution of the observed misalignment?

Of historical note: Around the same time as the advent of computers (our earliest attempt at creating technical collaborative extensions), general rumblings about systems demonstrating high complexity became common, particularly within the cybernetics community—an emerging discipline that heavily cross-pollinated between machine, man and society [44].[9]

[8]This *perception* should be disambiguated from the *measurement* of complexity. A *perception* of complexity intrinsically carries with it the notion of *significance to some end.*

[9]Prior to the 1940s, discussions of "complexity" seemed focused on mathematical systems.

More recently, consider the pattern of outcomes as a consequence of applying the prevailing viewpoint to late amplifying and early collaborative extension development paradigms:

- Cost overruns;
- Schedule overruns; and
- Value deficiencies (systems that do not resolve the need even as ill-defined as they are; includes instances of defective or missing features).

More specifically, consider the findings of a 2012 McKinsey & Company study concerning the project outcomes for 5,400 large IT projects whose initial budgets were greater than $15M, a rough stand-in for high-tech systems that are arguably nearing or entering the collaborative system paradigm [45]. The study concluded that, on average, these projects were: 45 % over budget, 7 % over time, and produced 56 % less value than predicted.

Perhaps more telling is a chart that was recently presented by DARPA, annotated to note the classification of the industries in terms of their technology development paradigms as discussed earlier in this section (Fig. 10). The chart depicts historical trends of resource expenditures (schedule and cost) as a function of measured "complexity" for three key industries:

- **Automobile**: A long-standing industry with well-crafted patterns that has historically been within the amplifying extension paradigm. With the advent of driverless cars, this industry has begun to rapidly shift into the collaborative-extension paradigm.

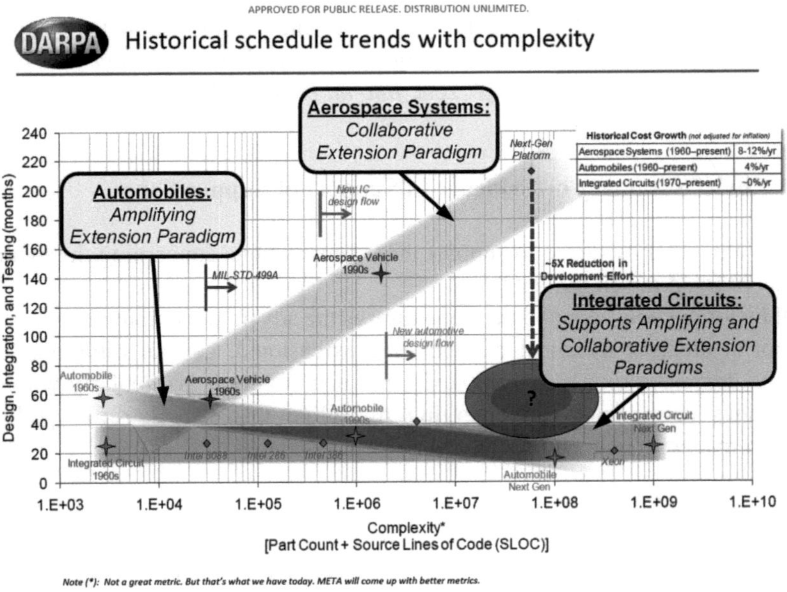

Fig. 10 Historical schedule (i.e., resource expenditure) trends with complexity (borrowed from DARPA slides [46]) annotated to highlight association with human technology development paradigms

- **Integrated circuit**: An industry within the amplifying-extension paradigm that supports both amplifying and collaborative extensions. Although the trends have been historically flat, they are also beginning to struggle with "complexity" in VLSI design [9].
- **Aerospace**: An industry that began in the amplifying-extension paradigm and has since moved within the collaborative-extension paradigm.

6.2 *On Two Potentially Harmful Mental Constructs for "Complex" System Design*

The whole is simpler than the sum of its parts.
—Willard Gibbs, renowned 19th century American scientist.

Evidence suggests we may be attempting to extend the *parts-centric* craftsman mentality into a "gap zone" associated with collaborative extensions. It is natural to continue to apply mental constructs that work well within the "grasp zone" of passive and amplifying extensions to a "gap zone" associated with collaborative extensions. However, if these mental constructs are *in conflict* with the nature of collaborative extensions, we might expect that their application could cause harm.

If a mental construct is used beyond its applicability, then we might expect a strong potential to observe discrepancies with that construct. When this happens, particularly when this mental construct is a core tenet of the mentality, we might expect a natural tendency to bring in *additional* mental constructs in an attempt to *explain these discrepancies* in observations.

For instance, the idea that systems are defined in terms of a boundary would have been a natural construct supporting the craftsman mentality working within the passive extension paradigm. Is this mental construct still consistent with the nature of collaborative extensions?

6.2.1 Systems Are Defined by Boundaries

Within the prevailing viewpoint, there appears to be a strong notion that a system is defined, in no small measure, by its boundary. The usage of boundaries to define what is part of a system versus what is not is prevalent and persistent throughout industry, academia and the various methodologies and tools that support them. The concept seems easily attributable to the underlying deep-rooted assumption of the viewpoint because interconnected parts must be cleanly separated and defined.

Thinking in terms of boundaries is an extremely useful simplifying assumption for the toolmaker. It bounds the problem space, breaks it into known components, and improves the tractability of the design of many kinds of systems. Furthermore, the application of a boundary is useful in analysis: to scale the flow of material, for instance.

However, it is important to recognize that definitions establish what is important and significant. A key quality of a definition is that it is *definitive*. In other words, it must be *objectively true*. If a critical component of that definition cannot hold objectively true over the scope and scale to which it is being applied, then the definition is at best *an approximation*, and should be noted as such. An important aspect of an approximation is to know when and to what degree it is valid.

Earlier, it was observed that if any part of the solution to resolve the need exists outside the system that fulfills that solution, then completeness and definitiveness are potentially unknowable. If completeness is unknowable, then there is a violation of one of the conditions of assurance required to meet the intent of system design.

To assess the validity of the notion that systems are defined by their boundaries, consider some example systems in the context of the three technology development paradigms and observe the definitiveness of the system boundary. By evaluating in the context of the paradigms, we may arrive at better understanding of the applicability of the assumption.

Passive Extensions A bicycle may be classified as a passive extension because it requires external actualization to extend the capability and capacity of the rider. We would consider it relatively simple to draw boundaries between the bicycle and the rider, and between the bicycle and the ground. From experience, we could then readily define boundaries around all of the parts and pieces required to construct the bike. If necessary, we could apply physics to predict the behavior in order to make improvements to the design of the bicycle's construction. In this case, the construct of the bicycle sufficiently encompasses everything necessary for the human to extend their capability and capacity. The interactions with the environment are within the human's capability to comprehend and appreciate because of our intrinsic familiarity with the interaction space. Therefore, completeness is within the capability and capacity of the human to comprehend.

For passive extensions, we can *usefully consider the system to be defined by the boundary.*

Amplifying Extensions Many cars may be classified as amplifying extensions. Here, we must support and resolve the needs of the motive force in addition to the needs of the tool user, including the fuel provisioning system—the collection of available gas stations that provision the fuel. The notion of what constitutes the system becomes more complicated: Is the fuel station part of the system, or not? What about the passengers? If the car is a plug-in electric vehicle, do we now include the electric grid? As the number of significant dependencies increase, we have more decisions to make about where to draw the boundary. Our sense of definitiveness is becoming less precise and neat, but not necessarily in a harmful way. We can compensate for the incompleteness by defining a loose set of external actors allowed to interact with the system in defined contexts, enabling us to contain the increasingly uncertain interactions and their consequential behaviors with approximations. The interactions of the system with its environment are generally within the human's capability to sufficiently comprehend and appreciate because of a combination of our intrinsic familiarity with the interaction space and the assistance of externalized constructs. Therefore, completeness is, *in general, sufficiently* within the capability and capacity of the *extended* human to comprehend.

For amplifying extensions, we can consider the boundary to be a *generally useful approximation* of the system definition.

Collaborative Extensions *Collaborative extensions emulate the social interactions of natural systems.* To illustrate the effect social behaviors have on the definitiveness of the system boundary, consider the behavior of fire ants. Suppose we observe two groups of fire ants; draw a boundary around each of the groups and call them system *alpha* and system *beta*. If *alpha* falls into a body of water, we will observe them interacting with the environment to form a raft. If *beta* likewise falls into the water, we will likewise observe them interacting with the environment to form a raft. Now suppose *beta* approaches *alpha* and they begin interacting with each other. Now, there is only one group of fire ants. We can draw a new boundary around this group and call this new system *gamma*. *Gamma* likewise interacts with the environment and forms *an entirely new raft*.

What happened to the original systems? The original members remain and are working together. Theoretically, we could continue to draw boundaries around the original members of *alpha* and *beta* separately, though these lines would be constantly shifting as the ants maneuver around in this new configuration; the systems would be inter-mix*ing*. Have *alpha* and *beta* become subsystems of *gamma*? Or is it more accurate to state that they have been subsumed and the distinctions of *alpha* and *beta* are no longer significant? Does it support the resolution of our needs—or improve our comprehension—to consider these boundaries? Or does it create confusion?

Now, suppose we take the viewpoint of *alpha*. Would we consider *beta* to have entered *alpha* as a set of inputs that simply have the net effect of growing the number of constituent ants in *alpha*? Thus, *alpha* remains, and therefore, *alpha* is also *gamma*. Or did *alpha* and *beta* collaborate to reorganize and form *gamma*? Is the boundary all of these or none of these? Are any of these potential definitions of the system relevant or significant?

To date, we lack sufficient externalized constructs to compensate for the lack of definitiveness and completeness observed in the social behaviors of natural collaborative systems. For this reason, we might expect that the prevailing viewpoint will always struggle with the construction of any system that requires us to *negotiate interactions amongst numerous collaborating systems, directed towards adapting the overall effect to the scale and scope of our objectives* (e.g., the internet, future transportation systems, and etc). The *adaptive* interactions of the system with its environment *and with itself* are *generally beyond* the human's capability to sufficiently comprehend and appreciate in order to successfully emulate,[10] even with the

[10]We have demonstrated the ability to emulate these behaviors on a small scale under controlled conditions—swarm robot demonstrations, for instance. But what happens when we attempt a system beyond our capability and capacity to comprehend its completeness?

assistance of externalized constructs. Therefore, completeness is *insufficiently* within the capability and capacity of even the *extended* human to comprehend.

For collaborative extensions, the *boundary is no longer a valid approximation* of the definition of the system.

Even when the systems should be within our capability and capacity to comprehend, our dependence on the practitioner and the questionable constructs provided by the viewpoint to support them can cause further issues with definitiveness and completeness. For instance, show two practitioners the same design problem and they will likely define the appropriate system boundary for the surrogate construction *differently*, based on their own experience. This effect introduces further ambiguity that only intensifies as the system is decomposed and the boundaries of the subsystems are defined based on arbitrary cleavage lines [46]. The process of system decomposition to determine subsystems and components is undertaken experientially with no grounded sense of order, structure or intent. The most valued outcome of the decomposition process is that the resultant system is described in the form of the expected artifact—one that is aligned with what is significant to the toolmaker—with no guarantee that it will align with what is significant to the system.

What is the inherent significance of a boundary? A boundary specifically separates two things from one another and values what and how much crosses that line. Is this separation truly useful for the design of the system or does it merely encourage us to compensate for incompleteness by adding yet more parts and connections all equally lacking in their sense of significance to the whole?

Have we created a definition for a system that when applied to collaborative extensions becomes *arbitrary and inconsistent*? We might expect that anything less than definitive could cause confusion in the construction and potentially aggravate the *perception of "complexity" as significant*. After all, the *resultant inconsistency of the behavior* of the system might encourage us to conclude that it is *the system itself* that is insufficiently knowable and predictable.

Since the concept of boundaries is a core tenet of the *parts-centric* view of systems, we might expect a natural tendency to bring in *additional* mental constructs in an attempt to *explain these inconsistencies* in observations.

6.2.2 Systems Have Emergent Behavior

In the prevailing viewpoint, the combination and interaction of disparate systems produces behavior that *cannot be inferred by the parts*. This *phenomenon is explained* as "emergence" [47]. Sometimes, emergent behavior is "deducible [47]." Deducibility enables us to predict these behaviors in our models.[11]

[11]Perhaps this apparent "deducibility" is a consequence of sufficient experiential knowledge and practitioner pattern-matching?

Emergent behavior can also manifest in undesirable and unpredictable ways during the course of the design process. Because this happens, we are encouraged to perform additional construction trial-and-error feedback cycles in an attempt to identify these behaviors earlier; and when found or feared, to build compensations into the system in an effort to manage these behaviors.

If emergent behavior is a generally unpredictable consequence of the interaction of parts, then the potential for undesirable emergent behavior should increase significantly with the number of the parts and their connections. Thus, as the "complexity" of a system increases, so might our propensity to introduce uncomfortable compensations into our process and design.

What do these compensations represent? They are additional burdens imposed in an effort to manage (limit) the effects of the "emergent behavior." When we *compensate by adding more parts into the system, have we not exacerbated the problem* by creating even more potential for emergent behavior?

Is it possible we are repeating history by *"adding epicycles" to compensate for discrepancies* between our model of systems and our observations of undesirable emergent behavior?

7 The Path Forward

Every block of stone has a statue inside it and it is the task of the sculptor to discover it.
—Michelangelo

This chapter has provided evidence suggesting the prevailing viewpoint becomes increasingly misaligned with the intent of system design as the system exceeds the capability and capacity of the practitioner to comprehend it. Rather, the prevailing viewpoint may be more appropriately aligned with the intent of construction and manufacturing. While this *parts-centric* viewpoint is critical for the construction of systems to resolve a need, it is arguably insufficient to handle the design of "complex" systems on its own—in part because of its inherent disconnect with the need. An alternative viewpoint has been proposed (see Fig. 6) that demonstrates theoretical agreement with the intent of system design, regardless of the scale and scope of the system. This alternative viewpoint appears to bridge the current gap between the formation of the need and the construction of the system. The existence of a viewpoint demonstrating improved agreement with the evaluation criteria suggests that its view of systems may be a more suitable basis for our comprehension of "complex" systems.

Unexpectedly, the analysis also uncovered the probable existence of at least two other viewpoints: one focused on forming the need; the other on the interactions among systems that performs the resolution of the need. These four viewpoints appear to work collaboratively to affect adaptive change within and between systems.

Until these three additional viewpoints are better understood and methods for their effective application and collaborative augmentation with the existing construction paradigm have been developed, we can expect to continue our struggles with "complexity." In the meantime, what can we do to improve our approximations?

7.1 Principles and Practices to Improve Alignment

A successful collaborative system is all about the *federation*: an alignment and unification towards fulfilling a common objective. Application of a properly federated approach in the space-time domain encourages minimal interfacing between the constituent systems and focuses the design only towards what is significant to resolve the perceived need.

What might we observe if we are constructing a well-federated system? What might we observe if we stray from that ideal? This section will briefly explore a set of principles and practices that seem to improve alignment with the underlying concepts of the alternative viewpoint and provide useful and actionable indicators of misalignment.

7.1.1 Coupling and Cohesion Principles

Coupling and cohesion principles are at the root of many design patterns that successfully guide the construction of passive and amplifying extensions. In general, coupling and cohesion states that elements with similar functional purposes should be organized together, and treated as a cohesive whole. The underlying assertion is that a system organized around these cohesive purposes is more likely to be stable and well-structured.

From the alternative viewpoint, coupling and cohesion principles begin to take the following form: At the scale and scope required to resolve the need, a well-federated system should exhibit distinct, non-overlapping *roles*, united toward fulfilling a common objective. For example, a race car pit crew demonstrates the following *core roles* that act to fulfill the overall objective of provisioning the car during the race: provision the tires, provision the fuel and provision the driver. Each role is distinct, cooperative and—once practiced—requires minimum communication for coordination. In addition, each *core role* represents its own common objective, which if insufficiently fulfilled by a single system, can be split into its own distinct, non-overlapping roles, united toward fulfilling that common objective. For instance, the objective of provisioning the tires cannot be fulfilled sufficiently by a single person, so it is instead split into four roles—one for each tire. In all cases, each role understands their contribution toward the common objective and works collaboratively toward fulfilling that common objective. The overall effect is accomplished with stunning speed and accuracy.

If the system is constructed with overlapping roles, in which at least one role acts in a way that exceeds its purview, we might expect to observe uncertainty, distrust and conflict. To compensate for the uncertainty, we might add interactions in an attempt to reconcile the overlap, which might manifest within the system as more coupling and less cohesion. To compensate for the distrust and conflict, we might add in protective barriers that also manifest within the system as more coupling and less cohesion.

7.1.2 Proper Sufficiency Principle

The basis for proper sufficiency is the observation that everything we find significant must be supported. When we stay focused on the true need, we support only that which is truly significant toward resolving the need—no more and no less. If we begin to support things that do not resolve the need, then we are *supporting systems that are not aligned with fulfilling the common objective*. Instead, we are supporting systems that are aligned toward fulfilling some other objective, and these tend to grow rapidly into systems that parasitically drain resources away from accomplishing our true intent.

The criteria for proper sufficiency states: If you add one more thing, the system does more than it must to resolve the need; if you remove one more thing, it does not do enough to resolve the need. What might we expect to observe if proper sufficiency is met? Returning to the pit crew example: Assigning four people to provision the tires demonstrates proper sufficiency—the resultant effect is the stunning speed and accuracy required to win the race.

What if it is not met? Add a person and you add the potential for unnecessary collisions in the pit, resulting in confusion and a reduction in timeliness and accuracy. Also, the system must sustain the increased resource burden of supporting that additional person. Would we observe that person focused toward fulfilling the common objective or more focused toward his or her own interests, such as avoiding collisions and collecting pay? Remove a person and now the remaining crew must correctly and completely coordinate to fill the role left vacant, resulting in the potential for overlapping roles (see coupling and cohesion principles), and a likely reduction of the timeliness and accuracy of the operation.

7.1.3 The Surgeon Principle

The "Surgeon Principle" states that the cuts we make to a system to expose its constituent elements (often referred to as "decomposition") ***should not*** make it "bleed." If we cut properly, then we have successfully maintained the integrity of the cohesive (correct and complete) whole. If we cut improperly, then we have created misalignments that spawn *undesired* systems around fulfilling some other objective that is not aligned with the whole. These other systems can be partially synergistic, parasitic (the primary objective is to persist itself) or antagonistic.

In any case, because we have created these secondary systems, we have made them significant to us. If they are not satisfied, we perceive that the whole of the system cannot be achieved. Thus, they must be supported. How are they supported? By creating new systems that may be as mild as a simple patch (i.e., a skin graft) or as massive as a full-blown outer control system (i.e., a heart bypass).

What are signs and symptoms that a system is bleeding? Perhaps if we hear a compensatory statement such as, "Because we found x, we must add in y to fix it." A bleeding system is a weakened system, leaving it highly sensitive to infection from outside influences; an *unhealthy sensitivity to small perturbations in its operating environment*, for instance.

A seemingly inevitable consequence of a bleeding system is a shift away from proper sufficiency, a substantial increase in coupling, and a corresponding reduction in cohesion (see earlier discussions). In other words, if we cut improperly and cause the system to bleed, this means we have created *unnecessary* problems we *cannot ignore* and the consequences are *detrimental to our success*.

The most effective way to stop the bleeding is to re-factor the system around the principles of successful collaboration and proper federation.

7.2 Concluding Remarks

> As a consequence of [emergent phenomena], we have to shift our attention from the components of our world to their interactions. In other words, we need a change from a component-oriented to an interaction-oriented, systemic view, which is at the heart of complexity science.[12] I claim that this change in perspective, once it becomes common wisdom, will be of similar importance as the transition from the geocentric to the heliocentric worldview.
>
> —Dirk Helbing, Complexity Time Bomb (2014) [48]

Thousands of years ago humans developed a craftsman mentality that encouraged us to view systems as constructions of things ("the parts give rise to the whole"), rather than as the space-time manifestation of elegantly patterned expressions of value, self-motivated to interact and federate to resolve their needs ("the whole before the parts"). As a consequence of the *human-scale* pervasiveness of our successful craftsman heritage, we have almost certainly applied our construction viewpoint to *all* things: How we comprehend and form our expression of needs. How we design solutions to resolve those needs. How we construct and manufacture those solutions as systems. How we view the space-time interaction between systems as they perform the resolution of needs. The problem with applying a single viewpoint to all things is that the intent of all of the other distinct, non-overlapping viewpoints must still be met. The single viewpoint approach violates proper

[12] Perhaps we should consider a more apt name for Helbing's referenced science?

sufficiency, creating the potential for confusion, uncertainty and distrust due to the overlap of roles and may *result in a system that cannot sufficiently affect adaptive change towards the resolution of need.* Once this occurs, compensations are applied that only serve to increase coupling and reduce cohesion. When excessive, these compensations can create a ***perception of complexity as significant to design***.

The seemingly insurmountable problems we experience with the burden of complexity may *have a path to resolution*, but only if we choose to accept the implications that our struggles may be *self-imposed*. The question for the system design community may no longer be "How do we manage complexity?" The question may have transitioned to: "Is it time we explored the path not taken?" It may make all the difference.

Acknowledgements We'd like to extend our appreciation to Draper for providing opportunities to ignore boundaries and question all things assumed and sacred; to Frank Serna for extending the invitation to propose our ideas for this volume and for being our advocate and advisor; to Dan Soares for his unwavering support and indispensable contributions; and to Spencer Lewis for volunteering to extend our research of the existing literature.

References

1. Parunak, H., & VanderBok, R. S. (1997). *Managing emergent behavior in distributed control systems.* Anaheim, CA: ISA-Tech. Retrieved from http://pdf.aminer.org/000/333/760/integrative_technology_engineering_emergent_behavior_into_materials_and_systems.pdf
2. Mogul, J. C. (2006). *Emergent (mis)behavior vs. complex software systems.* 1st ACM SIGOPS/EuroSys European Conference on Computer Systems 2006, New York, pp. 293–304. doi: 10.1145/1217935.1217964.
3. Kilner, S. *Complexity metrics and difference analysis for better application management* [whitepaper] [Online]. [Cited: January 2, 2015]. http://www.databorough.com/downloads/White_Papers/Complexity-Metrics-and-Difference-Analysis-White-Paper.pdf
4. Kreimeyer, M., & Lindenmann, U. (2011). *Complexity metrics in engineering design. Managing the structure of design processes* (Vol. XIII). New York: Springer.
5. Houghton Mifflin Company. (1985). *The American heritage dictionary* (2nd College ed.). Boston: Houghton Mifflin Company.
6. Calvano, C. N., & John, P. (2004). Systems engineering in an age of complexity. *Systems Engineering, 7*(1), 25–34.
7. Deshmukh, A., & Collopy, P. (2010). *Fundamental research into the design of large-scale complex systems (AIAA 2010–9320).* 13th AIAA/ISSMO multidisciplinary analysis optimization conference. AIAA 2010–9320. Fort Worth, TX: AIAA/ISSMO.
8. Collopy, P. (2012). *A research agenda for the coming renaissance in systems engineering (AIAA 2012–0799).* 50th AIAA aerospace sciences meeting including the new horizons forum and aerospace exposition. Nashville, TN: AIAA.
9. Sangiovanni-Vincentelli, A. (2010). Managing complexity in IC design. *Complexity workshop presentation preceding the DARPA Adaptive Vehicle Make (AVM) program.* Spring 2010.
10. Pughat, A. (2012). Power, interconnect and complexity crises in future VLSI: From a designer's point of view. *International Journal of Electrical Engineering and Technology (IJEET), 3*(2), 210–222.
11. Kubik, A. (2003). Toward a formalization of emergence. *Artificial Life, 9*(1), 41–65.

12. Newman, D. V. (1996). Emergence and strange attractors. *Philosophy of Science, 63*(2), 245–261.
13. Goldstein, J. (1999). Emergence as a construct: History and issues. *Emergence, 1*(1), 49–72.
14. Emmeche, C., Koppe, S., & Stjernfelt, F. (1997). Explaining emergence: Towards an ontology of levels. *Journal for General Philosophy of Science, 28*, 83–119.
15. Stepney, S., Polack, F. A. C., & Turner, H. R. (2006). *Engineering Emergence*. ICECCS 2006: 11th IEEE International Conference on Engineering of Complex Computer Systems (pp. 89–87). Stanford, CA: IEEE.
16. Fricke, E., & Schulz, A. P. (2005). *Design for changeability (DfC): Principles to enable changes in systems throughout their entire lifecycle*. Systems Engineering *4*(8).
17. de Neufville, R. (2004). *Uncertainty management for engineering systems planning and design*. MIT Engineering Systems Symposium.
18. Hastings, D. E., Weigel, A. L., & Walton, M. A. (2003). Incorporating uncertainty into conceptual design of space system architectures. INCOSE International Symposium, *13*(1), 1380–1392.
19. de Weck, O. L., & Jones, M. B. (2004). *Isoperformance: Analysis and design of complex systems with known or desired outcomes*. 14th Annual International Symposium of the International Council on Systems Engineering (INCOSE), Toulouse.
20. Project Performance International. Training Available On-Site. *Project Performance International Web site*. [Online] [Cited: May 12, 2015.] www.ppi-int.com/training/onsite-training.php
21. Applied Technology Institute. ATI Courses. *Applied Technology Institute Web site* [Online]. [Cited: May 12, 2015]. www.aticourses.com/index.htm
22. Kuhn, T. S. (1970). *The structure of scientific revolutions*. Chicago: The University of Chicago Press.
23. Elon Musk's secret weapon: A beginner's guide to first principles. *Evantostudio (blog article)* [Online] November 2013. [Cited: January 5, 2015]. http://studioblog.envato.com/elon-musks-greatest-weapon-laymans-guide-first-principles/
24. Boyle, R. (1661). *The skeptical chymist: Or chymico-physical doubts & paradoxes*. London: J. Cadwell.
25. Hill, J. H., & Petrucci, R. H. (1996). *General chemistry*. Upper Saddle River, NJ: Prentice Hall.
26. Cooper, G. R., & McGillem, C. D. (1967). *Methods of signal and system analysis*. New York: Holt, Rinehart and Winston.
27. Kegan, R., & Lahey, L. L. (2009). *Immunity to change: How to overcome it and unlock the potential in yourself and your organization (leadership for the common good)*. Boston: Harvard Business School Publishing Corporation.
28. Dolling, L. M., Gianelli, A. F., & Statile, G. N. (Eds.). (2003). *The tests of time: Readings in the development of physical theory*. Princeton, NJ: Princeton University Press.
29. Barnett, P. (2015, January/February). Measures that matter. *Corporate Finance Review*, 5–10.
30. Ackoff, R. L., Magidson, J., & Addison, H. J. (2006). *Idealized design: How to dissolve tomorrow's crisis. . .today*. Upper Saddle River, NJ: Prentice Hall.
31. Maier, M. W., & Rechtin, E. (2000). *The art of systems architecting* (2nd ed.). New York: CRC Press.
32. OK Go. (2014). The writing's on the wall. *You Tube* [Music Video].
33. Routio, P. Normative Analysis and Preparing the Proposal. *Arteology, the science of products and professions* [Online]. [Cited: May 12, 2015]. www2.uiah.fi/projecti/metodi/179.htm#compare
34. Crutchfield, J. P. (1994). *Is anything ever new? Considering emergence*. SFI Working Paper.
35. Estefan, J. A. (2008, May 23). *Survey of model-based systems engineering (MBSE) methodologies* [Online]. [Cited: January 5, 2015.] http://pdf.aminer.org/000/260/416/towards_a_unified_paradigm_for_verification_and_validation_of_systems.pdf
36. Office of the Under Secretary of Defense (Acquisitions, Technology and Logistics). (2008, August). *Systems engineering guide for systems of systems* [Online] 1.0. [Cited: January 5, 2015.] http://www.acq.osd.mil/se/docs/SE-Guide-for-SoS.pdf

37. Larson, W. J., & Wertz, J. R. (Eds.). (1999). *Space mission analysis and design (space technology library)* (3rd ed., Vol. 8). Boston: Microcosm Press.
38. Karwowski, W., Soares, M. M., & Stanton, N. A. (2011). *Human factors and ergonomics in consumer product design: Methods and techniques: Needs analysis: Or, how do you capture, represent, and validate user requirements in a formal manner/notation before design*. Boca Raton, FL: CRC Press.
39. Babou, S. (2008). *What is stakeholder analysis? The project management hut Web site* [Online]. http://www.pmhut.com/what-is-stakeholder-analysis
40. Wogalter, M. S., Dempsey, P. G., & Hancock, P. A. (2000). What's in a name? Using terms from definitions to examine the fundamental foundation of human factors and ergonomics science. *Theoretical Issues in Ergonomics Science, 1*, 1.
41. Hughes, T. P. (1983). *Networks of power: Electrification in Western society, 1880–1930*. Baltimore, MD: Johns Hopkins University Press.
42. Hughes, T. P. (2004). *Human-built world: How to think about technology and culture*. Chicago: University of Chicago Press.
43. Arthur, W. B., & Polak, W. (2006). The evolution of technology within a simple computer model. *Complexity, 11*(5), 23–31.
44. Wolfram, S. (2002). A new kind of science. *Wolfram Media*.
45. Bloch, M., Blumberg, S., & Laartz, J. (2012, October). Delivering large-scale IT projects on time, on budget, and on value. *McKinsey & Company Web Site* [Online]. [Cited: January 2, 2015.] http://www.mckinsey.com/insights/business_technology/delivering_large-scale_it_projects_on_time_on_budget_and_on_value
46. Eremenko, P. (2010, October 7). Adaptive vehicle make [proposer's day briefing to DARPA for the Adaptive Vehicle Make Program]. *DARPA Web Site* [Online]. www.darpa.mil/WorkArea/DownloadAsset.aspx?id=2659
47. Baas, N. A., & Emmeche, C. (1997). On emergence and explanation. *Intellectica, 2*(25), 67–83.
48. Helbing, D. (2014). Complexity time bomb—When systems get out of control. *Chapter 2, Digital Society, 2. Forthcoming*. SSRN: http://ssrn.com/abstract=2502559

Early Phase Estimation of Variety Induced Complexity Cost Effects: A Study on Industrial Cases in Germany

Sandra Eilmus, Thomas Gumpinger, Thomas Kipp, Olga Sankowski, and Dieter Krause

1 Variety Induced Complexity: Causes and Effects

In this contribution, trans-disciplinarity is understood from an industrial point of view where inter-corporate trans-disciplinarity refers to different branches and intra-corporate trans-disciplinarity refers to different corporate disciplines such as development or sales. Therefore, analyzing and reducing variety induced complexity is a trans-disciplinary challenge. The following sub-sections describe the causes and trans-disciplinary effects of variety induced complexity.

1.1 Complexity Is Induced by Different Causes

Complexity can be described and classified in various ways, e.g. as technical, organizational, cognitive complexity etc. (Chapter "So It's Complex, Why Do I Care?"). In this contribution, we distinguish between technical functional complexity and variety induced complexity. While technical functional complexity is caused by numerous interacting functions pertaining to a single product, variety induced complexity is caused by the presence of numerous product variants. The

S. Eilmus (✉) • T. Gumpinger
Odego GmbH, Hamburg, Germany
e-mail: eilmus@odego.de

T. Kipp
Knorr-Bremse Systeme für Schienenfahrzeuge GmbH, München, Germany

O. Sankowski • D. Krause
Institute for Product Development and Mechanical Engineering Design (PKT), Hamburg University of Technology, Hamburg, Germany

F.-J. Kahlen et al. (eds.), *Transdisciplinary Perspectives on Complex Systems*,
DOI 10.1007/978-3-319-38756-7_10

nature of this variety induced complexity as well as their challenges and effects on various corporate disciplines are described in this contribution. Nonetheless, variety induced complexity is especially hard to handle if the variant products include a certain technical functional complexity as well.

1.2 The Challenges of Individualization and Globalization in Mechanical and Plant Engineering

The megatrends of individualization and globalization determine the way products and services are demanded in market. Individualized customers expect products exactly fitting their individual needs and highly specific requirements. On the other hand, globalized markets imply globalized customers with country-specific needs, customs and laws as well as global competitors along with a global availability of a broad range of products. These trends highly affect companies in mechanical and plant engineering. According to a study among over 300 German companies [1], increasing demand for customized system and integration solutions, i.e. increasing complexity of systems due to growing number of demands from multiple domains, is the most important trend influencing German mechanical and plant engineering industries. Another example how increasing complexity leads to increasing cost in aerospace systems is given in Chapter "So It's Complex, Why Do I Care?".

These conditions lead mechanical and plant engineering companies to fulfill individual customer needs at globally marketable prices. Individual customer needs are fulfilled by a broad range of different product variants. Developing, producing and offering a broad range of product variants causes high cost, i.e., complexity cost as compared to mere mass production. In turn, this complexity cost reduces the marketability in terms of pricing. Furthermore, increased product variants result in a loss of transparency leading to high communication effort and limited-grounded decision making. This intensifies the cost effect of variety induced complexity (Fig. 1).

1.3 Trans-disciplinary Complexity Cost Effects

Variety induced complexity causes cost in various disciplines of a company. This implies that the handling of variety induced complexity is a trans-disciplinary challenge. In order to reduce complexity cost, finding a trans-disciplinary solution and its evaluation are indispensable. Overall cost reduction needs to be achieved, rather than mere partial improvements in single disciplines.

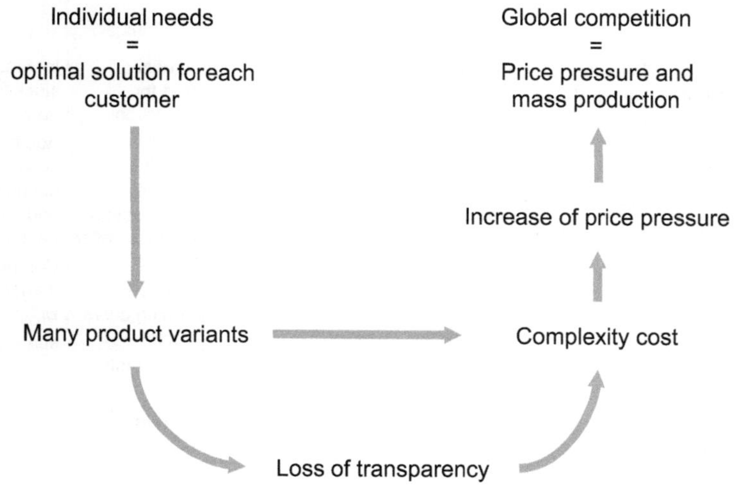

Fig. 1 The challenge of variety induced complexity

One of the models to describe the affected disciplines along the product's supply chain is the model of product life phases corresponding to the product life oriented perspective proposed in Chapter "Designer Systems of Systems: A Rational Integrated Approach of System Engineering to Tailored Aerodynamics, Aeroelasticity, Stability, Control, Geometry, Materials, Structures, Propulsion, Performance, Sizing, Weight, Cost". Figure 2 describes the complexity cost effects in different disciplines along the product life phases of the supply chain.

2 Scientific Contributions on Reducing Internal Variety

Several fields of research address the problem of reducing internal variety, namely, management approaches, such as complexity management and variant management, and product development approaches that involve the development of product families and product platforms. Furthermore, approaches to cost estimation may help in understanding complexity cost effects.

2.1 *Complexity Management and Variant Management*

It is important to clarify the definition of complexity in mechanical and plant engineering industry that suffices the needs of industrial practice. Brosch [3] proposed the following definition of complexity: Complexity is the property of a high number and variety of elements and relations causing additional cost in an

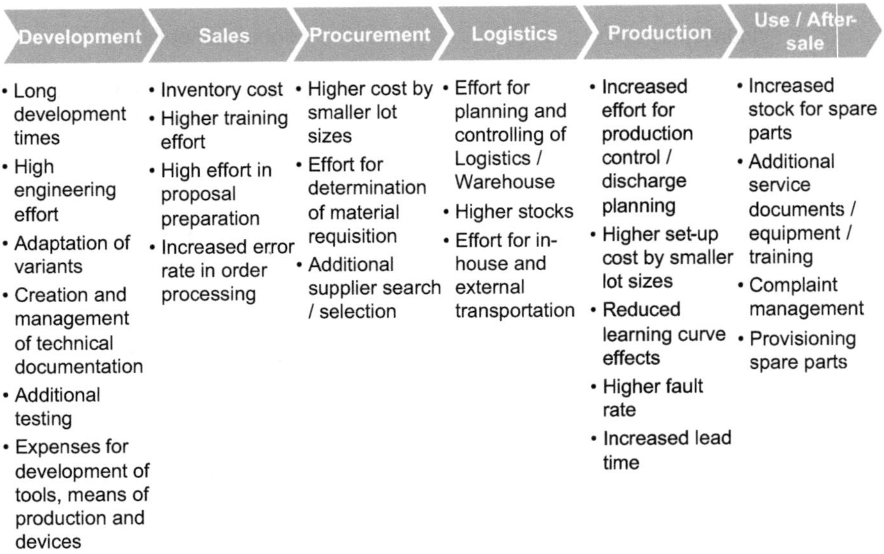

Fig. 2 Complexity cost effects in different disciplines along the product life phases of the supply chain based on a literature research done by Ripperda and Krause [2]

industrial firm. In industrial context, these elements include components, processes and persons.

Complexity management describes approaches to handle complexity more efficiently in a company. Design Structure Matrix (DSM) [4] is a matrix based approach to complexity management. It clusters components, processes and persons with strong relations to each other into modules to reduce complexity.

Variant management focuses on the handling of variety induced complexity. A prominent approach of variety management is Mass Customization suggested by Pine [5] that describes general management changes required in a company to industrially produce tailored products at competitive prices.

2.2 *Development of Modular Product Families and Platforms*

In order to enable effective complexity management as described earlier, modular product structures need to be developed. Modular product structures are product structures that can be characterized by commonality, combinability, function binding, decoupling and standardized interfaces to certain degrees [6]. Modular product families describe the whole of all the variants of a product that are configured from a set of modules. Product platforms can be understood as a special type of a product family that reuses a cluster of common assets across all product variants of the family. These common assets can be components, processes, knowledge, people and relationships [7].

Different approaches to the development of modular product families and platforms have been published [8]. There are approaches, such as by Jiao and Tseng [9], Du et al. [10], and Haug et al. [11], that support the visualization of product families in order to enhance their configurability. While some other approaches, such as of Simpson et al. [12], de Weck et al. [13] and Hölttä-Otto [14], focus on the optimization of the product variants, component variants or configurations to specific cost or performance targets. Integration of product strategic module drivers into the modular product structure is the core idea of Modular Function Deployment proposed by Erixon [15]. Krause et al. developed a holistic approach combining the ideas of modularization according to module drivers with the redesign of components and products to enhance their configurability, as well as other modular properties such as decoupling and function binding [16].

2.3 Complexity Cost Analysis

"Complexity cost denotes the cost of indirect functions at a company and its suppliers that are caused by component variety; complexity cost includes, for instance, the cost of designing, testing, and documenting a component variant" [17]. Traditional cost methods distinguish between direct cost and overhead cost. These cost methods focus on cost accounting for controlling the overall expenditure. For example, the differentiated overhead method [18] assigns overhead cost by overhead absorption rates relative to the direct cost. As overhead cost is assumed globally, it does not consider distinction between product variants with high or low lot sizes, or product variants causing additional processes. Resource based methods [19], such as activity-based costing (ABC) (e.g., used by Park and Simpson [20]), or the variety accounting method [21] show overhead cost more transparently. These methods are used as basis for strategic variant management decisions and need a proper data base as input. Tu et al. [22] searched and compared data of existing processes with processes of variant products to estimate cost. This approach supports optimization and decision making. More detailed literature research on complexity cost analysis has been conducted by Ripperda [23] with the conclusion that the traditional cost accounting systems are not sufficient to acquire the effects of modularity. So far, only some contributions studying these effects have been published, e.g., Fujita et al. [24] and Sedlmeier et al. [25]. Yet, there is a need for more detailed analysis and empirical research, especially on the trans-disciplinary nature of complexity cost effect.

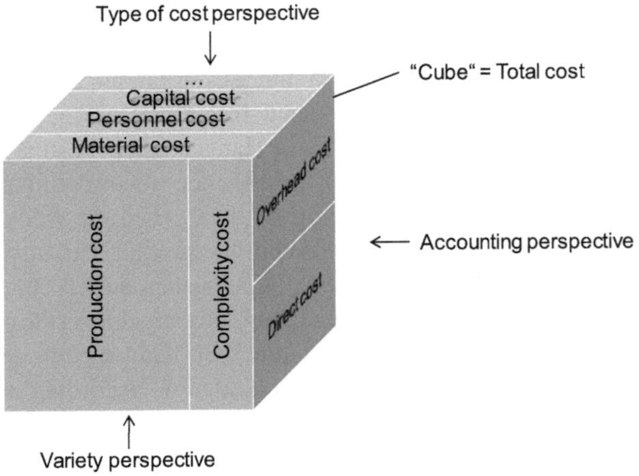

Fig. 3 Perspectives of cost modeling cf. [18] extended by variety perspective

3 Hypotheses on the Reduction of Variety Induced Complexity Cost

To gain new insights on the cost effects of variety induced complexity and its reduction by modularization, the following hypotheses are introduced. Consider the simple cost model shown in Fig. 3. This cost model does not replace other cost models but enables us to take into account the new perspective of variety induced complexity cost.

$$TC = PC + CC$$

Where

TC = Total cost
PC = Production cost
CC = Complexity cost

Hypothesis 1
Modularization is a useful approach to reduce complexity cost.

$$CC_{as\ is} > CC_{modular\ concept}$$

Hypothesis 2
Modularization can influence production cost positively or negatively.

Hypothesis 3
Modular concepts show different complexity cost effects in different disciplines.

Hypothesis 4

The cost effects of modularization depend on lot size and are thus different for different segments and branches.

Hypothesis 1 and 2 aim to give more insight on the cost effects of modularization. In industrial practice, the expectations on the cost effects of modularization projects are rather high and diffused. This is why clear relations between modularization and cost effects need to be described. The questions that need to be answered are: Does modularization lead to reduced complexity cost? Does modularization further lead to reduced production cost or does the effect of oversizing due to modularization lead to the production cost increase?

Hypothesis 3 is fundamental for developing methodical support to reduce complexity cost. If the complexity cost effects of modularization differ over different disciplines, it is crucial to develop a methodical support that enables a differentiated trans-disciplinary evaluation of modular concepts.

On basis of hypothesis 4, the transferability to different branches and segments of the findings of this contribution is investigated. Different branches, among others things, are characterized by different annual lot sizes. Because of this, the consistency of findings is analyzed based on lot size dependency.

4 Industrial Application of the Integrated PKT-Approach for Developing Modular Product Families

Industrial modularization projects are analyzed to gain insights into whether their data support the above hypotheses. These modularization projects were performed on subassemblies of products from German mechanical and plant engineering industry. The Integrated PKT-Approach for Developing modular Product Families was used [16] for modularization.

Detailed data and insights are gained from a previously published case study on a family of electrical device industrial vehicles [26]. Especially, the part of cost estimation is now provided with more branch specific data to gain further detailed insights into the topic pursued in this contribution. Yet, this case is simplified and generalized for publication due to confidentiality reasons. In the following sub-sections, the PKT-approach will be first described in general and then demonstrated through its methods and tools that are applied on the electrical device family.

4.1 General Means to Enable Industrial Application

The integrated PKT-Approach aims to realize a customer-required external variety, based on possible low internal product and process variety [16] (Fig. 4). It was

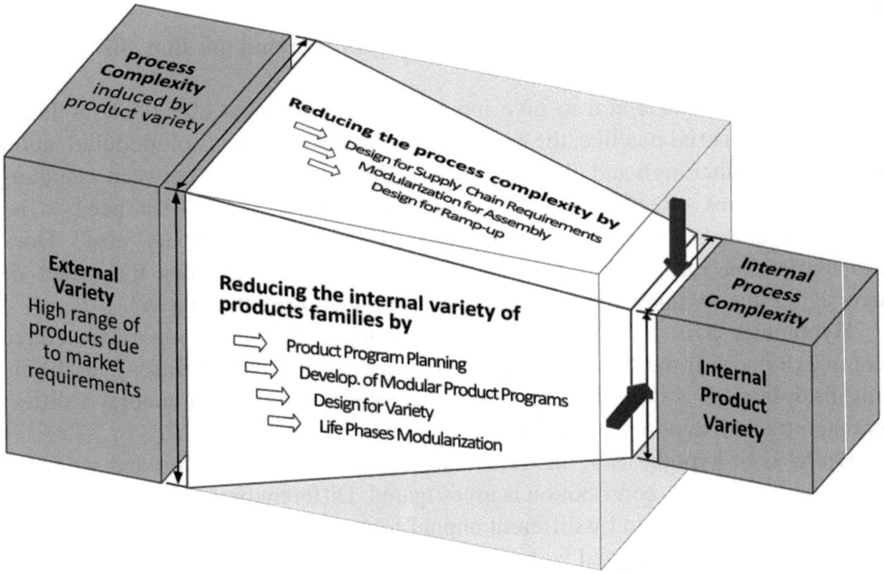

Fig. 4 Method units of the Integrated PKT-approach for developing modular product families [16]

developed at the PKT institute, Hamburg University of Technology, in order to provide methodical support for developing modular product families for industrial application.

This is fulfilled by the following means:

- A configurable methodical toolkit with different method units and tools to tailor the methodical support to different corporate situations.
- Visualizations that foster solution finding in trans-disciplinary teams.
- Actual redesign of components to enable their configurability and to enhance their commonality.

The method units applied for modularization of the electrical device family are presented below in detail.

4.2 As Is Analysis

The starting point of modularization is an "as is" analysis of the existing external and internal variety in order to understand the causes of variety induced complexity. External market variety is analyzed with the Tree of Variety (ToV) [27]. This tool illustrates the properties that differentiate the individual product or module variants from customers' point of view. It can even be used for demonstrating internal variety. The electrical device family provides handling support for the driver of the

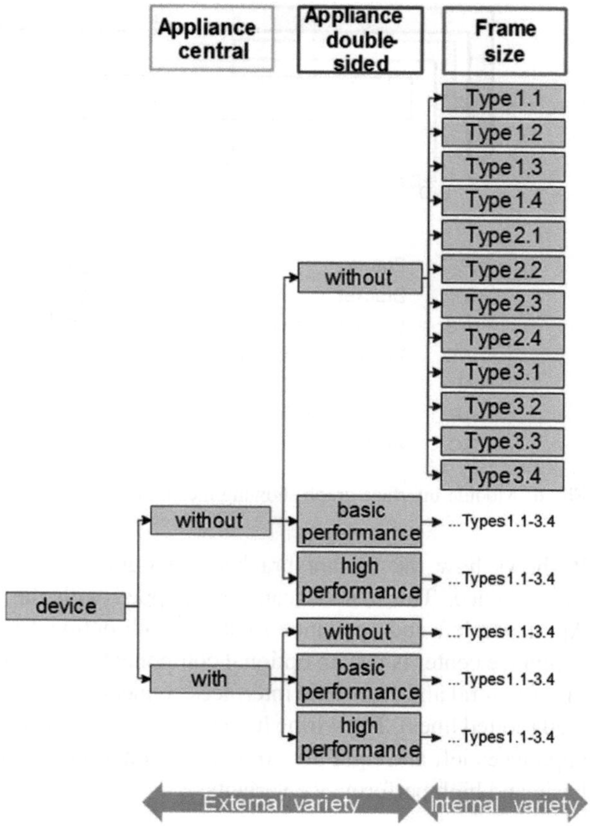

Fig. 5 Tree of variety showing the variety of the devices mounted on the associated frame [26]

industrial vehicle. Depending on the specific use case of a particular customer, the device can be individualized (Fig. 5).

The customer may choose between six different device types equipped with none or three different appliances and their combinations. One appliance can be attached in the middle and/or the two other appliances at both the sides of the bracket. This double-sided appliance is available as a high and basic performance variant. As this device is handled in internal processes mounted to a variant frame, its internal variety even multiplies with the variety of the frame. A tool for analyzing the internal variety on a technical level is the Module Interface Graph (MIG) [28], which is shown in Fig. 6.

It visualizes rough shapes and geometrical layout of the device as well as technical functional interfaces. The variety of components is symbolized with white for standard components, grey for variant components and dotted lines for optional components. A variant component is a component that is generally applied in each variant but with different specifications, e.g., high or basic performance. Optional components are those components that are only applied in some variants. The symbols enable us to summarize whole product or module family in a single visualization.

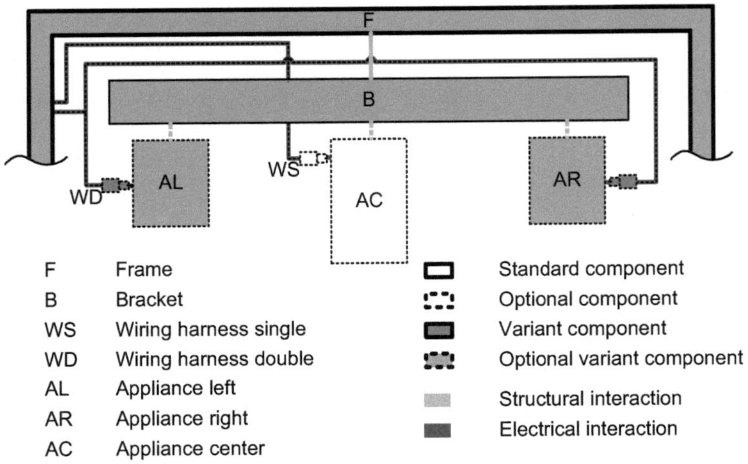

Fig. 6 Module interface graph showing the components of the device [26]

It shows how the variant bracket is mounted to the variant frame in the electrical device. The bracket can be equipped with the following optional devices: Appliance left and appliance right are both optional variant components while the appliance center is just an optional component. Since the appliances are all optional, the structural and electronic interfaces connecting these to the bracket are optional as well (dotted lines). The wiring harness double as well as the electrical interfaces to the appliances left and right are not only optional but also have variant adapters for the basic and high performance variants.

4.3 Product Program Integration

As is and pursued carryover of components are visualized using the Carryover Chart (CoC), which is shown in Fig. 7.

The rows of this figure demonstrate the product lines and families that use the same modules. Similarly, the columns depict the individual modules comprising the whole machine. This overview is especially important for huge machines and plants that are difficult to modularize in one step and instead require multiple steps to handle the complexity of the modularization task.

For the case of the electrical device of industrial vehicles, the device shows no carryover across product families as is. However, with modularization, carryover of this device over the whole Product Line A is pursued.

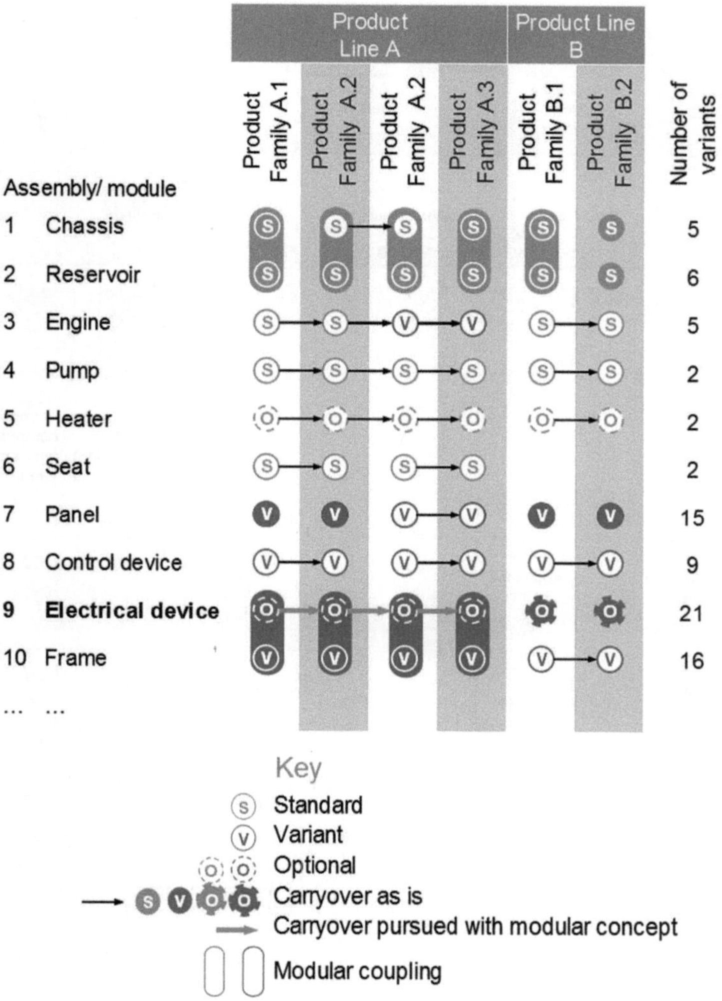

Fig. 7 Carryover chart with electric device [29]

4.4 Design for Variety

Design for Variety aims to redesign components to enhance configurability and commonality. This is supported by the Variety Allocation Model (VAM) [27, 28], which is shown in Fig. 8.

Filling the layers of differentiating properties, technical characteristics and variant and optional components, causal connections can be drawn showing how customer related properties cause variety of components. If a customer related property spreads over several components, then the product's configurability is restrained by these couplings. If a component is influenced by several properties,

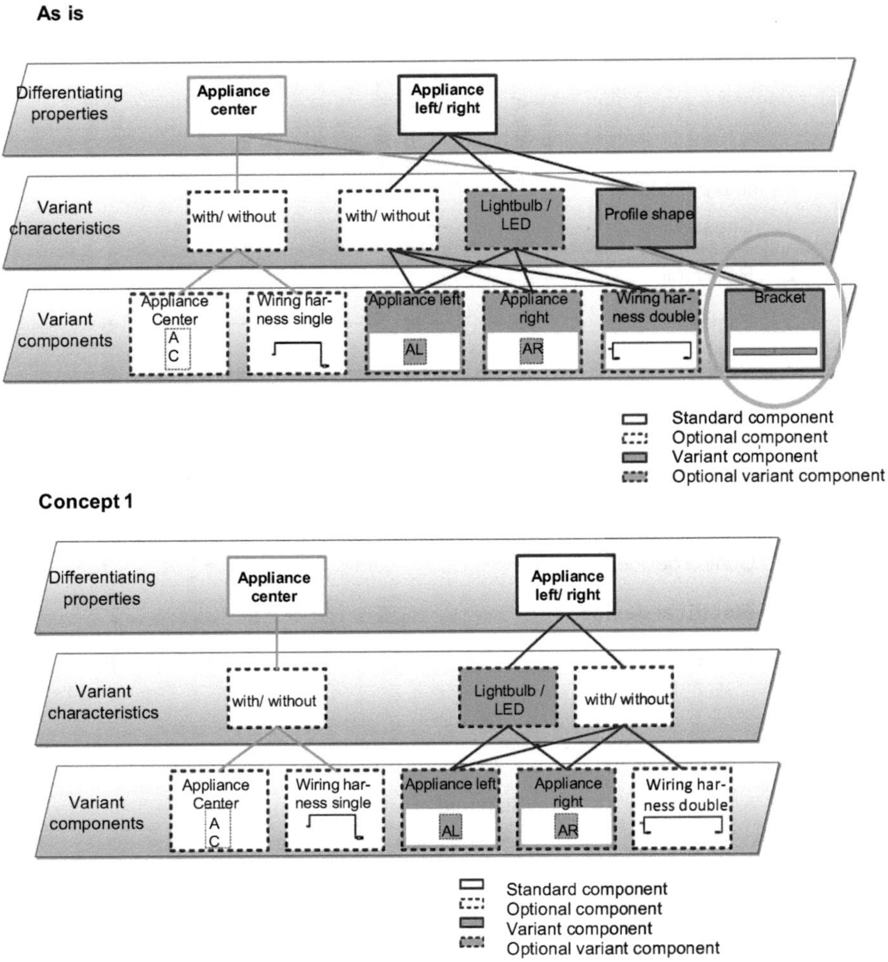

Fig. 8 VAM of the electrical device as is and of concept 1

the component is variant often with multiplication effects among the influencing properties. These influences need to be reduced in order to enhance commonality. The VAM shows how far the as is components are not designed for variety. The aim is to derive (based on this visualization) technical solutions regarding how to design the components for variety, i.e., a clear 1: n-mapping of properties to components.

The VAM of the electrical device shows that mainly the bracket is not designed for variety. As is, this bracket has another design for the device without any appliances. It is suggested to standardize this bracket for all devices and thereby reduce the influence of the customer related properties.

Other solutions like standardizing the wiring harness double by an integrated adapter or oversizing the appliances left/right are combined into three concepts as depicted in Fig. 9.

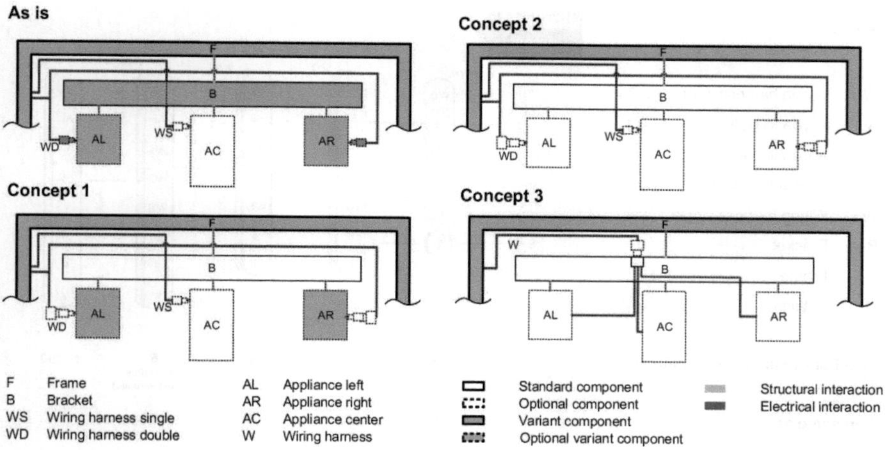

Fig. 9 MIG of as is and three new concepts [26]

4.5 *Life Phases Modularization*

The method unit Life Phases Modularization aims to provide a modular structure that enables cost benefits in all relevant life phases. Firstly, module drivers are defined. Module drivers are reasons as to why components should be handled as one module in a life phase. For example, a module driver for the life phase procurement is "modular sourcing", i.e.; that components shall be procured in a preassembled state together as one module from a supplier. Assigning the module drivers to the corresponding components each life phase derives its own optimal modular structure. Visualizing those in the Module Process Chart (MPC) shown in Fig. 10, trans-disciplinary alignment is performed to derive a modular structure that is beneficial over all life phases. Small numbers symbolize the variety in which the components and modules need to be handled in each life phase. Hence, early multiplication of variants in the process can be detected and reduced.

From left to right the chart showing the electrical device demonstrates that product development has to develop six component types, two of them optional-standard, four of them variant or optional variant, ten code numbers in total. When sales is handling an order, the order reflects how the customer choices lead to one optional module including three components in two variants, another optional module including two components and a single component in two variants.

During procurement, this order is passed to the supplier as one whole module where the device in its six variants is mounted to a frame of 12 variants. Due to the multiplication of variant components, this bigger module is ordered in 72 variants. Counting only the code numbers of the device, this effect cannot be detected. This is why the line "perceived as" counts actual variant processes caused by the product variety. Logistics needs to handle this module in 72 variants and supply the assembly line with the right variant at the right time in the right order. As the vehicle itself can be individualized in up to 800 variants, the assembly with the

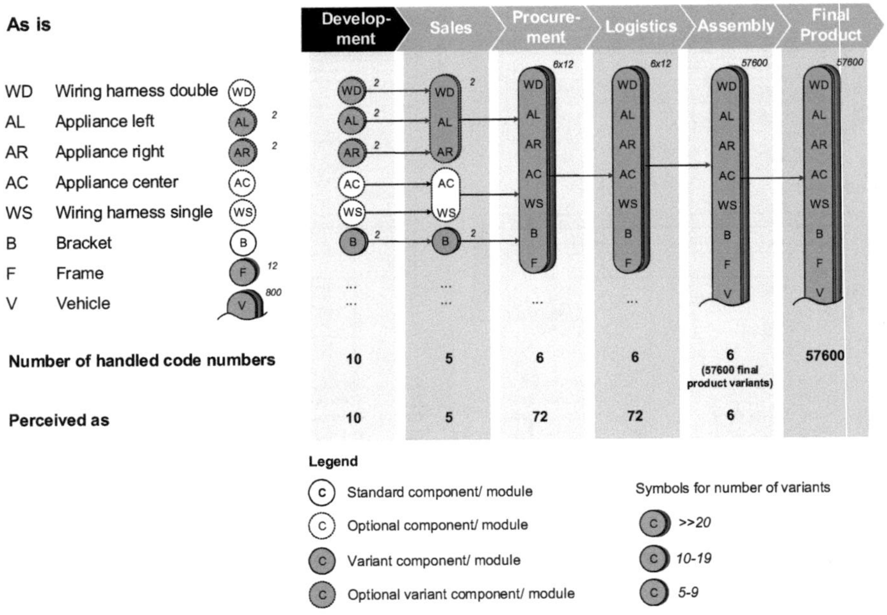

Fig. 10 Modular concepts over the product life phases in the MPC as is [26]

considered module causes further multiplication so that products for customers are available in 57,600 variants.

Development and sales handle small number of modules with small number of variants that enable the building of 72 variants of the appliance and contribute to offering 57,600 vehicle variants to the customer. The effects of commonality can be utilized to a great extent. Procurement and logistics need to handle only one module but its 72 variants need to be handled separately. Both procurement and logistics life phases need to create individual code numbers for each of the 72 variants that must be administered and serviced. Logistics also need to store these 72 variants separately and supply the assembly line with the right variant at the right time in the right order, which causes high cost as this process requires an in-sequence supply-chain (Fig. 10).

By Design for Variety and Life Phases Modularization, the three modular concepts are derived. They all show reduction of code numbers by Design for Variety as well as reduction of handled code numbers in the processes by postponement in the product life phases (Fig. 11).

4.6 Complexity Cost Effect Estimation

The modular concepts are evaluated by estimating their production cost and extrapolating them to the complexity cost. The extrapolation is based on the assumption

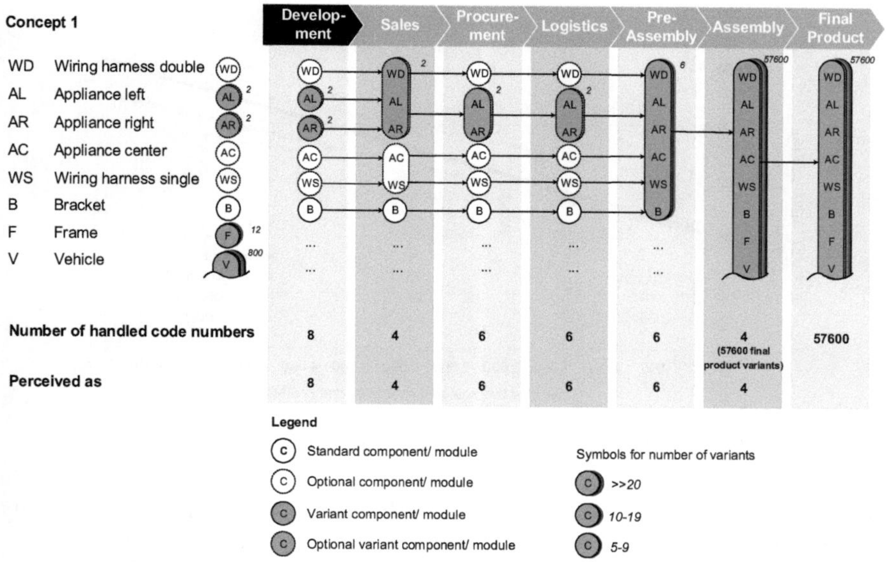

Fig. 11 Modular concepts over the product life phases in the MPC for concept 1 [26]

that a huge part of the complexity cost is caused by the number of different code numbers to handle. The extrapolation is defined by the following equation [26, 29]:

$$\text{TCCI} = 1 - \frac{d-1}{\sum_{j=1}^{d} \phi - 1}$$

$$TC(CNC) = PC + n \cdot CNC$$

Where

TC = Total cost
N = Number of code numbers
PC = Production cost
CNC = Average code number cost per year

As many companies do not have a process to accurately calculate the code number dependent cost, they use a given factor CNC representing each code number cost per year. The extrapolation provides the range in which this factor needs to be in order to choose a particular concept. The code number of the life phase development is used as the number of code numbers ($n_{as\ is} = 10$, $n_{concept\ 1} = 8$, $n_{concept\ 2} = 6$, $n_{concept\ 2} = 7$). The cost extrapolation in Fig. 12 shows that the concept 1 has similar production cost as the as is analysis, but shows better complexity cost effects. If CNC is estimated to be much higher than 2000€ per code number per year, concept 2 is to be preferred. In any case, concept 3 does not

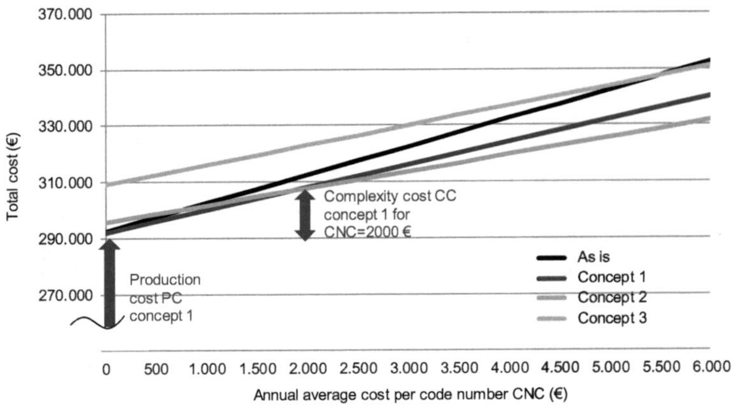

Fig. 12 Breakeven analysis to evaluate modular concepts [29]

show any benefit as the production cost is too high to make up for good complexity cost effects.

5 Insights on the Nature of Variety Induced Complexity Cost: Industrial Cases

The treatment of Hypothesis 1–4 from the industrial case on an electrical device family is presented above and published by Eilmus et al. [26]. These hypotheses are evaluated in this section. They are compared with the results from another industrial case study on a vehicle frame family, which has not been published yet.

5.1 Hypothesis 1: Modularization Is a Useful Approach to Reduce Complexity Cost

It is assumed that the complexity cost factors presented in Fig. 2 would increase with a rise in the number of code numbers to be handled in the product life phases. Therefore, as long as modularization leads to reduction of code numbers by increasing commonality, it would result in reduced complexity cost.

$$CC_{as\ is} > CC_{modular\ concept} \text{ if } n_{as\ is} > n_{modular\ concept}$$

Where

CC = Complexity cost
N = Number of code numbers

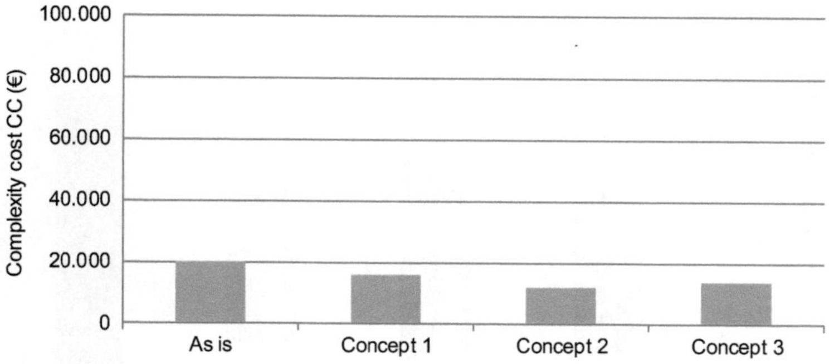

Fig. 13 Estimation of annual complexity cost for the industrial case of electrical device family

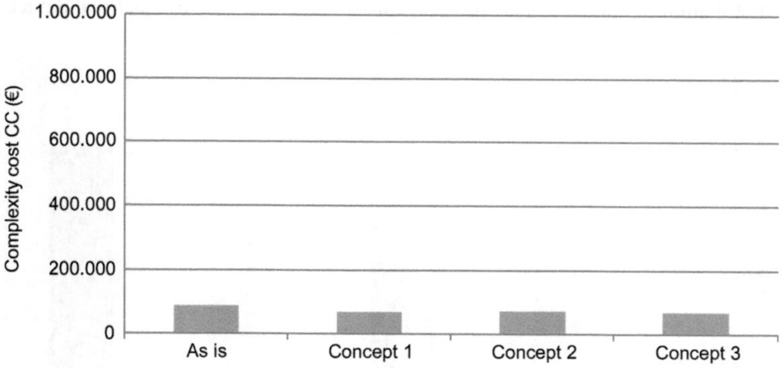

Fig. 14 Estimation of annual complexity cost for the industrial case of vehicle frame family

Hypothesis 1 can be supported by the evaluation of the two industrial cases summarized in Figs. 13 and 14 based on the above assumption. In the case of the electrical devices, complexity cost can be reduced by up to 40 %. A reduction of up to 23 % can be achieved with modular concepts for the industrial case of vehicle frames.

5.2 *Hypothesis 2: Modularization Can Influence Production Cost Positively or Negatively*

The effect of modularization on production cost is less explicit than the effect on complexity cost. On the one hand, modularization leads to clear modules with integrated and coupled components within a module as well as simple and standardized interfaces between modules. This can reduce cost in assembly and interfaces. Commonality can reduce tool cost and raise lot size effects. On the other hand, commonality can lead to oversizing thereby raising the overall production cost.

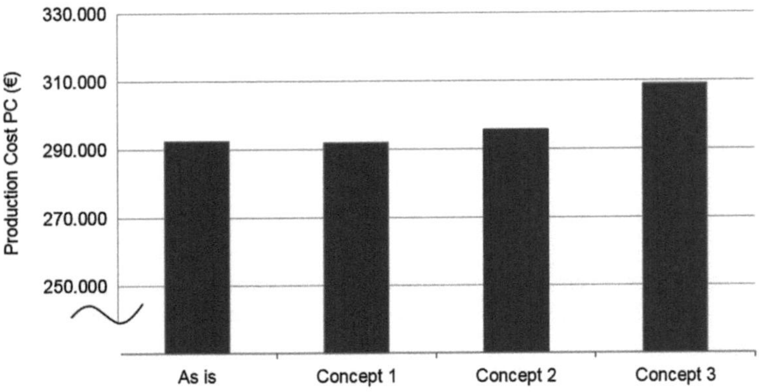

Fig. 15 Estimation of annual production cost for the industrial case of electrical device family

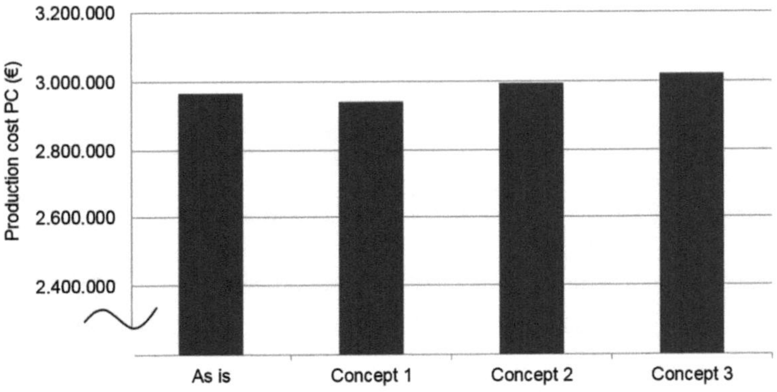

Fig. 16 Estimation of annual production cost for the industrial case of vehicle frame family

The industrial cases show different effects. For the electrical device case study, one modular concept resulted in a similar production cost as the as is situation. Due to oversizing, the other concepts would cause higher production cost (Fig. 15). In the case of vehicle frame family, a concept could be developed to reduce the production cost by 1 % as compared to the as is situation, which amounted to a sum of more than 25,000€ (Fig. 16). However, the other concepts would cause higher production cost.

The conclusion is that modularization causes different effects on production cost—increasing, decreasing or leaving it unchanged depending on the situation. Yet, in many cases oversizing in order to enhance commonality leads to higher production cost. Thus, redesigning and finding more effective new solutions to enhance commonality rather than oversizing need to be investigated, e.g., completely new working principles that foster common technologies for different product variants. Furthermore, accepted methods for estimating complexity cost effects become increasingly important as companies often make a decision merely

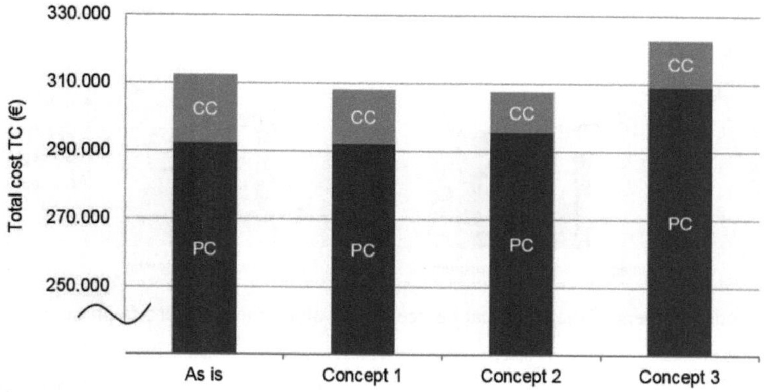

Fig. 17 Estimation of annual total cost for the industrial case of electrical device family

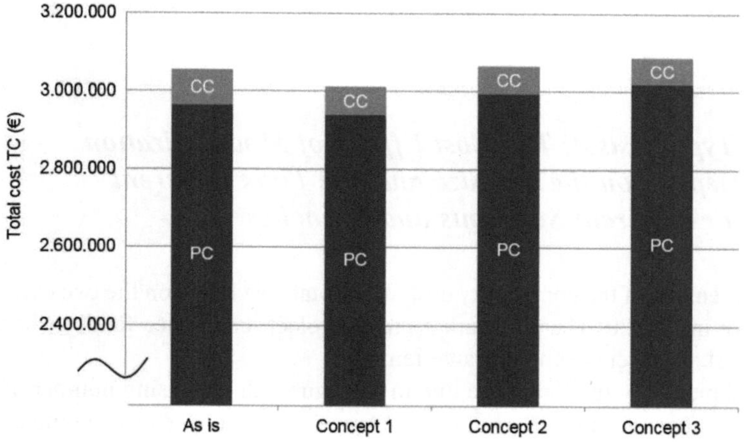

Fig. 18 Estimation of annual total cost for the industrial case of vehicle frame family

based on production cost without knowing about the total cost a modular concept may entail (Figs. 17 and 18).

5.3 *Hypothesis 3: Modular Concepts Show Different Cost Effects in Different Disciplines*

Assuming that complexity cost rises with the number of code numbers, this hypothesis can be supported by showing that the number of code numbers differs in the various disciplines. Taking the modular structure of the different disciplines along with the product life phases (Figs. 10 and 11) into account, the code numbers of the disciplines vary with the differing modular structures (Fig. 19). Thus, the complexity cost might also differ trans-disciplinarily along the product life.

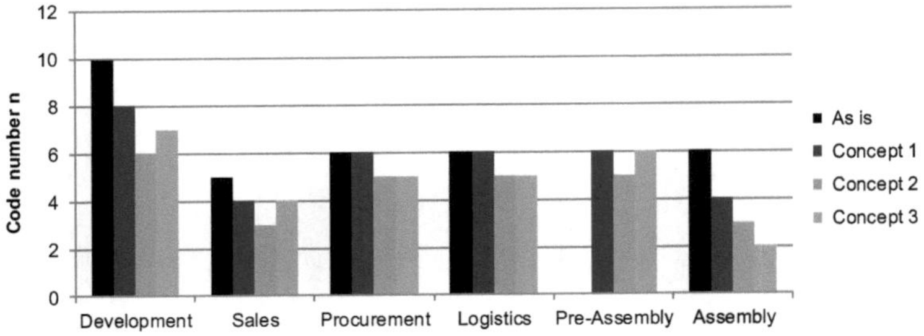

Fig. 19 Code numbers of the electrical device family along the product life phases

Based on this analysis, a generalized cost estimation like the breakeven analysis of complexity cost effect presented above in Fig. 12 needs to be enhanced via a more differentiated approach by taking all the relevant disciplines into account separately.

5.4 *Hypothesis 4: The Cost Effects of Modularization Depend on the Lot Size and Are Thus Different for Different Segments and Branches*

The percentage of the complexity cost to the total cost based on the two case studies is shown in Fig. 20. The left mark represents electrical device family whereas the right mark represents vehicle frame family.

As depicted by the light blue line in the figure, an increasing number of pieces per variant cause the percentage of the complexity cost to reduce significantly.

In the case of electrical device family, a 1 % increase of production cost can be compensated by 15 % reduction of the complexity cost. The declining share of the complexity cost implies that in the case of the vehicle frame family, a reduction of complexity cost by 34 % is required to compensate 1 % increase in production cost. As a result, in the first case study, concept 2 with the lowest complexity cost shows the lowest total cost. In contrast, in the second case study, concept 1 with the lowest production cost has the lowest total cost (refer to Figs. 17 and 18).

This shows that a positive cost effect of modularization can be achieved per variant for small lot sizes by reducing the complexity cost, while large lot sizes require a reduction of production cost.

In addition to the lot sizes, other segment and branch specific aspects determine the percentage of cost of complexity in the total cost. Examples of such aspects are distribution cost or specific cost-intensive production processes. The focus of

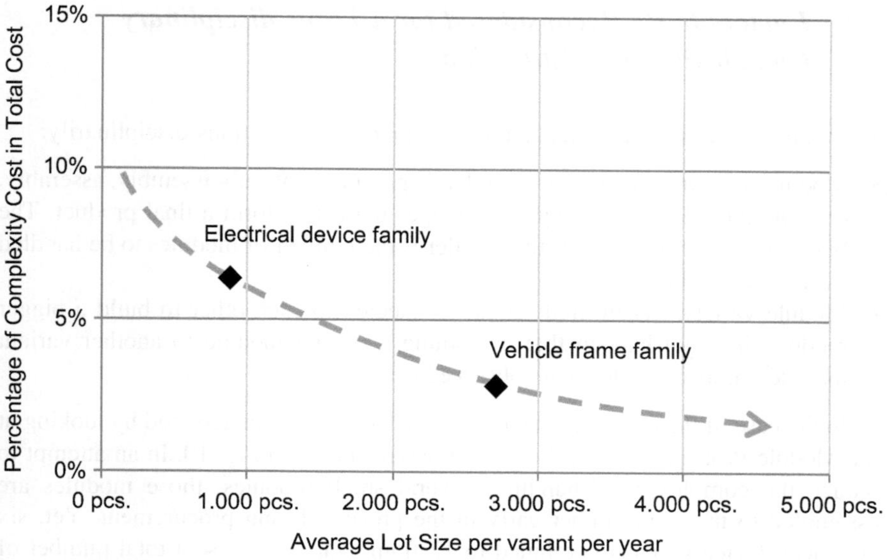

Fig. 20 Percentage of complexity cost versus average lot sizes per variant

modularization has to be therefore adopted to industry, segment and lot sizes to achieve a positive impact on total cost.

5.5 Resulting Need for Improvement

The findings elaborated in the context of these four hypotheses lead to the following needs for further improvement.

1. Development of trans-disciplinary complexity cost estimation taking the different product life phases into account.
2. Description of distinct approaches for branches and segments with low and high lot sizes.

Initial approaches to address these needs are presented in the following sections.

6 Proposal for a Method to Estimate Trans-disciplinary Complexity Cost of Modular Concepts

The aim of the method proposed in this section is to reconsider the complexity cost estimation taking into account the diverse handling of a product family's modules in different disciplines along the product life phases.

6.1 Factors to Be Reconsidered for a Trans-disciplinary Complexity Cost Estimation

There are two reasons how the handling of modules differ trans-disciplinarily:

- Module size: In the course of a product's procurement, sub-assembly, assembly, etc., the modules grow bigger and bigger until they form a final product. The bigger a product's modules, the smaller is the number of modules to be handled (refer to Fig. 11).
- Module variety: When modules are connected to each other to build a bigger module, it might happen that by joining a variant module to another variant module, multiplier effects are observed (Fig. 10).

In the case of the electrical device these effects can be understood by looking at the Module Process Chart of the as is situation shown in Fig. 10. In an attempt to reduce the complexity of handling several small modules, those modules are assembled to the frame rather early in the process during procurement. Yet, six electrical device variants combined to 12 frame variants cause a total number of 72 variants of this bigger module thereby increasing the overall complexity. In order to understand and evaluate the complexity caused by a modular structure, it is necessary to analyze the processes in the different disciplines across the product life phases. Note that it is important to understand a module's interaction with other variant modules as only taking the analyzed module into consideration is not sufficient. While the assembly of electrical device and frame causes multiplier effects in the procurement and logistics product life phases (Fig. 10), the assembly of the highly variant electrical device frame module of the vehicle results in very lean processes. This is realized by standardized interfaces that vary based on the number of wiring harnesses connected to the vehicle.

In order to take these effects of multiplication or standardization into account, it is proposed to not only count the number of handled code numbers, but also how they are perceived in the various processes of the product life phases (shown in Fig. 21). This is evaluated through process analysis. The diversity in the handled code numbers and perceived handled code numbers requiring a trans-disciplinary complexity cost estimation but also diverse shares on the total cost. Performing complexity cost estimation with a breakeven analysis as shown above in Fig. 12, total cost is estimated globally across all the product life phases from a development point of view. However, cost is differently induced in separate product life phases as established by Wildemann [30] (Table 1). Taking this data into account, a product life specific view on complexity cost is enabled.

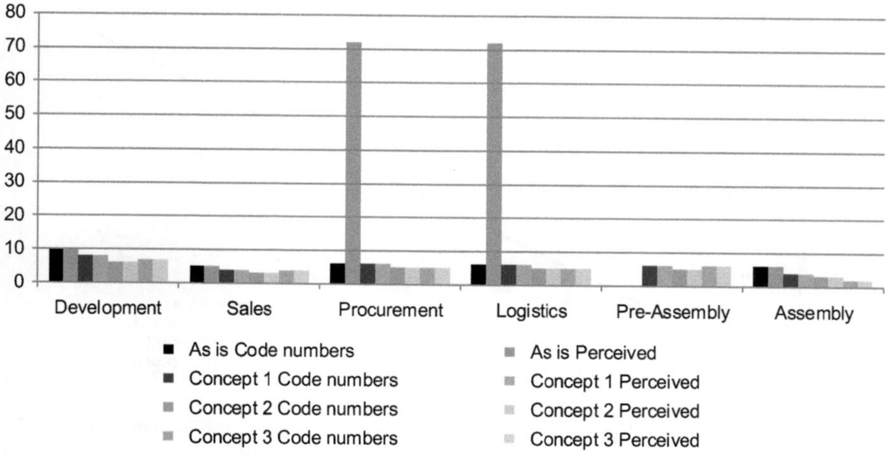

Fig. 21 Code numbers compared to perceived code numbers of the electrical device family along the product life phases

Table 1 Share of the various product life phases on total complexity cost [30]

	Development	Sales	Procurement	Logistics	Production
DCF_l	Min (%)	20	10	10	30
	Max (%)	40	20	20	40
	Avg (%)	30	15	15	35

6.2 *Enhanced Breakeven Analysis of Complexity Cost*

The existing approach of breakeven analysis is enhanced by the aspects discussed below in order to achieve a trans-disciplinary life phase specific cost estimation:

- Analysis of perceived handled modules in each product life phase.
- Breakdown of total complexity cost to life phase specific shares.

The enhanced Breakeven Analysis of Complexity Cost chart (BACC) can be described using the following equation by including the above aspects.

$$CC(CNC) = CNC \cdot \sum_{l=1}^{6} DCF_l \cdot n_{perceived}$$

$$TC(CNC) = PC + CNC \cdot \sum_{l=1}^{6} DCF_l \cdot n_{perceived}$$

where

CC = Complexity cost

$n_{perceived}$ = Number of perceived code numbers per life phase (l)

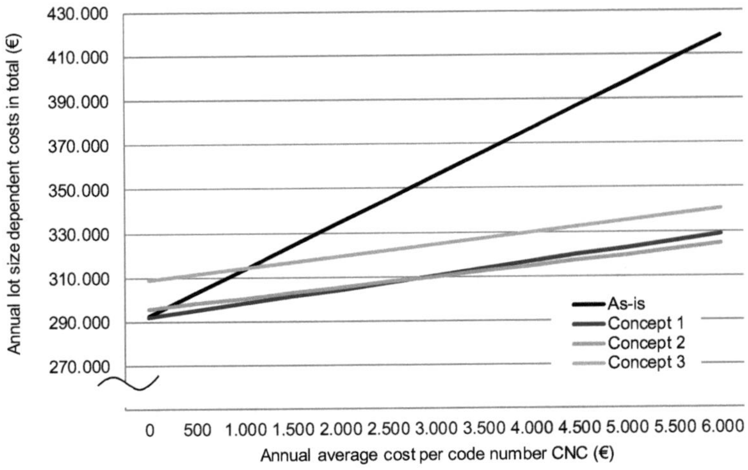

Fig. 22 Enhanced Breakeven Analysis of Complexity Cost chart (BACC) of the electrical device family

CNC = Average code number cost per year
DCF_l = Distributed cost factor per life phase (l)
TC = Total cost
PC = Production cost

The BACC (Fig. 22) gives an answer to the question of how far the complexity cost effects of different concepts are dependent on the assumption of the annual average cost per code number (CNC). In this example, it was assumed that CNC = 2000€. The BACC chart shows that even if this assumption is vague, the corresponding result would be the same, i.e., concept 1 would still be slightly cheaper than concept 2 as far as total cost is concerned. Furthermore, this chart shows that implementing concept 1 would bring about total cost reduction without taking the complexity cost reduction into account. The breakeven of concept 2 is at CNC < 500€ and the breakeven of concept 3 is at CNC < 1500€.

When compared to the breakeven analysis presented earlier in Fig. 12, the gradient of the as is modular structure is higher so that all new concepts have their breakeven point at a smaller average code number cost per year.

6.3 *Product Life Induced Complexity Cost*

In order to understand the discipline specific complexity cost effects in each product life phase, the Product Life Induced Complexity Cost (PLICC) is visualized. The PLICC is based on the same equation as the BACC, but now a constant value is

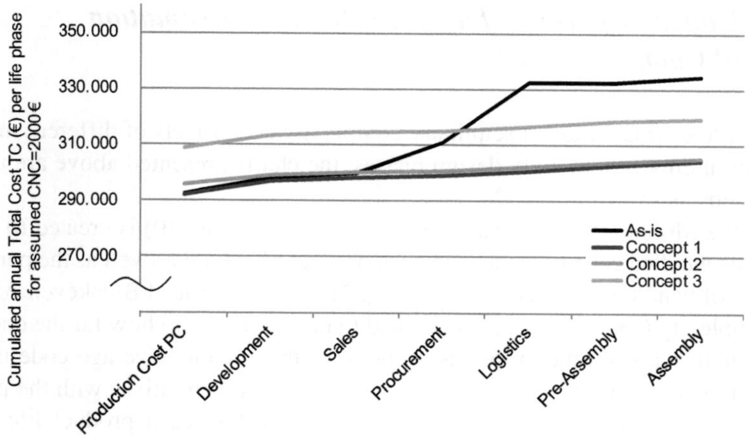

Fig. 23 Product Life Induced Complexity Cost (PLICC) of the electrical device family

assumed for CNC. In the case of the electrical device, it is assumed that $CNC_{assumed} = 2000€$. Based on this assumption, the PLICC is visualized by a line with a section for each product life phase thus cumulating the cost induced in the course of the product life phases (Fig. 23).

$$TC(l) = PC + CNC_{assumed} \cdot \sum_{1}^{1} DCF_l \cdot n_{perceived}$$

where

$n_{perceived}$ = Number of perceived code numbers
$CNC_{assumed}$ = Assumed average code number cost per year, specifically $CNC_{assumed} = 2000€$
DCF_l = Distributed cost factor per life phase (l)
TC = Total cost
PC = Production cost

The resulting chart shows that in the case of the electrical device family, most of the complexity cost induced is estimated to be in the procurement and logistics phases, which corresponds to the actual corporate situation of this case study. So, the aim is now to choose a new concept with the following attributes:

- Lower cumulated sum than the as is situation,
- Less high gradients,
- Postponement of high gradients to the right to prevent a high tied cost over the lead time.

6.4 *Holistic Approach for an Early Phase Estimation of Complexity Cost*

In order to estimate trans-disciplinary complexity cost effects of different modular concept alternatives in early design phases, the charts presented above are used in three steps as shown in Fig. 24.

First, a Module Process Chart (MPC) (as shown in Fig. 10) is created to understand the handled code numbers n in each product life phase as well as the perceived number of handled modules, i.e., $n_{perceived}$. Then, the enhanced Breakeven Analysis of Complexity Cost (BACC) is performed to understand as to how far the choice of concepts depends on the proper assumption of the assumed average code number cost per year, i.e., $CNC_{assumed}$. To discuss the concept alternatives with the product life phases and to understand how they induce cost in each product life phase, the Product Life Induced Complexity Cost (PLICC) is applied.

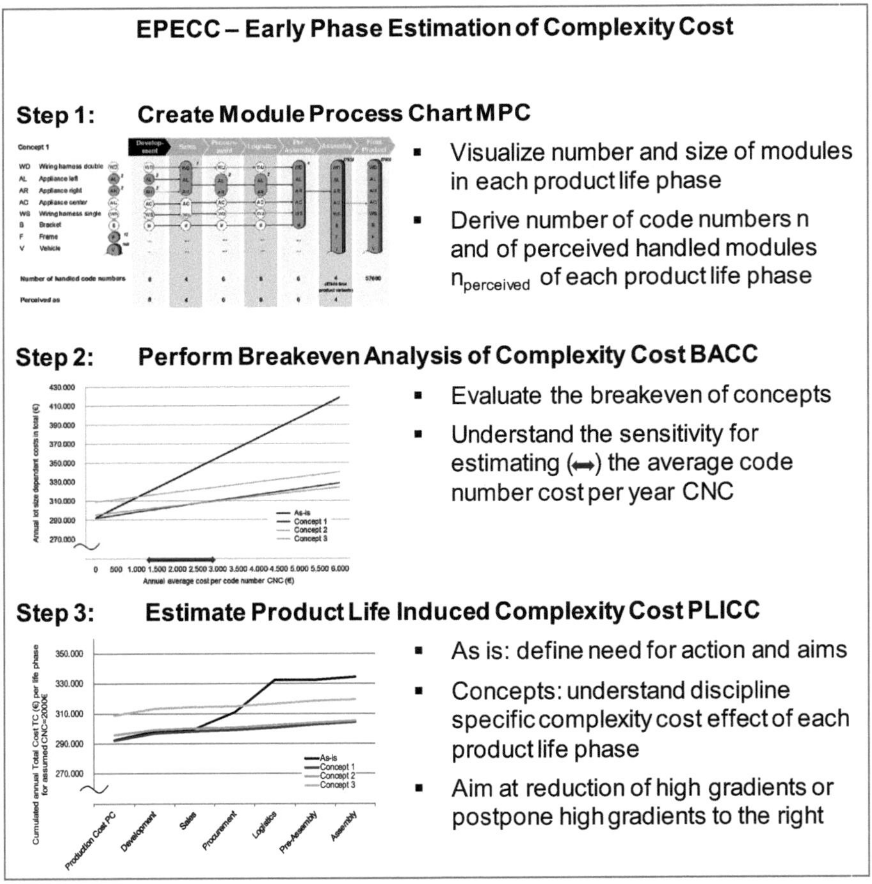

Fig. 24 Steps of the holistic approach for Early Phase Estimation of Complexity Cost (EPECC)

6.5 Discussion of the Proposed EPECC Approach

EPECC was developed to enhance the breakeven analysis to evaluate modular concepts (Fig. 12) by reconsidering cost differentiation for single product life phases. This differentiated approach enables a trans-disciplinary evaluation of modular concept alternatives integrating all product life phases. There is evidence that the enhanced estimation procedure EPECC complies better with reality than the breakeven analysis used before. First, especially the corporate experts of the case study of electrical devices reported high cost in procurement and logistics as represented by the PLICC. Second, the complexity cost share to total cost was estimated 6.4 % earlier for the electrical device family. The enhanced EPECC approach leads to a share of 12.6 %. In the literature, a share of 15–20 % is published by Wildemann [30], which is more than the share calculated by EPECC.

In order to get more insights on the influence of lot sizes on the share complexity cost to total cost, the curve of Fig. 20 is updated with the complexity cost derived by EPECC (Fig. 25).

While the general finding is the same, the values are higher and their difference to each other is bigger with the EPECC procedure (green) compared to before (blue). In order to confirm this curve, more empirical data needs to be gathered in the future.

However, the curve shows that cost reduction efforts by modularization need to be addressed differently for branches and segments with high lot sizes and those with low lot sizes. In the following sections, success factors from industrial and consultancy practice are summarized for both.

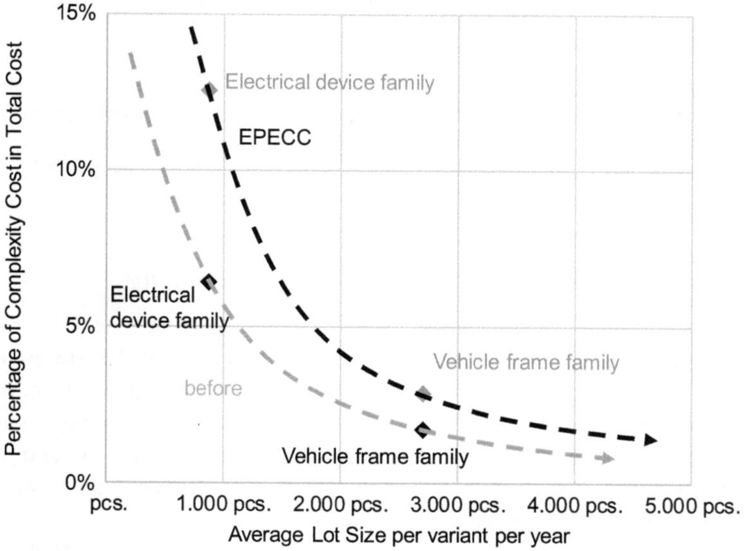

Fig. 25 Percentage of Complexity Cost versus Average Lot Sizes per variant derived by EPECC (*green*) and evaluated before (*blue*)

7 Success Factors of Modularization in Branches and Segments with High Lot Sizes

Branches and segments with high lot sizes typically produce a few thousand units of one variant per year. Such quantities are typical for the consumer goods industry, e.g., household appliances. This case is located on the very right side of Fig. 25.

These sectors and segments are characterized by the fact that complexity cost is increasing, but continues to contribute only a comparatively small part of the total cost. Due to this comparatively small proportion, higher production cost is not acceptable to reduce complexity cost. This is especially true for those variants of a product family that occupy the lowest price points and often reach the highest quantities.

In this case, modularization is promising if both complexity cost and production cost are reduced. To achieve this, the important success factors are:

- A cost controlling accompanying modularization in particular for those variants that occupy the lowest price points within the product family.
- Early and intensive involvement of the purchasing and production department in modularization to identify potential for cost reduction.
- Focus on components and production processes with high economics of scale when standardizing components or modules.
- Avoidance of any kind of over-design during modularization.

8 Success Factors of Modularization in Branches and Segments with Low Lot Sizes

In German industry, low lot sizes are typical for plant engineering as well as for many small and medium-sized enterprises (SMEs).

8.1 Modularization Factors for Plant Engineering

In plant engineering, highly application-specific individual plants are engineered and produced, often with lot size 1. In Fig. 25, companies in plant engineering would be located at the very left side of the chart, having a high share of complexity cost to production cost. That is why efforts to raise component and module commonality by modularization are very effective to reduce total cost, even if these efforts raise production cost.

In the context of plant engineering, success factors for cost reduction by modularization are:

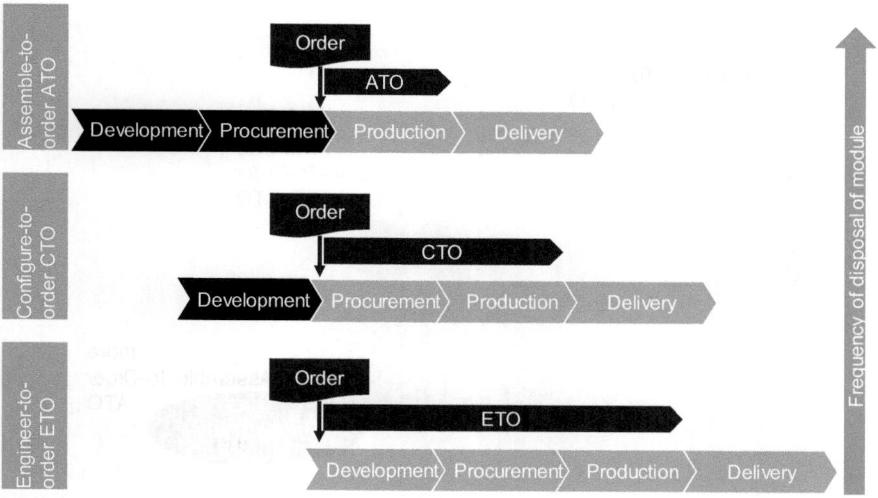

Fig. 26 Supply chain strategies

- Definition and pre-development of areas of plants that can be equipped with the same common modules across different plant variants.
- Identification and standardization of common modules that can be reused within the same plant order.
- Shift from order specific processes to order independent processes to save storage and tooling cost.
- Distinguish supply chain strategies for different modules according to frequency of disposal (Fig. 26).

8.2 *Modularization Factors for Small and Medium-Sized Enterprises*

Small and medium-sized enterprises (SMEs) are enterprises with less than 250 employees. SMEs often produce in low lot sizes, so that even here, raising component and module commonality by modularization is very effective. Furthermore, in many cases SMEs are strongly driven by customer requirements and special request. This is why they need a certain degree of flexibility, which might hinder the implementation of modular product structures as well as the implementation of standard processes and modules coming along with a modular strategy.

Modularization success factors for SMEs are:

- Clear definition of the product program (refer to Fig. 27). What part of the product program should be covered by modular product structures? What other parts of the product program need to be kept outside in order to enable, e.g., order specific engineering?

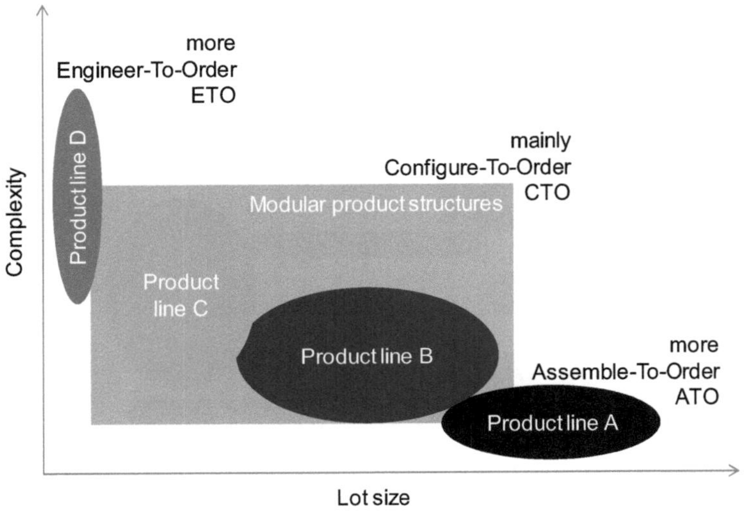

Fig. 27 Clear structuring of the product program and definition of modularization focus

- Structuring of product lines as modular product families with high commonality and clear configuration rules.
- Carryover of single modules to product lines outside the modularization focus.
- Training of the permanent staff engineers on variety management and modularization as there are no resources to establish an extra role or department for variety management and modularization.

9 Summary and Conclusion

Variant products lead to complexity thereby increasing cost. The development of modular product structures enables cost reduction. Yet, the effect of modular concept alternatives is dependent on lot sizes and varies in the different product life phases. This is why a differentiated cost estimation approach is developed. The Early Phase Estimation of Complexity Cost (EPECC) provides a trans-disciplinary cost estimation for modular concept alternatives showing the breakeven of each alternative compared to the as is situation as well as its effect in each product life phase. This cost estimation can be used during a methodical modularization project as summarized in Fig. 28. To develop modular concepts, the different lot sizes of the specific branch or segment need to be reconsidered. While modularization efforts in branches and segments with high lot sizes need to address reduction of production cost, modular concepts reducing complexity cost by commonality are effective in branches and segments with low lot sizes.

Further empirical data are needed to gain even more insights on the influence of lot size on cost effects of modularization and to confirm these conclusions.

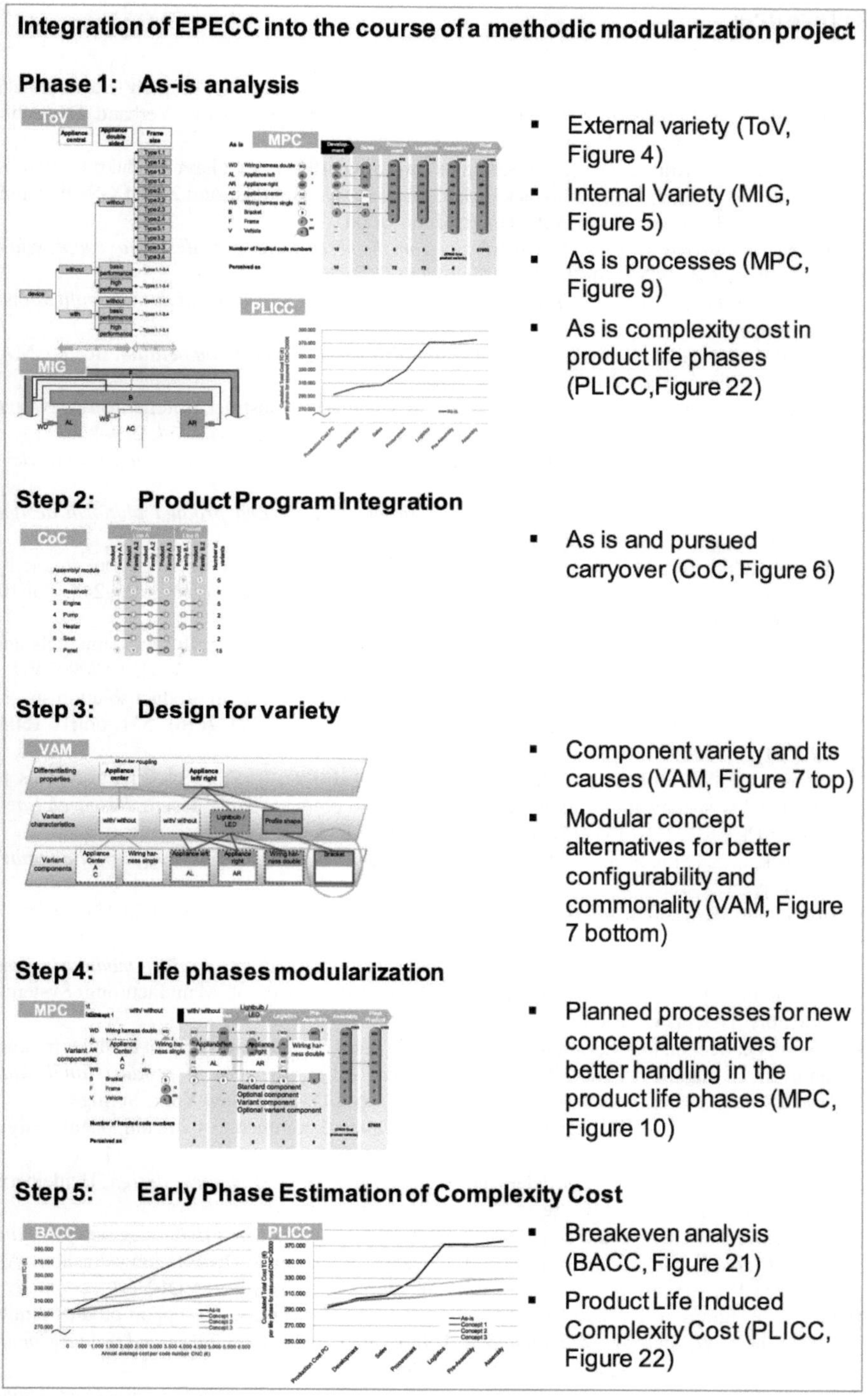

Fig. 28 Integration of EPECC into the course of a methodical modularization project

References

1. Verband Deutscher Maschinen- und Anlagenbau e.V., & McKinsey & Company (Eds.). (2014). *The future of German mechanical engineering*. Frankfurt, Germany: Verband Deutscher Maschinen- und Anlagenbau e.V
2. Ripperda, S., & Krause, D. (2013). Komplexitätskosteneffekte modularer Produktfamilien. In D. Krause, K. Paetzhold, S. Wartzack (Eds.), *Design for X*. Beiträge zum 24. DfX-Symposium Oktober 2013, 24. Hamburg: Aufl. TuTech Innovation.
3. Brosch, M., & Krause, D. (2011). *Complexity from the perspective of the design for supply chain requirements* (pp. 102–116). Hamburg, Germany: TuTech Verlag
4. Eppinger, S. D., & Browning, T. R. (2012). *Design structure matrix methods and applications*. Cambridge, MA: MIT Press.
5. Pine, B. J. (1993). *Mass customization. The new frontier in business competition*. Boston, MA: Harvard Business School Press.
6. Salvador, F. (2007). Toward a product system modularity construct: Literature review and reconceptualization. *IEEE Transaction on Engineering in Management, 54*, 219–240.
7. Robertson, D., & Ulrich, K. (1998). Planning for product platforms. *Sloan Management Review, 39*(4), 19–31.
8. Simpson, T. W. (Ed.). (2013). *Advances in product family and product platform design. Methods & applications*. New York: Springer.
9. Jiao, J., & Tseng, M. (2000). Understanding product family for mass customization by developing commonality indices. *Journal of Engineering Design, 11*(3), 225–243. doi:10.1080/09544820075002l003.
10. Du, X., Jiao, J., & Tseng, M. M. (2001). Architecture of product family: Fundamentals and methodology. *Concurrent Engineering, 9*(4), 309–325. doi:10.1177/1063293X0100900407.
11. Haug, A., Hvam, L., & Mortensen, N. H. (2013). Reducing variety in product solution spaces of engineer-to-order companies: The case of Novenco A/S. *IJPD, 18*(6), 531. doi:10.1504/IJPD.2013.058556.
12. Simpson, T. W., Bobuk, A., Slingerland, L. A., et al. (2012). From user requirements to commonality specifications: an integrated approach to product family design. *Research Engineering Design, 23*(2), 141–153. doi:10.1007/s00163-011-0119-4.
13. de Weck, O. L., Suh, E. S., & Chang, D. (2003). *Product family and platform portfolio optimization* (pp. 175–185).
14. Hölttä-Otto, K. (2005). *Modular product platform design*. Helsinki: Helsinki University of Technology.
15. Erixon, G. (1998). *Modular function deployment. A method for product modularisation*. Stockholm: The Royal Institute of Technology, Department of Manufacturing Systems, Assembly Systems Division.
16. Krause, D., Beckmann, G., Eilmus, S., et al. (2013). Integrated development of modular product families: A methods toolkit. In T. W. Simpson (Ed.), *Advances in product family and product platform design. Methods & applications* (pp. 245–269). New York: Springer.
17. Thonemann, U. W., & Brandeau, M. L. (2000). Optimal commonality in component design. *Operations Research, 48*(1), 1–19.
18. Ehrlenspiel, K., Kiewert, A., Lindemann, U., et al. (2007). *Cost-efficient design*. Heidelberg, New York: Springer, ASME Press.
19. Schuh, G. (1989). *Gestaltung und Bewertung von Produktvarianten. Ein Beitrag zur systematischen Planung von Serienprodukten*. Technische Hochschule Dissertation—Aachen, Als Ms. gedr. Berichte aus der Produktionstechnik, Vol. 177. Düsseldorf: VDI-Verl.
20. Park, J., & Simpson, T. W. (2006). An activity-based costing method to support product family design. In T. W. Simpson, Z. Siddique, & J. Jiao (Eds.), *Product platform and product family design. Methods and applications*. New York: Springer.
21. Pfeiffer, W. (1992). Variantenkostenrechnung. In W. Männel (Ed.), *Handbuch Kostenrechnung* (pp. 861–877). Wiesbaden, Germany: Gabler-Verlag.

22. Tu, Y. L., Xie, S. Q., & Fung, R. (2007). Product development cost estimation in mass customization. *IEEE Transactions on Engineering Management, 54*(1), 29–40. doi:10.1109/TEM.2006.889065.
23. Ripperda, S., & Krause, D. (2014). *Costs in modularization approaches: A co-citation analysis*. In 13th International Design Conference—Design 2014, pp. 965–974.
24. Fujita, K., Akagi, S., Yoneda, T., et al. (1998). Simultaneous optimization of product family sharing system structure and configuration. In: *Proceedings of DETC '98 Design Engineering Technical Conferences*. Atlanta, Georgia, USA: ASME.
25. Sedlmeier, L., Möller, K., Schultze, W., et al. (2013). *Kostenwirkung der Modularisierung (KosMo): Entwicklung eines Prognose- und Entscheidungsmodells für Markt- und Kosteneffekte von Modularisierungskonzepten.* In: *Schriftenreihe Forschungshefte der FVA, Nr. 1081*. Frankfurt, Germany: Forschungsvereinigung Antriebstechnik e.V.
26. Eilmus, S., & Krause, D. (2013). Product life-oriented development of component commonality and variety. In: *Proceedings of the ASME 2013 International Mechanical Engineering Congress and Exposition (IMECE2013), San Diego.*
27. Kipp, T. (2012). *Methodische Unterstützung der variantengerechten Produktgestaltung* (Hamburger Schriftenreihe Produktentwicklung und Konstruktionstechnik 1st ed., Vol. 4). Hamburg: TuTech Verl.
28. Blees, C., Kipp, T., Beckmann, G., et al. (2010). Development of modular product families: Integration of design for variety and modularization. In: A. Dagman & R. Söderberg (Eds.), *NordDesign 2010. Proceedings of the 8th biannual conference NordDesign 2010. Product and Production Development* (pp. 159–170). Göteborg: Chalmers University of Technology.
29. Eilmus, S., Ripperda, S., & Krause, D. (2013, August, 19–22). Towards the development of commonal product programs. In *Proceedings of iced13. 19th International Conference on Engineering Design (ICED13), Design for Harmonies* (Vol. 4, pp. 209–218). Seoul: Design Society.
30. Wildemann, H. (2000). *Komplexitätsmanagement. Vertrieb, Produkte, Beschaffung, F & E, Produktion und Administration* (TCW-Report Nr. 20). München: Verl. TCW Transfer-Centrum.

Problem Solving and Increase of Ideality of Complex Systems

Helena V.G. Navas

1 Introduction

In the last decades has been developed a new system thinking called self-organization theory or complexity theory. The complexity can be related to several quadrants: physics, mathematics, biology, statistics, social service, sociology, linguistics, psychology, etc. The definition of complexity varies significantly by area of expertise.

In the popular perception, a system is considered as complex when it contains a large number of components that interact with one another in multiple ways. However, for many researchers, a system can be considered as a complex system, when its properties are not a natural consequence of their constituent elements viewed in isolation, but when various parts interacting with the ability to generate new qualities on the collective behavior.

The interactions between the parts of a complex system create a collective pattern called emergent property. The emergent properties of a complex system stems much of the nonlinear relationship between the parts. Nonlinearity is considered to be essential for complexity. Thus, to characterize a system is necessary to know not only the parts but also the modes of relationship between them.

The main focus of the complexity theory is the manipulation of information and how it is transformed from a given representation to another desired representation. The study of complex systems is a new challenge for the science that investigates how the relationship between the elements influence the collective behavior of a system and how the system interacts and how it relates to the environment in which it operates.

H.V.G. Navas (✉)
Faculdade de Ciências e Tecnologia, Departamento de Engenharia Mecânica e Industrial, UNIDEMI, Universidade Nova de Lisboa, Quinta da Torre, 2829-516 Caparica, Portugal
e-mail: hvgn@fct.unl.pt

F.-J. Kahlen et al. (eds.), *Transdisciplinary Perspectives on Complex Systems*,
DOI 10.1007/978-3-319-38756-7_11

The complexity of a system is not an absolute characteristic is a relative characteristic. Two forms of complexity can be distinguished: a disorganized complexity and an organized complexity. The interactions of the elements in a disorganized complexity system can be seen as almost random. The behavior of the whole system can be described by using probability and statistical methods. Organized complexity resides in non-random, or correlated, interaction between the components.

The thematic of complex systems is transversal crossing the traditional disciplines of science and engineering, business and medicine. A trans-disciplinary approach can be very useful in the study of complex systems. A number of philosophical concepts and analytical, modeling and simulation tools has been developed to assist the study of complex systems.

The state of equality or congruence between organizational intelligence and performance management is very important. Flumerfelt, Alves, Calvo-Amodio, Hoyle and Kahlen explore a new approach to systems management through an adaptive method that informs systems competency (chapter "Managing Systems Complexity Through Congruence").

The current market demands require innovation in products, processes, or business models that enables the company to achieve higher levels of sustainability. Innovation does not only demand the identification, implementation, and monitoring of new ideas aimed at improving the performance of the organization but must also have an impact on the overall level of competitiveness.

Enterprises need to invest in systematic innovation if they plan to win or at least survive. Innovation can no longer be seen as the product of occasional inspiration. It has to be transformed into a capacity, not a gift. Innovation has to be learned and managed. Unexpected occurrences, inconsistencies, process requirements, changes in the market and industry, demographic change, and changes in perception or new knowledge can give rise to innovation opportunities.

Systematic innovation can be understood as a concept that includes the instruments necessary to develop the right inventions needed at that point in time and incorporate them into new products and processes.

Incremental innovation is not always sufficient to prevent impact of economic activities, thus more radical innovation initiatives are needed. Radical innovation presupposes a profound shift in the use of resources. More radical forms of innovation can cause a sustainable transition to be difficult to deploy [1].

Radical solutions are very important, especially considering the long-term gains. Traditional engineering and management practices can become insufficient and inefficient for the implementation of new scientific principles or for vast improvements of existing systems.

2 Theory of Inventive Problem Solving (TRIZ)

Organizations need new management paradigms in order to become more competitive. The gains in resource efficiency generally result from process improvements. The traditional incremental improvement of existing technologies is no longer sufficient. All economic activities need to radically increase the efficiency of resource utilization.

Organizations need methodologies and analytical tools to implement radical changes and very new techniques. Thus, sporadic initiatives of innovation must evolve to the continuously planned and scheduled activities; the innovation must become systematic. Innovation has to be served at the right time in the right dose. The Theory of Inventive Problem Solving (TRIZ), Brainstorming, Collateral Thinking, Mind Maps, and other methodologies can stimulate individual and collective ability to innovate.

The Theory of Inventive Problem Solving, better known by its acronym (TRIZ), was developed by Genrich Altshuller in 1946 [2]. TRIZ is a theory that can assist in the inventing process.

The TRIZ methodology can be considered on several levels. At the highest level, the TRIZ can be seen as a science, as a philosophy or a way to be in life (a creative mode and a permanent search of continuous improvement). In more terms, the TRIZ can be seen as a set of analytical tools that assist both in the detection of contradictions on systems and in formulating and solving problems through the elimination or mitigation of contradictions [3].

The TRIZ methodology is based on the following grounds:

- Systems,
- Levels of innovation,
- Law of ideality,
- Contradictions.

Every system can contain one or more subsystems. The hierarchy of systems can be complex with many interactions. When a system produces harmful or inadequate effects, the system needs to be improved. Systems emerge, ripen to maturity, and die (they are replaced with new systems).

TRIZ systematizes solutions that can be used for different fields and activities. Over the past decades, TRIZ has developed into a set of different practical tools that can be used collectively or individually for problem solving and failure analysis.

The integral development of TRIZ consists of a set of concepts [4]:

- Problem formulation system;
- Contradictions solving;
- Concept of the ideal state of a design;
- "Substance-Field" Analysis;
- Algorithm of Inventive Problem Solving (ARIZ).

There are different methods to assess the levels of innovation. Different methods take into account different aspects of innovation initiatives. However, innovation requires analysis on different levels; therefore, the evaluation systems must be organized accordingly. There is a need for approaches to capture complexities of a multi-level innovation analysis.

The measurement of innovation is necessary; it contributes to the establishment of longer-term policies for innovation and for evaluation and comparison of different solutions of problems in complex systems.

Altshuller's analysis of a large number of patents reveals that the inventive value of different inventions is not equal. Altshuller systematized the solutions described in patent applications by dividing them into five levels [5]:

- Level 1: routine solutions using methods well known in their area of specialty. The Level 1 is not highly innovative. This category constitutes about 30 % of the total.
- Level 2: small corrections in existing systems using methods known in the industry. This level makes up approximately 45 % of the total.
- Level 3: major improvements that solve contradictions in typical systems of a particular branch of industry. About 20 % of the total are on this level, which is where creative design solutions appear.
- Level 4: solutions based on the application of new scientific principles. They solve the problem by replacing the original technology with new technology. Nearly 4 % of the total are classified to be on this level.
- Level 5: innovative solutions based on scientific discoveries not previously explored. This level consists of less than 1 % of the total.

The five-level TRIZ classification can be used for analysis of innovation initiatives and problem solutions. The same problem can be solved by solutions with different levels of innovation. The five-level classification can be applied for the evaluation of innovation level of different innovation solutions, and it can be used during the development process of new proposals.

The development of a new solution can follow different procedures:

- conventional improvement of existing system (levels 1 and 2);
- new forms but with existing principles of operation (levels 2 and 3);
- creation of new system generation with new principles of operation (levels 4 and 5).

The creative solutions classified in levels 4 and 5 (and especially the solutions at level 5) can be considered radical.

The five levels of innovation also can be useful for the prognosis of evolution of a specific system [6].

Contradictions constitute one of the most important terms of the TRIZ vocabulary.

A system conflict or contradiction occurs when the improvement of certain attributes results in the deterioration of others. The typical conflicts are:

reliability/complexity, productivity/precision, strength/ductility, etc. A problem is defined by contradictions.

The traditional way of contradiction solving is through the search of possible compromise between contradicting factors, whereas TRIZ aims to remove contradictions and compromises. The inconsistencies are eliminated by modification of the entire system or by modification of one or more subsystems.

Altshuller distinguished three types of contradictions [7]:

1. Administrative contradiction—contradiction between the needs and abilities;
2. Technical contradiction—the classical engineering "trade-off," an inverse dependence between parameters/characteristics of a machine or technology;
3. Physical or inherent contradiction—opposite/contradictory physical requirements of an object.

A traditional approach is based on trade-offs through preferable combinations of characteristics in conflict. TRIZ aspires to solve the contradictions by modification of systems to avoid deterioration of any characteristic in case of improvement of other characteristics.

3 Complex Systems Problem Solving in Trans-disciplinary Context

The problem solving is a very pertinent issue in complex systems management.

To solve problems, some people consult books, others feel that they only can solve problems doing, others prefer to devote to the analysis and synthesis of problems, others usually seek solutions by consulting of experts, some people seek existing solutions... We need new skills to really be systemic and creative. The need to learn to solve problems and be able to improve this capacity is placed. It is necessary to find new approaches to solve complex system problems.

Through the complementarity, a transdisciplinary perspective can be adopted to modern complex problems. Hester, Akpinar-Elci, James and Margaret Shaeffer explore the concept of complementarity as a motivation for seeking transdisciplinary solutions to complex problems (chapter "On Complementarity and the Need for a Transdisciplinary Perspective in Addressing Infectious Diseases").

A dynamic complex problem leads us to think about the whole system. We cannot look at the problem alone without considering the entire system. We need to work with the various players directly.

Complex systems present multifaceted problems with multiple players and various interests involved. There are many problems to predict and to control complex systems. In truly complex problems, there will hardly be a unique and unerring resolution.

Companies need new approaches to expect the unexpected and to be able to react quickly to emergent risks or to opportunities. Holt, Collopy and Deturris propose transdisciplinary strategies for engineering complex systems, above all a structured approach with experimentation (chapter "So It's Complex, Why Do I Care?").

The first step in the TRIZ's conflict solving process is drawing up a statement of the problem in order to reveal the contradictions contained in the system. Then, the parameters that affect and improve system performance are identified.

It is important to identify and to understand the contradiction that is causing the problem as soon as possible. TRIZ can help to identify contradictions and to formalize problems to be solved. The identification and the formalizing of problems is one of the most important and difficult tasks, with numerous impediments. The situation is often obscured.

The problem can be generalized by selecting one of the TRIZ problem solving tools. The generic solutions available within TRIZ can be of great benefit when choosing corrective actions.

Generally, the TRIZ's problem solving process is to define a specific problem, formalize it, identify the contradictions, find examples of how others have solved the contradiction or utilized the principles, and finally, apply those general solutions to the particular problem.

Figure 1 shows the steps of the TRIZ's problem solving process.

In TRIZ, the problems are divided into local and global problems [2].

The problem is considered as local when it can be mitigated or eliminated by modifying a subsystem, keeping the remaining unchanged.

The problem is classified as global when it can be solved only by the development of a new system based on a different principle of operation.

An identification and analysis of contradictions should be included in any process of TRIZ inventive solving of problems. When a contradiction is identified it becomes easier to find creative and effective solutions for the problem. A contradiction does not solve a problem, but it gives direction for a solution.

We can use two ways of problem solution. One way is to resolve all contradictions by applying one of the appropriate TRIZ analytical tools for solving contradictions (for example, 40 Inventive Principles, Contradiction Matrix, or others). The other way is to transform the technical contradictions into physical contradictions resolving all contradictions at the physical level (for example, with databases of physical phenomena and effects).

All technical contradictions can be transformed into a corresponding physical contradiction. Any physical contradiction can be resolved using one of the four principles (Principles of Physical Contradiction Solving):

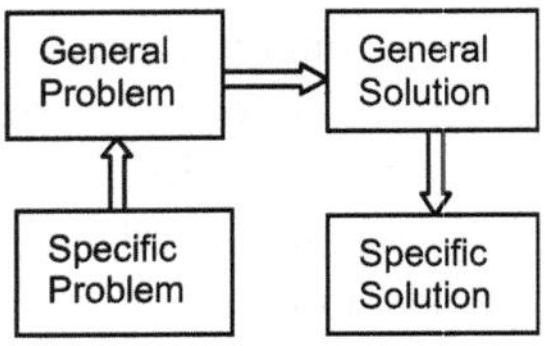

Fig. 1 Steps of the TRIZ's algorithm for problem solving [8]

- Separation of contradictory characteristics in time;
- Separation of contradictory properties in space;
- System transformation;
- Phase transformation (physical and/or chemical transformation of substances).

Altshuller [5] found that, despite the great technological diversity, there are only 1250 typical system conflicts. He also identified 39 engineering parameters or product attributes that engineers usually try to improve. All 1250 conflicts can be solved through the application of only 40 principles of invention [5], often called Techniques for Overcoming System Conflicts.

However, most of the principles of invention have a specific meaning introduced by Altshuller [9].

The inventive principles are simple analytical tools to solve contradictions and finally resolve the problem. In practical activities, various methods are used to apply the 40 principles in the process of problem solving. The simplest method is to examine each contradiction and try to apply the principles of each of them or their combinations to solve the contradiction technique on the specific problem.

Another method is the development of a contradiction and use of the contradiction matrix in order to determine the set of recommended principles to solve the problem (usually between two to four principles).

Substance-Field Analysis is one of TRIZ analytical tools. It can be used in the solution of problems through functional models building [2]. Substance-Field Analysis is a useful tool for identifying problems in a system and finding innovative solutions to these identified problems.

Recognized as one of the most valuable contributions of TRIZ, Substance-Field Analysis is able to model a system in a simple graphical approach, to identify problems and also to offer standard solutions for system improvement [10].

The process of functional models construction comprehends the following stages [11]:

1. Survey of available information.
2. Construction of Substance-Field diagram.
3. Identification of problematic situation.
4. Choice of a generic solution (standard solution).
5. Development of a specific solution for the problem.

There are mainly five types of relationships among the substances: useful impact, harmful impact, excessive impact, insufficient impact and transformation [11].

Substance-Field Analysis has 76 standard solutions categorized into five classes [12]:

Class 1: Construct or destroy a substance-field (13 standard solutions)

Class 2: Develop a substance-field (23 standard solutions)

Class 3: Transition from a base system to a super-system or to a subsystem (6 standard solutions)

Class 4: Measure or detect anything within a system (17 standard solutions)

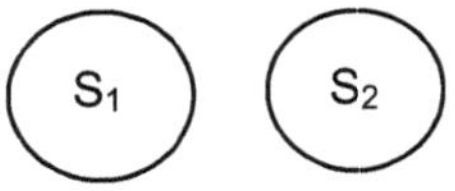

Fig. 2 Problematic Situation 1—incomplete model

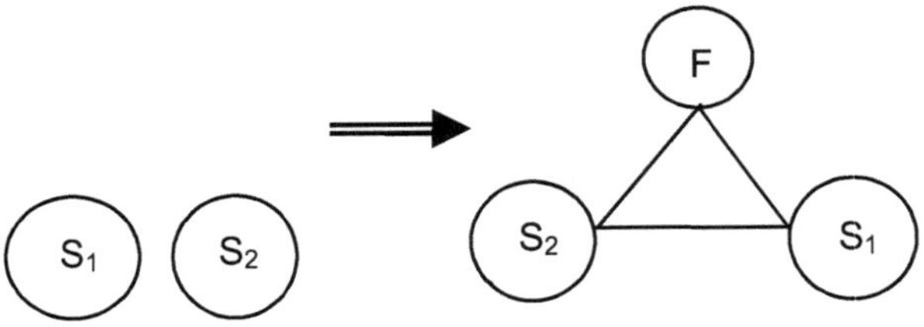

Fig. 3 General Solution 1 for Problematic Situation 1

Class 5: Introduce substances or fields into a system (17 standard solutions)

These 76 solutions can be condensed and generalized into seven standard solutions.

Figure 2 shows the Problematic Situation 1—Incomplete Model [11].

The Substance-Field Model is incomplete, a field is missing. The problem corresponds to Problematic Situation 1 and can be solved resorting to General Solution 1.

Figure 3 shows the solution.

Then the model becomes complete.

Figure 4 shows the problem (Problematic Situation 2—Harmful Interactions between the Substances).

The Substance-Field Model is complete however the interaction between the substances is harmful. The problem corresponds to Problematic Situation 2 and can be solved resorting to General Solution 2.

Figure 5 shows the general solution.

Then the harmful interaction is reduced or eliminated.

General Solution 3: Modify S1 to be Insensitive or Less Sensitive to Harmful Impact

The problematic situation is the same (see Fig. 4).

General Solution 3 is similar to General Solution 2, but instead of substance S2 modification, the substance S1 is modified. The characteristics (physical, chemical and/or other) of substance S1 are changed in order to become it less sensitive or insensitive to a harmful impact. The changes can be internal and/or external, can be temporary or permanent.

The physical and/or chemical characteristics of substance S1 may be altered internally or externally, so that it becomes less sensitive or insensitive to a harmful impact, as seen in Fig. 4. The modification may be either temporary or permanent. Additives may be needed in the modification.

Figure 6 shows the general solution.

Then the harmful interaction is reduced or eliminated.

General Solution 4: Change Existing Field to Reduce or Eliminate Harmful Impact

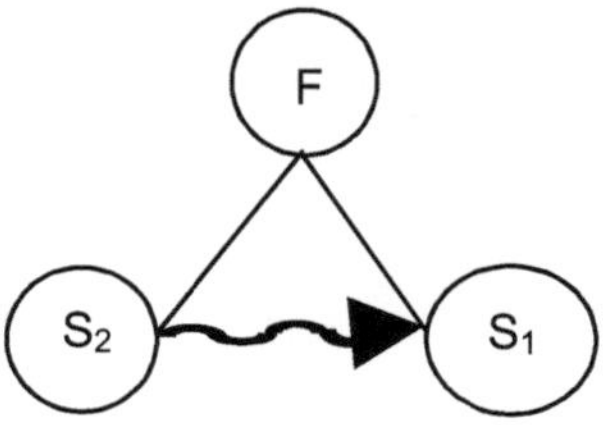

Fig. 4 Problematic Situation 2—harmful interactions between the substances

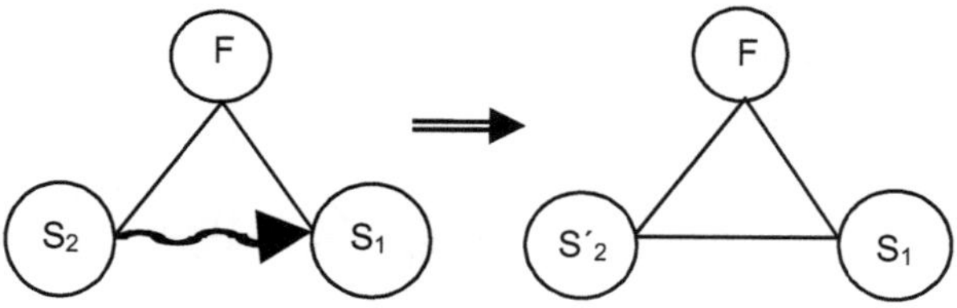

Fig. 5 General Solution 2 for Problematic Situation 2

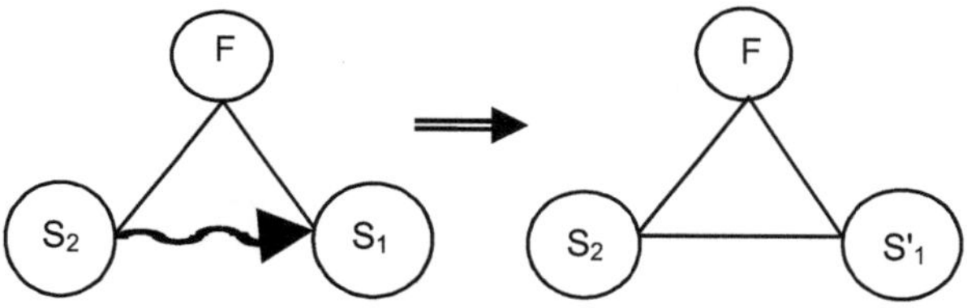

Fig. 6 General Solution 3 for Problematic Situation 2

The problematic situation is the same (see Fig. 4).

General Solution 4 is similar to General Solutions 2 and 3, but instead of substances modification, the field F is modified.

Changing the existing field while keeping the same substances may be a choice to reduce or removing the harmful impact. The existing field can be increased, decreasing, or completely removed and replaced by another one.

Figure 7 shows the general solution.

General Solution 5: Eliminate, Neutralize or Isolate Harmful Impact Using Another Counteractive Field Fx

The problematic situation is the same (see Fig. 4).

General Solution 5 presupposes introduction of a counteractive field FX in order to remove, neutralize or isolate the harmful impact. The substances S2 and S1 and the field F will not change its characteristics in this solution.

Figure 8 shows the general solution.

General Solution 6: Introduce a Positive Field

The problematic situation is the same (see Fig. 4).

General Solution 6 is very similar to General Solution 5.

Another field is added to work with the current field in order to increase the useful effect and reduce the negative effect of the existing system keeping all elements without change.

Figure 9 shows the general solution.

General Solution 7: Expand Existing Substance-Field Model to a Chain

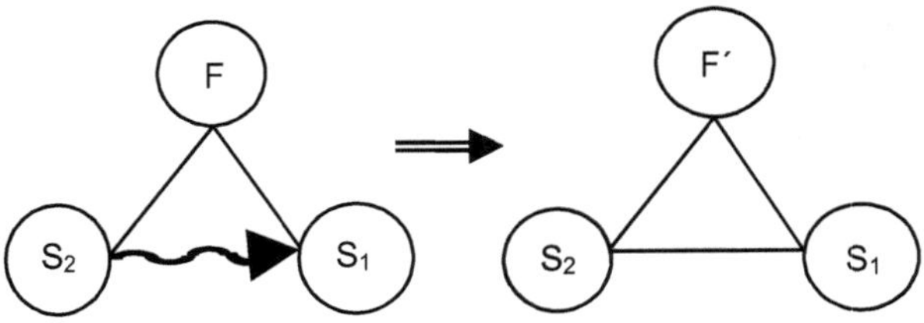

Fig. 7 General Solution 4 for Problematic Situation 2

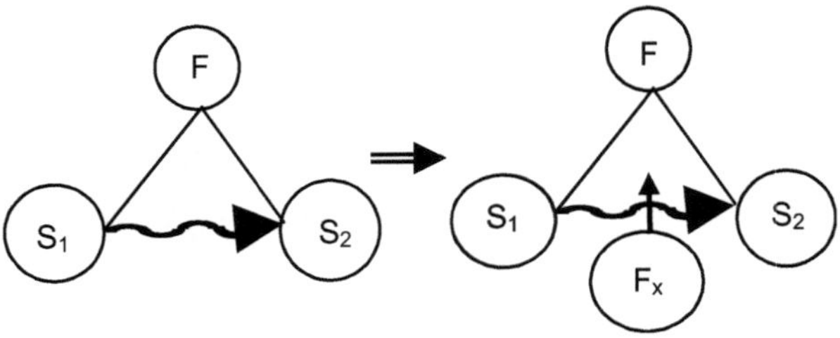

Fig. 8 General Solution 5 for Problematic Situation 2

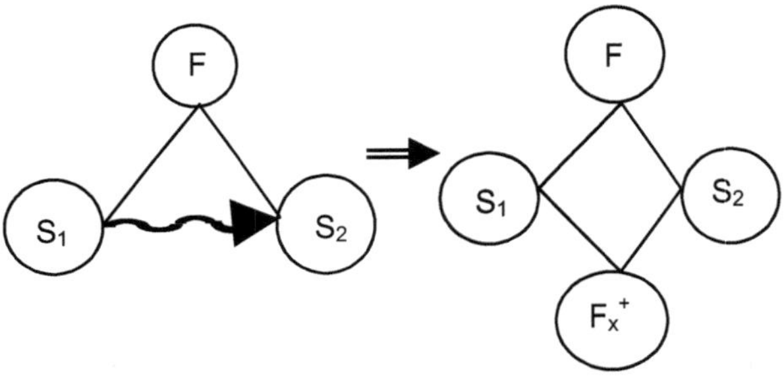

Fig. 9 General Solution 6 for Problematic Situation 2

The problematic situation is the same (see Fig. 4).

The existing Substance-Field Model can be expanded to a chain by introducing a new substance S3 to the system. Instead of directly acting upon S1, S2 will interact indirectly with S1 via another medium, substance S3.

Figure 10 shows the general solution.

Beyond the Problematic Situation 1 (incomplete model) and the Problematic Situation 2 (harmful or undesirable interactions between the substances), also the Problematic Situation 3 (insufficient or inefficient impact) can occur.

Figure 11 shows the Problematic Situation 3.

The general solutions used for the Problematic Situation 2 can be used for the Problematic Situation 3. Figs. 12, 13, 14, 15, 16 and 17 show the general solutions.

TRIZ includes a database of scientific effects structured according to technological functions. According to TRIZ, the scientific effects are one of the principles for contradiction solving by transformation of an action or field to another with the application of physical, chemical, biological, geometric, or another phenomena.

Currently, there are over 8000 known effects and different phenomena, and 400–500 of them are most applicable in the practice of engineering activity.

Fig. 10 General Solution 7 for Problematic Situation 2

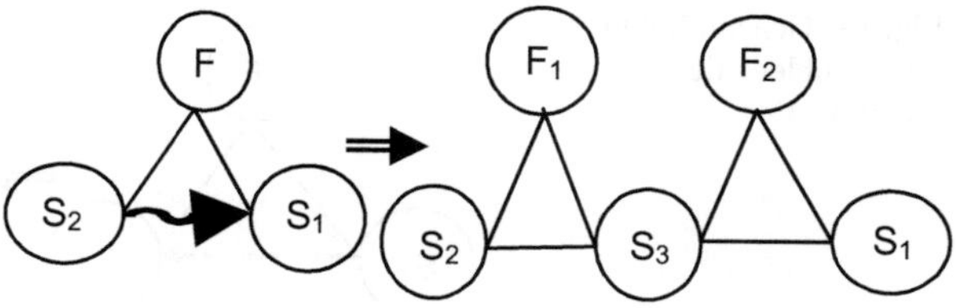

Fig. 11 Problematic Situation 3—insufficient or inefficient impact between the substances

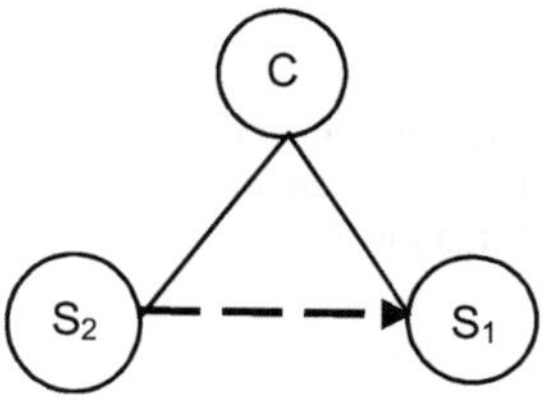

Fig. 12 General Solution 2 for Problematic Situation 3

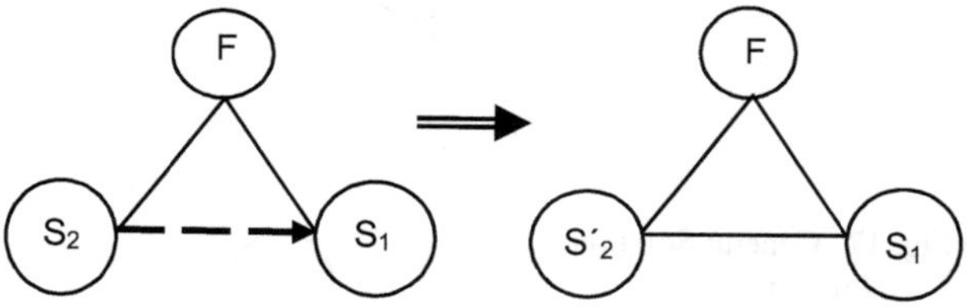

Fig. 13 General Solution 3 for Problematic Situation 3

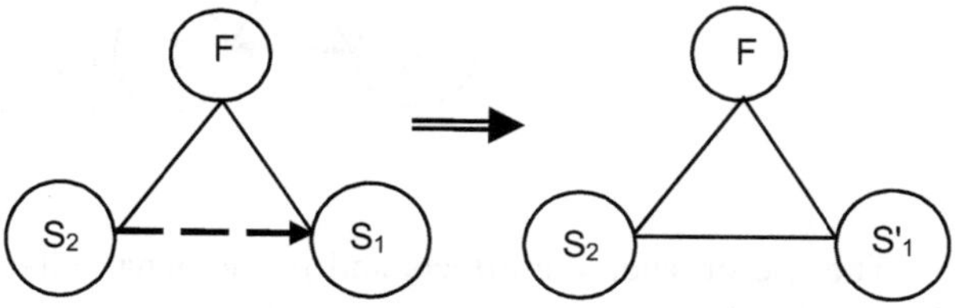

Fig. 14 General Solution 4 for Problematic Situation 3

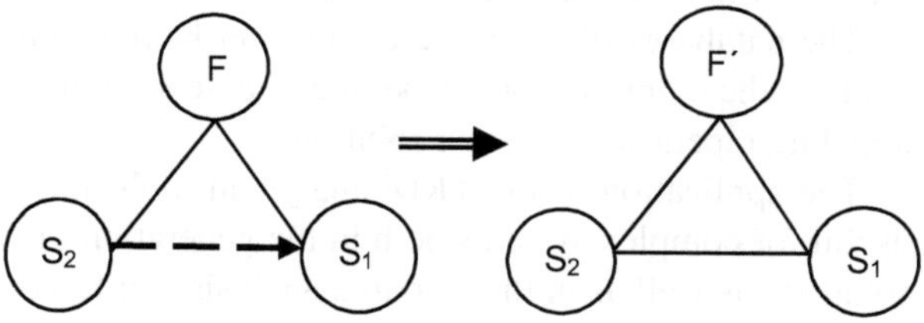

Special tables and descriptions of scientific phenomena exist, which give us the opportunity to define the required effect of an output action or function that should be performed according to the identified problem. Also, there are special software with databases of scientific and engineering phenomena. These programs allow effective selection based on the desired function. Some software provides access to more than 4500 engineering and scientific effects, theorems, laws, and phenomena.

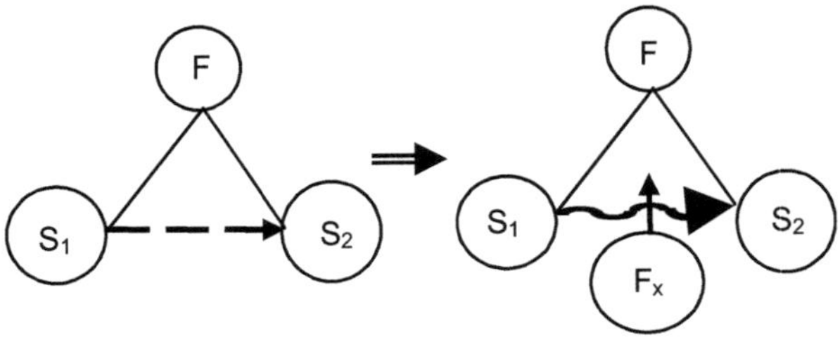

Fig. 15 General Solution 5 for Problematic Situation 3

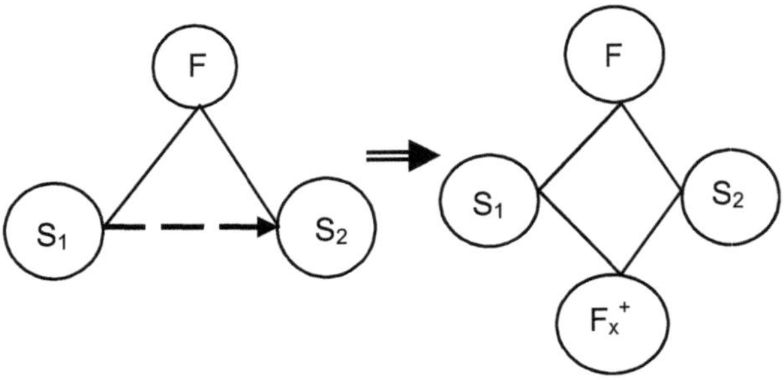

Fig. 16 General Solution 6 for Problematic Situation 3

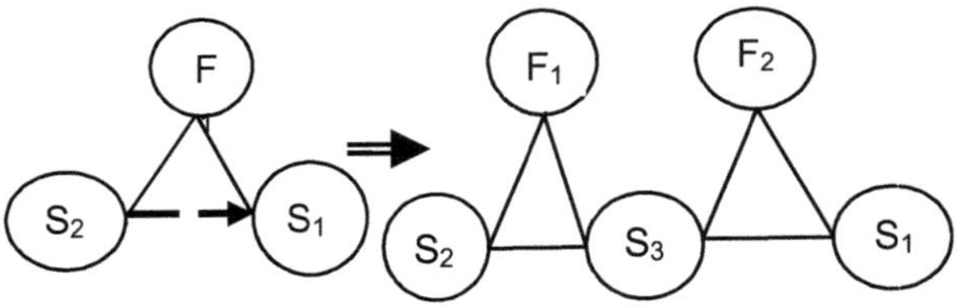

Fig. 17 General Solution 7 for Problematic Situation 3

The use of scientific effects and phenomena helps to develop solutions at the highest level of innovation. Scientific effects can be used to solve problems outside the field where the original problem was found.

The databases of scientific effects can help all initiatives of innovation, especially radical innovation dispelling the fear of using new techniques but also avoiding repetition of wrong solutions.

The application of the TRIZ analytical tools and techniques can be especially useful for complex systems both in the generation of innovative and revolutionary solutions as well as in the resource analysis and forecast [13].

The identification of available resources around a problem and the maximization of their use is important for finding cost-effective and environmentally friendly solutions. TRIZ demands that the analysis of resources take into account the negative as well as the positive resources in a system [13].

The improvements must continue until the resources are fully utilized.

Resources can be grouped according to the following [3]:

1. Natural or environmental resources
2. System resources

3. Functional resources
4. Substance resources
5. Energy and field resources
6. Time resources
7. Space resources
8. Information resources

Altshuller also grouped resources in the following categories:

1. Based on accessibility:
   (a) Internal (limited to the main elements of the system).
   (b) External, including resources from the general environment and those which are specific for the given system.
   (c) Resources from the super-system or other accessible, inexpensive resources (including waste).

2. Based on readiness for utilization
   (a) Readily-available resources.
   (b) Derived (modified readily-available resources).

The key for sustainability is resources productivity.

TRIZ analysis of resources can be useful in innovation initiatives related to more efficient and responsible usage of resources, including energy use.

The gains in resource efficiency generally result from process improvements. Thus, sporadic initiatives of innovation must evolve to the continuously planned and scheduled activities; the innovation must become systematic.

The traditional incremental improvement of existing technologies is no longer sufficient. All economic activities need to radically increase the efficiency of resource utilization.

The application of the TRIZ analytical tools and techniques can be especially useful for complex systems both in the generation of innovative and revolutionary solutions as well as in the resource analysis and forecast.

The resources can be used to solve problems according to the following workflow:

1. Formulate the problem;
2. Build the list of resources in the following order: internal, external, byproducts, and complex resources;
3. Define what kind of resources are needed to solve the problem;
4. Estimate each of the existing resources and the effects of its use;
5. Propose how to use the resource.

4 Law of Systems Evolution and Ideality of Complex Systems

The concept of ideality refers to the observation of the evolution of the system over time, leading to increases of useful functions and features of decreasing useless, harmful, and even of neutral function. The occurrence of this fact can be observed both through incremental improvements as the radical product innovations.

The TRIZ Axiom of Evolution reveals that, during the evolution of a system, improvement of any part of that system can lead to conflict with another part.

The Law of Ideality states that any system tends to reduce costs, energy wastes, and space and dimensional requirements as well as become more effective, more reliable, and simpler. Any system, during its lifetime, tends to become more ideal.

We can evaluate an inventive level of a system by its degree of ideality.

The ideality can be calculated as the ratio of a system's useful functions to its harmful functions with the formula:

$$\text{Ideality} = \text{Useful Functions}/(\text{Harmful Functions} + \text{Cost}) \qquad (1)$$

Useful functions include the following:

- Primary useful functions—the purpose for which the system was designed;
- Secondary functions—other useful outputs;
- Auxiliary functions—functions that support the primary useful functions, such as corrective functions, control functions, housing functions, transport functions, etc.

Harmful functions include all harmful factors associated to the system (for example, costs, area that it occupies, emission of noises, expended energy, resources needed for system maintenance, etc.).

The level of ideality increases with the increase of useful functions and reduces with the increase of harmful functions.

There are three ways to increase an ideality of a system:

- Increasing the useful functions;
- Reducing any harmful or expensive function;
- A combination of the first two paths.

According to the TRIZ methodology, an Ideal System does not exist. An absolute ideality is impossible to achieve, but other relative levels of ideality are achievable. A real system can approximate to the Ideal System by increasing the useful functions and eliminating the harmful functions through contradiction solving, more efficient use of resources, and the reduction of system complexity and number of components.

The ideality can be used both to improve existing systems and also for the creation of new technologies or new systems to fulfill specific functions.

The concept of increasing the degree of ideality is crucial for predicting the evolution of the system.

There are several concepts derived from the same concept of ideality, such as ideal final result, ideal final goal, ideal solution, ideal product, ideal process, etc.

Ideal final result (IFR) is the ultimate idealistic solution of a problem when the desired result is achieved by itself.

Systematic innovation in complex systems can be supported by TRIZ ideality, whereas the planned and continuous improvement can be made in terms of increasing the level of ideality of a given system.

The initiatives of a radical innovation can be seen as actions aimed at dramatically increasing the level of ideality. The ideal final result can be very useful, especially for a radical innovation in complex systems.

The ideality matrix helps identify the interactions between the requirements and distinguish the positive and negative effects of iterations. For example, weight reduction can lead to reduction in volume but may lead to reduction of the tank capacity.

Based on this analysis, measures can then be established to increase the level of ideality by increasing the useful functions, reducing any harmful or expensive function, or by combination of the first two paths.

Previously, the storage capacity of computers was very limited, requiring the development of hardware capable of storing more. The size of the devices was also a serious problem. Nowadays, these same devices have a much higher storage capacity and their size decreased significantly (Fig. 18), thus enabling their easy transport and its adaptability.

It has been observed that the systems and processes generally follow certain regularities in their development. These regularities were translated into evolutionary patterns and are useful for the construction solutions for problems. They are also very good to predict the future development of a technique [15]. Savransky tells us [3] that it is possible to express the idea of a development technique through the concept of ideality, using the notion that any development technique brings the increase of its ideality.

An another example of the evolution of the ideality can be seen in the evolution of portable music playback devices. One can do a little comparative analysis between the Walkman whose first version was released in 1979 and the iPod which had its first release in 2001. The Walkman features are quite dissimilar from the iPod, which in terms of size are much larger, have no memory storage, operate from batteries or connected to electric current and only play music via cassettes.

Since the first version of the iPod presents a considerably small size, its own battery with a battery life of 10 h and the possibility to download 1000 songs, among other features. Thus, it can be said that the iPod is closer to the ideal solution than the walkman.

According to Savransky [3], a technique evolves towards increasing its ideality in two ways:

- Evolution over its useful life to increase local ideality;
- Evolution by the transition into another technique.

The evolution over the life to increase local ideality is described as the evolution α by Savransky [3]. Here, the operation mode technique (the way it performs its

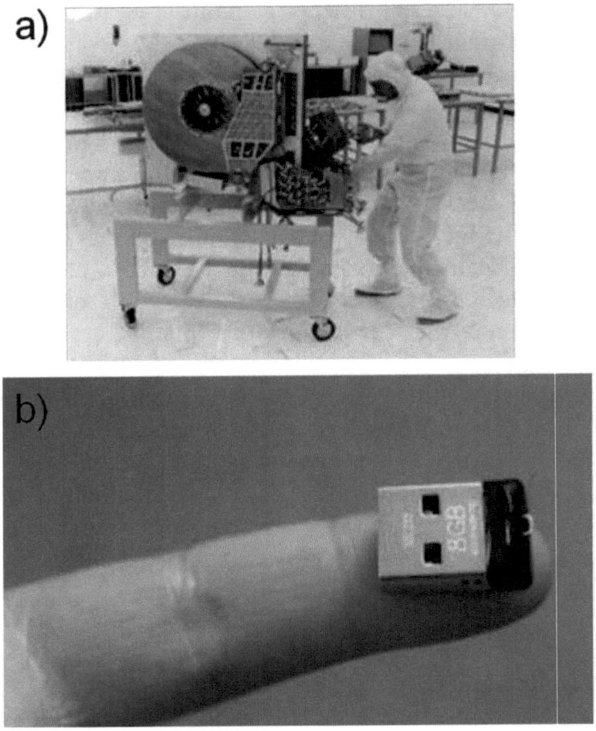

Fig. 18 Evolution of data storage devices (**a**) Hard disk with 1 GB in 1980 (**b**) USB data storage device with 8 GB in 2010 (adapted from [14])

primary function) is not changed, but its parameters are improved. It increases its useful function and reduces their harmful functions and resource costs, thus increasing its ideality.

When an ideality of one technique is placed in graph with time along the stages of technique development (birth, childhood growth, maturity, and decline), an S curve is usually produced. For the end of the useful life, the ideality is close to the limits as it becomes increasingly difficult to improve any further.

The S curve may be combined with other curves showing the equivalent of steps in terms of level of creativity, innovation and profitability associated with the development of the technique (Fig. 19).

A given system has a life cycle consisting of four phases, among them, the period of infancy (childhood), growth, maturity, and decline as shown in Fig. 19.

The childhood phase is characterized by the appearance of a new system, embryonic and resulting renewal of a previous system. This phase is considered delayed due to lack of human and financial resources to implement the new system.

The growth of the technological system will only begin when it is recognized by society. At this stage, most system problems have been resolved and most of its main features have been improved, thus captivating the investment by individuals or organizations.

The technological system reaches maturity when the interest of the society stabilizes and although still a stage where the investment is something high, the result is not as satisfactory as in the childhood stages and growth.

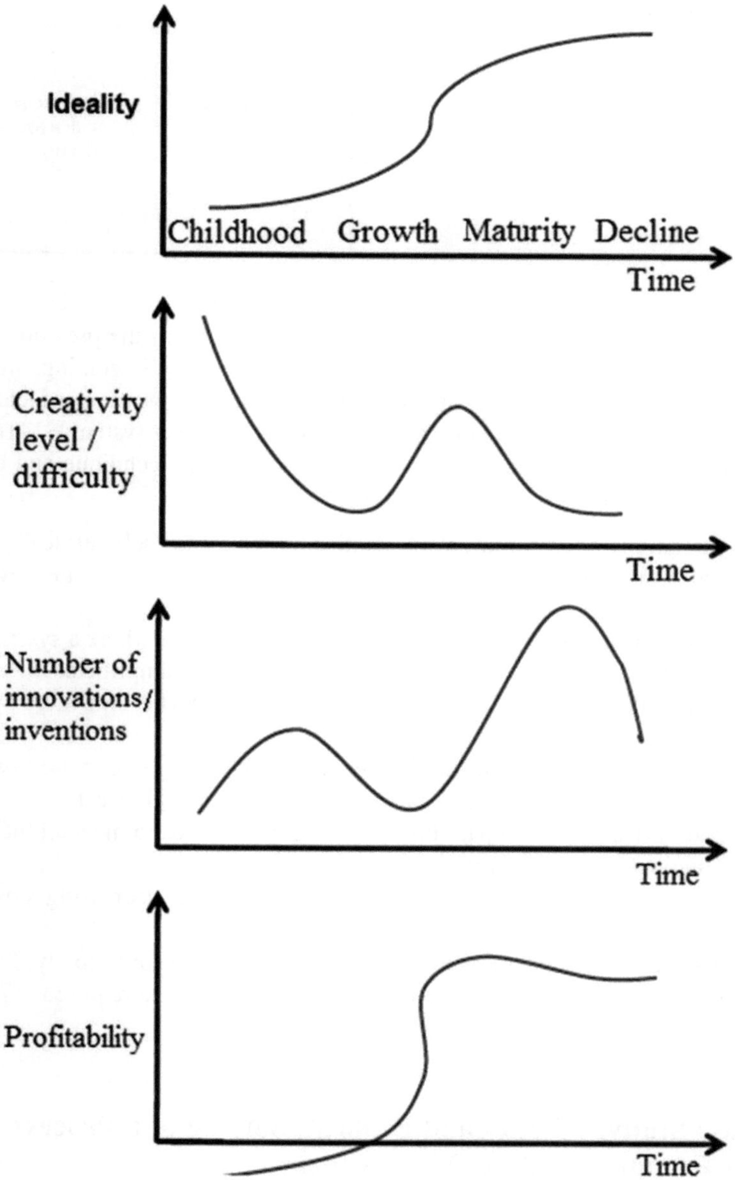

Fig. 19 Evolution α of a system (adapted from [3, 16])

On the decline phase the technological system can no longer provide any kind of innovation and fails to captivate society.

The evolution by transition to another technique is described as the β-evolution by Savransky [3]. It occurs as we approach to technique that is at the end of its useful life, and the potential for improving its ideality reaches its limits. As shown in Fig. 20, the transition to a new technique can be performed by an inventive solution.

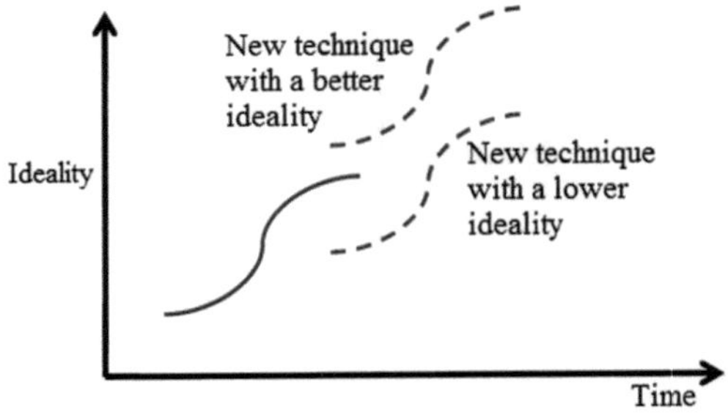

Fig. 20 Evolution β of a system (adapted from [3, 16])

The primary function of the new technique is the same as the previous but the way in which it is achieved is different. Right from the beginning, this new technique can have a better ideality of the prior technique, or a lower ideality but with the potential to improve more quickly than the previous system [3] [16].

There are eight different trends guiding a development technique and each of them is divided into lines of evolution [16]:

- Less human involvement: more automation systems and "self-service".
- Non-uniform development of parts: some parts of the system are developed more quickly than others.
- Simplicity—Complexity—Simplicity: a repeating pattern, where a system starts out as simple, and increases in complexity and is again simplified.
- Increasing dynamism, flexibility and control: the systems become more dynamic and flexible.
- Increasing the segmentation and the use of all fields: progressive use of smaller parts until the parts are so small that, together, have a field effect.
- Compatibilities and incompatibilities: the system evolves to meet all the necessary functions more effectively.
- Increased ideality: more benefits are achieved while decreasing costs and damages.
- Stages of development: the newly invented systems improve slowly, followed by a rapid increase in ideality up to where new systems are required [17].

5 Case Study: Ideality and Problem Solving of a Process for Complex Projects Management

Having in mind the TRIZ principles and based on the article reviews and individual knowledge of project management process from project management practitioners in a brainstorming session [18] eight main parameters were identified, namely:

- Psychological pressure
- Emotional stability
- Information loss

- Time waste
- Reliability
- Harmful factors acting on the process
- Adaptability
- Productivity

The Table 1 contains the ideality matrix built based on the established parameters.

In this case, the ideality is:

$$I = 22/32 \approx 0.6875$$

In case of the contradiction between Emotional Stability (2.) and Information Loss (3.), the principle to be applied to solve the problem would be the number 8 (Balance). The resolution of this contradiction has an impact on all processes of project management, to ensure support of top management and delivery of reports to management prior to implementation of new initiatives.

In case of the contradiction between Information Loss (3.) and Adaptability (7.), the principle to be applied to solve the problem would be the number 6 (Universality). The resolution of this contradiction has impact on development planning by diversification of skills in a matrix structure. It has also implications in all processes of project management in the use of standard specifications.

In case of the contradiction between Time Waste (4.) and Reliability (5.), principles to be applied to remedy the situation would be the number 5 (Combination) and 9 (Neutralizing). The resolution of this contradiction has impact on the process of Initiation, by creating a network of alliances to achieve unity among all stakeholders and / or conduct collective approaches to problems or opportunities in project meetings. It also has implications in the process of Planning and Monitoring and Control in the use of PERT and GANTT charts, promoting for example collaborative scheduling and crisis management reduction or elimination.

In case of the contradiction between Reliability (5.) and Harmful Factors Acting on the Process (6.), principles to be applied to remedy the situation would also be

Table 1 Project management ideality matrix [19]

Parameter	1.	2.	3.	4.	5.	6.	7.	8.
1. Psychological pressure		−	+	B	−	+	−	−
2. Emotional stability	−		−	−	+	−	+	+
3. Information loss	+	−		+	−	+	−	−
4. Time Waste	B	−	+		−	+	−	−
5. Reliability	−	+	−	−		−	+	+
6. Harmful factors	+	−	+	+	−		−	−
7. Adaptability	−	+	−	−	+	−		+
8. Productivity	−	+	−	−	+	−	+	

−Harmful iteration
+ Useful iteration
B—Both (harmful and useful) iteration

the number 5 (Combination) and 9 (Neutralizing). The resolution of this contradiction has impact on the process of Initiation. It is important to create a network of alliances to achieve unity among all stakeholders and / or conduct collective approaches to problems or opportunities in meetings. It also has implications in the process of Planning, Monitoring and Control, in the use of PERT and GANTT charts and in the crisis management reduction or elimination.

In case of the contradiction between Harmful Factors Acting on the Process (6.) and Emotional Stability (2.), principles to be applied to remedy the situation would also be the number 5 (Combination) and 9 (Neutralizing). The resolution of this contradiction has impact on the process of Initiation, by performing an action to create a network of alliances to achieve unity among all stakeholders and / or conduct collective approaches to problems or opportunities in meetings. It also has implications in the process of Planning, Monitoring and Control, in the use of PERT and GANTT charts and the elimination of crisis management.

There are cases in which no contradiction was identified, and therefore, according to TRIZ, would not be treated. One such case would be the Information Loss vs. Time Waste. This can be harmful in Project Management. So in the near future, it should be necessary to consider solutions to the full adaptability of TRIZ methodology in Project Management.

Table 2 contains the TRIZ principles applied to Project Management Processes.

Table 2 TRIZ principles applied to Project Management Processes [19]

Principles	Applied to engineering	Applied to project management	Process group
5. Merging	Join or merge identical or similar objects, assemble identical or similar parts, in order to perform operations in parallel	Create a network of alliances to achieve unity among all stakeholders of the project. Perform collective approaches to problems or opportunities in meetings	Initiating
	Link operations in sequential or parallel, in view of the surrounding temporal	Use of PERT and GANTT charts	Planning
6. Universality	Make a system or object perform multiple functions, eliminating the need for additional parts	Facilitate the diversification of skills in a matrix structure	Planning
	Use of standard methods	Use of forms to the specifications of products or processes	All
8. Anti-weight (Balance)	To compensate for the weight (a negative trend) of an object or system, associating other object or system that provides a positive trend	Ensure support from Top Management, Delivery reports to management prior to implementation of new initiatives	All
9. Preliminary anti-action (Neutralizing)	Need to perform an action with both effects: beneficial and harmful, the same action should be replaced by one that neutralizes the harmful effects	Use of PERT; Eliminate the need for crisis management	Planning, Monitoring and Controlling

Table 3 Matrix of ideality of the camping stove [20]

Parameter	1.	2.	3.	4.	5.	6.	7.	8.
1. Volume		+			−	−	−	
2. Weight	+				−	−	−	
3. Firing time					+			
4. Noise level								
5. Time required to boil water	−	−	+				−	+
6. Tank capacity	−	−	+		+		+	+
7. Burning time at maximum flame	−	−			−	−		−
8. Boiled water per unit of gas	−	−			+	+	−	

− Harmful iteration
+ Useful iteration

6 Case Study: Ideality and Application of Ideality Matrix to a Complex Product Problem Solving

Consider the case of a camping stove design.

Customer requirements were collected, pooled and prepared by an affinity diagram, yielding the following list:

- Volume;
- Weight;
- Firing time;
- Noise level;
- Time required to boil water;
- Tank capacity;
- Burning time at maximum flame;
- Boiled water per unit of gas.

Table 3 contains the Matrix of Ideality built for the camping stove.

The Ideality Matrix helps identify the interactions between the requirements and distinguish the positive and negative effects of iterations. For example, weight reduction can lead to reduction in volume, but may lead to reduction of the tank capacity.

In this case, the level of ideality is:

$$I = 11 / 30 \approx 0.367$$

7 Conclusions

The need for problem solving is a current issue in complex systems. The complex systems need achieve its objectives. Complex systems managers need powerful and highly efficient analytical tools to solve problems, generate creative ideas and innovation. The lack of ideas can lead to the failure of objectives. Creativity is crucial for competitiveness.

The novelty of this study is to survey the applicability of the key pillars of TRIZ to complex systems, especially to ideality and problem solving in complex systems.

TRIZ methodology, with its strong theme of problem solving, can contribute to accelerating the resolution of problems in innovation activities of complex systems. The TRIZ analytical tools would be very useful for schematization of innovation tasks, system analysis, identification, and formalization of contradictions and problematical situations and their solving processes.

TRIZ is a methodology especially suited for supporting the development of radical solutions where traditional techniques usually do not yield positive results.

Moreover, TRIZ is considered the scientific basis of systematic innovation. Thus, the extension of TRIZ for complex systems seems to be a logical step of evolution.

Acknowledgements The present author would like to thank the Faculty of Science and Technology of the New University of Lisbon (UNL) and the Portuguese Foundation for Science and Technology (FCT) through the Strategic Project no. UID/EMS/00667/2013. Their support helps make our research work possible.

References

1. Hellström, T. (2007). Dimensions of environmentally sustainable innovation: The structure of eco-innovation concepts, sustainable development. *Sustainable Development, 15*, 148–159.
2. Altshuller, G. S. (1995). *Creativity as an exact science: The theory of the solution of inventive problems*. New York: Gordon and Breach Publishers.
3. Savransky, S. D. (2000). *Engineering of creativity: Introduction to TRIZ methodology of inventive problem solving*. Boca Raton, FL: CRC Press.
4. Radeka, K. (2007). *TRIZ for lean innovation: Increase your ability to leverage innovation across the enterprise and beyond*. Camas, CA: Whittier Consulting Group.
5. Altshuller, G. S. (2001). *40 principles: TRIZ keys to technical innovation*. Worcester, MA: Technical Innovation Center.
6. Kraev, V. (2006, November). Kraev's Korner: Levels of innovations. *The TRIZ Journal*.
7. Altshuller, G. S. (1986). *To Find an Idea*. Novosibirsk: Nauka [in Russian].
8. Fey, V. R., & Rivin, E. I. (1997). *The science of innovation: A managerial overview of the TRIZ methodology*. Southfield, MI: TRIZ Group.
9. Terninko, J., Zusman, A., & Zlotin, B. (1998). *Systematic innovation: An Introduction to TRIZ (Theory of inventing problem solving)*. Boca Raton, FL: CRC Press LLC.
10. Mao, X., Zhang, X., & Rizk, S. (2007, August). Generalized solutions for Su-field analysis. *The TRIZ Journal*.
11. Altshuller, G. S. (1999). *Tools of classical TRIZ*. Southfield, MI: Ideation International.

12. Terninko, J., Domb, E., &, Miller, J. (2007, February, March, May, June, July). The seventy-six standard solutions, with examples. *The TRIZ Journal*.
13. Mann, D. L. (2000, June). *The four pillars of TRIZ*. Invited paper at Engineering Design Conference, Brunel.
14. Tech, M. H. (2010). *A evolução dos computadores* [cited 2013 09/04]. Retrieved from http://russ4n0.wordpress.com/2010/06/15/a-evolucao-dos-computadores/
15. Rantanen, K., & Domb, E. (2010). *Simplified TRIZ: New problem solving applications for engineers and manufacturing professionals*. Hoboken, NJ: Taylor & Francis.
16. Gadd, K. (2011). Inventing with TRIZ. In *TRIZ for engineers: Enabling inventive problem solving*. Chichester: Wiley.
17. Ilevbare, I., et al. (2011). *Integration of TRIZ and roadmapping for innovation, strategy, and problem solving*.
18. Triep, B., & Caldeira, H. (2010, December 5). *A Metodologia Triz na Gestão de Projectos*. Technical Report. Faculdade de Ciências e Técnologia da Universidade Nova de Lisboa.
19. Navas, H. V. G., Alexandra, M. B. R. T., & Cruz-Machado, V. A. (2011, November, 2–4). *Integrating TRIZ in project management processes: An ARIZ contribution*. TFC 2011—11th ETRIA World TRIZ Future Conference 2011, ISBN 978-0-9551218-2-1, Dublin, pp. 387–396.
20. Navas, H. V. G. (2013). TRIZ: Design problem solving with systematic innovation. In *Advances in industrial design engineering* (pp. 75–97). Rijeka: InTech. doi: 10.5772/55979.

MIX
Papier aus verantwortungsvollen Quellen
Paper from responsible sources
FSC® C105338
FSC
www.fsc.org

If you have any concerns about our products,
you can contact us on
ProductSafety@springernature.com

In case Publisher is established outside the EU,
the EU authorized representative is:
Springer Nature Customer Service Center GmbH
Europaplatz 3, 69115 Heidelberg, Germany

Printed by Libri Plureos GmbH
in Hamburg, Germany